Climate Change in Santos Brazil: Projections, Impacts and Adaptation Options

Lucí Hidalgo Nunes • Roberto Greco
José A. Marengo
Editors

Climate Change in Santos Brazil: Projections, Impacts and Adaptation Options

Springer

Editors
Lucí Hidalgo Nunes
Santos, São Paulo, Brazil

José A. Marengo
National Center for Monitoring and Early
Warning of Natural Disasters (CEMADEN)
São José dos Campos, São Paulo, Brazil

Roberto Greco
Department of Geography
UNICAMP – Instituto de Geociências
Campinas, São Paulo, Brazil

ISBN 978-3-319-96534-5 ISBN 978-3-319-96535-2 (eBook)
https://doi.org/10.1007/978-3-319-96535-2

Library of Congress Control Number: 2018959106

This Springer imprint is published by the registered company Springer Nature Switzerland AG
The registered company address is: Gewerbestrasse 11, 6330 Cham, Switzerland

To my mother Maria Carolina, to my father José (in memoriam) and to my sister and brother, Lení and Marcos.
Lucí H. Nunes
To my mother Deanna (in memoriam), to my father Remo and to my sisters Cristina and Francesca.
Roberto Greco
This book is dedicated to my beloved wife Angela Cristina and son José Antonio, without whose
love, encouragement, support and inspiration I would never have made it this far. I also wish to dedicate this book to my mother and brothers for their continuous love and support all the way since the beginning.
José A. Marengo

Foreword

The entire world is challenged to find answers regarding adaptation to ongoing global environmental modifications, among them, climate change. At the same time, it has increased the perception that the processes observed at local scale are controlled by regional and global processes. Therefore, increasing knowledge at local level of processes arising from global changes, providing adaptive responses and finding solutions to the social and economic impacts associated is timely and needed.

Climate change operates throughout the world in different manners and at different rates and thus affects the areas differently. Therefore, understanding how the physical processes of a given place can respond to large-scale environmental modifications and knowing the capacity of society to adapt to these transformations are of utmost importance for local development.

These aspects were central issues in the Project Metropole, an international consortium involving Brazil, the UK and the USA, linked to the Belmont Forum: Call Coastal Vulnerability. Based upon an integrated multidimensional framework encompassing the social, cultural, political, economic, environmental and physical dimensions, the project evaluated local decision-making processes and provided feedback to local policy makers and society on possible actions towards adaption to sea level rise.

This book is rooted in the results of Project Metropole for Brazilian case study: the dynamic city of Santos, state of São Paulo. The initial step of the Project was to establish sea level rise scenarios for Santos. Prior to the project, the municipality counted on the IPCC projections only, which proved more alarming than the more realistic local evaluation provided by Project Metropole.

Having the projections of sea level rise, their impacts associated with storm surges were simulated by using an interactive computer-based scenario through the platform CoAST (Coastal Adaptation to Sea level rise Tool). The evaluation was performed for two contrasting sectors of Santos: the Southeastern Zone, with upscale neighbourhoods closer to the shoreline, and the Northwestern Zone that clusters poor quarters encroached in mangrove areas.

It is important to highlight that the Project Metropole is anchored in some assumptions. One of them is that the partnership with the local government facilitates the internalisation of the results and the implementation of public policies and appropriate legislation, allowing a better management of the area. Another important premise is that the involvement of population in the process of adaptation is crucial. Thus, several meetings were organised by the scientific community of Metropole together with the city council, engaging policy decision-makers, militaries, civil defence, NGOs, other academics and representatives of different sectors of civil society. In the first part of the first stakeholder engagement workshop, on 30 September 2015, the damages to assets from expected 100-year storms for 2050 and 2100 was presented. In the second part of the meeting, attendees discussed possible adaptation measures feasible for Santos and voted to model two of them for each zone. Between the two meetings, platform CoAST run again using the two more voted actions for each zone. These results were presented in the second stakeholder engagement on 1 December 2015 and showed how much building damage might be avoided over time if the actions chosen by population are implemented. During both meetings, attendees respond to a survey elaborated to define values, attitudes and perceptions related to climate change and adaptive responses. The questions evaluated aspects such as individual and community experience with coastal hazards, preferences for several potential adaptation actions and for public finance mechanisms for adaptation and perceptions about barriers to implementation. In a third general meeting, on 17 August 2017, a comprehensive overview of the Project Metropole was provided. Representatives of the federal government also showed efforts towards adaptive measures at the national level, underscoring the importance and novelty of the Project Metropole, which may inspire other coastal municipalities in Brazil.

The involvement of the civil society in the process of adaptation was also promoted within the Project Metropole by the use of the Adaptive Capacity Index (ACI), which reveals the performance of risk reduction related to adaptation to climate change. The index is established through a series of interviews with people representing key institutions from the public, private and nongovernmental organisation sectors of the municipality, with questions that allow a choice of responses. The performance of a given aspect is compared over time, in order to observe the evolution concerned to the adaptive capacity of a given institution. In a final meeting, held on 15 December 2016, the results were presented to the interviewees, who once again could discuss, this time together, issues related to the adaptive capacity of institutions based in Santos.

The importance of the results of the Project Metropole is amplified by the fact that the case study was the municipality of Santos. One of the oldest cities of Brazil, Santos is inserted in the humid tropics, with rich ecosystems like the Atlantic Rainforest and mangroves, and extensive topographic diversity. However, the entire region underwent transformations since the beginnings of the discovery of Brazil, which affected and compromised its ecosystem and increased its natural fragility.

Connected with the main consumer markets in the country by a multimodal transport system and located just 60 km away from São Paulo, the largest Brazilian city, the region has an undeniable strategic position that contributed to the installation in the 1950s of an important petrochemical and steel industrial complex in Cubatão – a nearby city located in a valley surrounded by the Serra do Mar mountain range, a quite unsuitable area for the dispersal of pollutants. The industrial park brought both benefits to the economy and deficits in the infrastructure: part of the labour force attracted by the industries began to live on the unstable hillslopes of Serra do Mar and mangrove areas, and the pollution generated by the industrial complex, for decades without control, caused damages to vegetation, speeded up mass movements and reduced the water retention capacity of the soil, contributing to urban floods.

In the decade of 1980, Cubatão became the most pollutant city of the world. It is of note that this informal land occupation with low-income settlements affects the ecological balance of the area and the life quality of the inhabitants. Problems such as the direct discharge of sewage into the water bodies and thus contamination of water resources, which affect the health of residents, especially children, and clearance of the Atlantic Rainforest and destruction of mangroves, which affect the local stability, are but some examples of the deleterious problems caused by this pattern of occupation, which is outside legal standards and beyond the local environmental capacity.

This situation of little or no concern for the environment began to change due to reasons of different natures. One of them was the end of the dictatorial period that had been in force in Brazil for more than two decades, which allowed for greater mobilisation of society in favour of a safer and healthier environment. Another aspect that contributed to this change was a severe precipitation event in late January 1985 that caused numerous landslides in the Serra do Mar, near the industrial park. If industries had been affected and toxic material had been escaped, an unprecedented tragedy could have occurred and hit a large area, fact that has influenced the adoption of measures to reduce pollution by the government of the state of São Paulo, which launched the first initiatives to pollution control in the same year.

At the same time, pressures imposed by urbanisation driven by speculative interests of the higher social classes led to a clear socio-spatial pattern of land use and land occupation in Santos, with low-income population installed in hillslopes and mangrove areas, while the wealthiest classes settled closer to the beach. The rapid transformation of the space and the disregard of land use and land occupation laws contributed to increase the physical susceptibility of the area and consequently to enlarge the vulnerability of all sectors of the population in the face of problems caused by storm surges, floods and landslides.

More recently, the discovery of oil in the ocean in a super deep layer called pre-salt has attracted new investments, including the expansion of the Port of Santos, and led to an exacerbated land valuation in the urban area, which in a few years changed the urban configuration of the city, with vertiginous increase of verticalisation that has altered the local climate and consequently the thermal comfort in a city that is naturally hot and humid throughout the year.

This scenario of economic vitality contrasts with the current environmental disruption and affects all sectors of the population of Santos albeit in a different way, as demonstrated by the results of the Project Metropole for two contrasting areas: the Northwestern and the Southeastern sectors of Santos.

Additionally, Brazil is experiencing a profound political, economic and value crisis, which must be seen by the local government as an opportunity to change various behaviours towards a fairer society and a healthier environment and, therefore, a sustainable development. This can only be achieved with scientific knowledge and popular participation. In this sense, the results of the Project Metropole can contribute for valuing and consolidating the important role of the municipality of Santos in the country.

The next chapters present the results obtained by this project, which was developed over 4 years, with the participation of researchers from different institutions in Brazil, as well as the valuable contribution of the municipal government and the population of Santos. Some chapters are contributions from other researches developed in the area that addressed additional issues and therefore help to compose a more general framework.

São Paulo, Brazil Lucí H. Nunes

Preface

The book is divided in three parts. Part I contains an introductory chapter, with information of the Project Metropole (An Integrated Framework to Analyse Local Decision Making and Adaptive Capacity to Large-Scale Environmental Change), which aimed to identify the factors that facilitate a shift in knowledge, attitudes, values and decision-making about local climate risks and adaptation strategies among decision-makers and stakeholders in case study of coastal communities in Brazil, the UK and the USA. The goal was to combine expertise from diverse scientific backgrounds to develop an integrated framework to analyse local decision-making and the adaptive capacity to local environmental change driven by large-scale processes like climate.

Part II has six chapters which evaluate the current environmental pattern and projections in Santos, Brazil. Chapter 2 explores the characteristics, detection and frequency of frontal systems in South and Southeast Brazil (pressure, temperature and wind fields) and response of the ocean to the atmospheric forcing. It also explores simulation and prediction of frontal systems with focus in Santos and surroundings. Chapter 3 evaluates the current patterns of extreme precipitation events at different temporal scales in Santos, whether the patterns are systemic and, in particular, whether they present some regularity. Complementing the analysis of extreme events, it also evaluates parameters such as standard deviation, coefficient of variation and number of rainy days, as well as the degree of concentration of daily rainfall. The objective of Chap. 4 is to assess the trend of storm surges that hit the coast of the Baixada Santista in the future climate. The assessment is based on the observed trend of the storms and the future change of the storm surges reproduced by the downscaling of global climate model simulations using the Eta Regional Climate Model at 20-km and 5-km resolutions. The downscaling reproduces, at higher resolution, the simulations from the HadGEM2-ES and MIROC5, under RCP4.5 and RCP8.5 scenarios. The number of storm surge detected in the present climate simulations is compared against observations. The extratropical cyclone located off the coast of South Brazil is simulated generally weaker and less frequent than reanalysis data. Chapter 5 provides a general overview of the causes of sea level variations, discussing how these variations are measured and analysed, the most important periodicities and long-term trends in sea level, with emphasis in the region of Santos. Aspects such as accuracy and space– time coverage of measurements and decadal and secular projections of the mean sea level are presented along the discussion. Chapter 6 presents and discusses patterns and trends of surfaces and "high tides" events in the region of Baixada Santista, where Santos is located, between 1961 and 2016. The evaluation of meteorological and oceanographic events such as storm surges and positive meteorological tides is of great importance for understanding coastal processes and the effects of current sea level rise and ongoing climate change, for the understanding of the vulnerability of coastal cities. Chapter 7 presents the first approach of a storm to the early warning system for Santos region. The system is based on the AQUASAFE platform and provides daily meteo-oceanographic forecasts for up to 5 days, detailed for the Bay and estuary scale of Santos, as well as providing meteo-oceanographic information in real time of the sensor network installed in the port region.

The four chapters of Part III analyse the impacts of sea level rise in Santos and surroundings. Chapter 8 evaluates the recent patterns of temporal distribution of extreme rainfall events and their impacts in the municipality of Santos and discuss the correlation between the occurrence of events (landslides) from the application of a technique which associates daily accumulated rainfall with the total amount that triggered landslides in the past and estimates the local stability thresholds due to accumulated precipitation. Chapter 9 offers an analysis of the Metropolitan Region of Baixada Santista, where Santos is inserted, from the perspective of global environmental changes, with emphasis on changes in land cover and land use. Images of the LANDSAT satellite were used for the years 1996, 2005 and 2011, from which it was possible to detect changes in the land cover, as well as projection of a trend scenario for the year 2021. The study also evaluates socioeconomic, as well as the carbon emission estimates. From this analysis, it is possible to identify the main transforming agents, impacts and challenges of the territorial management of the metropolitan region in the face of a challenging scenario of global environmental changes. Chapter 10 presents an overview of environmental risk management, associating the concepts with case studies of how dengue fever outbreaks respond to climate variability. It also discusses the main factors that can influence future patterns of dengue fever in São Paulo State and the macro region of Baixada Santista and the city of Santos. Study cases are focused on the period from 2008 to 2015 to show how these patterns are associated to weather fluctuations.

The focus of Part IV is adaptation, discussed in the three final chapters. Chapter 11 examines the impact of the construction of terminals between 2006 and 2014 along the Port of Santos navigation channel, located in the complex of Santos Estuary, in the stability of Port of Santos inlet. The Port of Santos is an estuarine port in Southeast Brazil which has received a great deal of investments for deepening dredging and for construction and expansion of port terminals in the past two decades. The port is under constant pressure for receiving larger vessels, one of the reasons to deepen the channel and to construct new terminals. Yet, land reclamation and basin reduction in bays or estuarine area decreases the tidal prism (the amount of water entering and leaving the inlet, so it works as a flow that controls sand deposit on bars). Chapter 12 carries out vulnerability studies for the Santos/SP region, applying the vulnerability index for coastal areas (SEVICA) under scenarios 4.5 and 8.5 of the IPCC AR5 report (IPCC, 2013). In the light of the vulnerability scenarios, critical infrastructure for the municipality, such as hospitals, highways and port structures, was analysed for situations of extreme events. Evaluations of Chap. 13 are based on the results of the Adaptive Capacity Index that incorporates the influence of structure and agency as defining characteristics in an attempt to move the discussion away from simply measurement into a more practical niche, generating an actor-identified solution mechanism through social learning upon which to create pro-active change in how climate change issues are addressed. Chapter 14 analyses the results of a survey applied for attendees in two engagement workshops in which projections of sea level rise and storm surges for 2050 and 2100 were presented for Santos, in view of identifying links between risk experiences, beliefs, values and attitudes about local government priorities for possible adaptation actions and public finance. Chapter 15 explores the creation of the Municipal Committee for Adaptation to Climate Change (MCACC) by the local government, in 2015, analysing the creation of local public policy in Santos coastal municipality focused on climate change adaptation. An initial evaluation of the decision-makers and local community perception were developed by interviews and participation in MCACC meetings. Local community perception was focused on current intense storm surges in the region. The community perception was about structural damages and economic losses in events related flooding and sea level rise.

In sum, the different chapters of this book offers insights to connect the knowledge on climate change – in special, the impacts due to sea level rise – in a coastal city under constant pressure of urbanisation, globalisation and land use changes. Because the problems faced for Santos are common to other coastal cities, this book aims to inspire other studies that, like Project Metropole, intend to strength the linkages among science, community and decision-makers.

In this book, the authors take this risk analysis approach to analyse in particular the impacts of climate change in the environment and in built-up areas, as well as in the health in a coastal city of Brazil. Its importance cannot be underestimated. The editors and authors have brought together a full range of appropriate experts to evaluate the risks that Brazil faces unless actions are taken quickly across the whole world to manage the challenges.

The pledge of the Paris Agreement, to keep the average global temperature rise to less than 2 °C and to aim for no more than 1.5 °C rise, is now a common commitment of 195 signatory countries. No other issue has brought such universal agreement for action. But this agreement is not matched by universal actions to meet the challenge. This book underlines the importance for all of us, as individuals, as urban, regional and national citizens, as city mayors and regional governors and as politicians and heads of government, to act now to help to avoid the severe risks set out here.

Therefore, the aim of this book is to explain how changes in the physical climate can lead to multiple complex changes in human systems, many of which are hard to predict and which tend to have adverse consequences when the changes fall far outside the normal range of variability.

São Paulo, Brazil

Lucí H. Nunes
Roberto Greco
José A. Marengo

Acknowledgements

The book, *Climate Change in Santos Brazil: Projections, Impacts and Adaptation Options*, is a result of contributions from a wide range of experts from across Brazil, the USA and the UK. We thank everyone who contributed to its richness and multidisciplinary outlook.

This book has been written by a team from the University of Campinas (UNICAMP), the National Center for Monitoring and Early Warning of Natural Disasters (CEMADEN), the National Institute for Space Research (INPE), the Municipal Government of Santos, the University of São Paulo (USP), the Institute of Geology – Secretariat for the Environment of the State of São Paulo (IG-SMA/SP), the Universidade Estadual Paulista (UNESP), the Rio de Janeiro State University (UERJ), the Universidade Santa Cecília (NPH-UNISANTA), the Catholic University of Santos (UNISANTOS), the Technological Institute of Aeronautics (ITA), the University of Southern Florida, King's College London and the Future Earth Coasts. The chapters received insightful comments from Frank Muller-Karger, C.J. Reynolds, Dr. Samuel Merril, Dr. Jack Kartez, Jonathan Lockman, Karen Langbehn and Alexander Gray and, from the UK branch, Prof. Dr. Mark Pelling, Dr. Shona Paterson and Kristen Guida. Additional thanks are due to the Civil Defence of Santos for providing with the data from the Saboó Station to Eduardo Garcia Rosa and Environmental Analyst Johann Constantino Lima for conducting part of the work of compiling and treating the data (until 2011). We also thank BNDO – National Bank of Oceanographic Data (Captain Vladimir C. Maluf, in particular), the Santos Civil Defence (Eng. Ernesto K. Tabuchi and Geologist Marcos P. Bandini, in particular), the Centre for Hydrodynamic Research at the University Santa Cecília (Prof. Renan B. Ribeiro, MSc., and Dr. Alexandra Canastra) for kindly making their respective databases available, and LECLIG/UNICAMP for making available their disaster registry for coastal São Paulo, *Cadastro de Desastres para o Litoral Paulista*.

This work was supported by the Belmont Forum-G8 Initiative Collaborative Research, METROPOLE: An Integrated Framework to Analyse Local Decision Making and Adaptive Capacity to Large-Scale Environmental Change, through the São Paulo State Research Foundation (FAPESP) (2012/51876-0), to the US National

Science Foundation (Grant No. NSF ICER-1342969) to the National Environmental Research Council (NERC (NE/L008963/1). Some of the chapters are a contribution of the Brazilian Research Network on Global Climate Change FINEP/Rede CLIMA Grant Number 01.13.0353-00, the National Institute of Science and Technology for Climate Change Phase 1 under the Brazilian National Council for Scientific and Technological Development (CNPq) Grant 573797/2008-0 and the São Paulo Research Support Foundation (FAPESP) Grant 2008/57719-9; and Phase 2 under CNPq Grant 465501/2014-1 and FAPESP Grants 2014/50848-9 and 2015/50122-0; the National Coordination for High Level Education and Training (CAPES) Grant 16/2014, and the Deutsche Forschungsgemeinschaft Grant DFG-GRTK 1740/2; FAPESP Grants 2012/51876-00, 2015/08192-1, 12/51876-0 and 14/14598-8, the *Companhia Docas do Estado de São Paulo* (2010–2011), Institute of Geology - Secretariat for the Environment of the State of São Paulo (IG-SMA/SP), the FEHIDRO (Fundo Estadual de Recursos Hídricos) Project nº2013-BS_COB-5, called "Implantação de sistema de monitoramento e previsão da qualidade da água por meio de modelagem numérica ambiental e desenvolvimento de base de dados na Bacia Hidrográfica do Estuário de Santos – SV", and DHI for providing a time-limited license for thesis of the Mike 21 FM software. Additional thanks to Espaço da Escrita provided by the UNICAMP for the translation of some chapters of the book.

São Paulo, Brazil

Lucí H. Nunes
Roberto Greco
José A. Marengo

Contents

Contributors

Lincoln Muniz Alves National Institute for Space Research (INPE), São José dos Campos, SP, Brazil

Luiz E. O. C. Aragão Remote Sensing Division, National Institute for Space Research (INPE), São José dos Campos, SP, Brazil

Marcos Pellegrini Bandini Civil Defense of Santos, Santos, São Paulo, Brazil

José Marques Carriço Municipality of Santos, Municipal Secretary of Government/Post-graduation Program in Law and the Architecture and Urbanism Course, Catholic University of Santos (UNISANTOS), Santos, SP, Brazil

Fernanda R. F. Carvalho Remote Sensing Division, National Institute for Space Research (INPE), São José dos Campos, SP, Brazil

Sin-Chan Chou National Institute for Space Research (INPE), São José dos Campos, SP, Brazil

Thiago Bezerra Corrêa Ramboll Environment & Health, São Paulo, SP, Brazil

Antonio Carlos da Silva Oscar Júnior Geography Institute, Rio de Janeiro State University (UERJ), Rio de Janeiro (RJ), Brazil

Pamela Pereira de Araújo Municipality of Santos, Trainee in Municipal Secretary of Environment, Santos, SP, Brazil

Ricardo de Camargo Institute of Astronomy, Geophysics and Atmospheric Sciences, University of Sao Paulo, São Paulo, SP, Brazil

Danielle Almeida de Carvalho São Paulo State University (UNESP), Institute of Biosciences, Coastal Campus, São Vicente, SP, Brazil

Débora M. de Freitas São Paulo State University (UNESP), Institute of Biosciences, Coastal Campus, São Vicente, SP, Brazil

Tiago Zenker Gireli School of Civil Engineering, Architecture and Urban Design, University of Campinas (Unicamp), Campinas, SP, Brazil

Roberto Greco Institute of Geosciences, University of Campinas (Unicamp), Campinas, SP, Brazil

Joseph Harari Institute of Oceanography, University of São Paulo, São Paulo, SP, Brazil

Eduardo Kimoto Hosokawa Municipal Government of Santos, Secretariat of Urban Development, Santos, SP, Brazil

José Chambel Leitão HIDROMOD, Porto Salvo, Portugal

Paulo Chambel Leitão HIDROMOD, Porto Salvo, Portugal

André Antonio Lyra National Institute for Space Research (INPE), São José dos Campos, SP, Brazil

José A. Marengo National Center for Monitoring and Early Warning of Natural Disasters (CEMADEN), São José dos Campos, São Paulo, Brazil

Fabiano de Araujo Moreira Department of Geography, University of Campinas (Unicamp), Campinas, SP, Brazil

Frank Muller-Karger Institute for Marine Remote Sensing/IMaRS, College of Marine Science, University of South Florida, St. Petersburg, FL, USA

Lucí H. Nunes Santos, São Paulo, Brazil

Beatriz Barbi de Oliveira Santos Azul Brazilian Airline, São Paulo, Brazil

Shona Paterson Future Earth Coasts, Cork, Ireland

Mark Pelling Department of Geography, King's College London, London, UK

Vinícius de Carvalho Neiva Pinheiro São Paulo State University (UNESP), Institute of Biosciences, Coastal Campus, São Vicente, SP, Brazil

Catherine J. Reynolds International Ocean Institute, College of Marine Science, University of South Florida, St. Petersburg, FL, USA

Renan Braga Ribeiro Núcleo de Pesquisas Hidrodinâmicas – Universidade Santa Cecília (NPH-UNISANTA), Santos, SP, Brazil

Wilson Cabral de Sousa Júnior Water and Environments Department, Civil Engineer Division, Technological Institute of Aeronautics, São José dos Campos, Brazil

Matheus Souza Ruiz Núcleo de Pesquisas Hidrodinâmicas – Universidade Santa Cecília (NPH-UNISANTA), Santos, SP, Brazil

Alexandra Franciscatto Penteado Sampaio Núcleo de Pesquisas Hidrodinâmicas – Universidade Santa Cecília (NPH-UNISANTA), Santos, SP, Brazil

Adan J. Silva National Institute for Space Research (INPE), São José dos Campos, SP, Brazil

Celia Regina de Gouveia Souza Institute of Geology – Secretariat for the Environment of the State of São Paulo (IG-SMA/SP), São Paulo, SP, Brazil

Post-Graduate Programme on Physical Geography – Faculty of Philosophy, Languages and Human Sciences, University of São Paulo (FFLCH-USP), São Paulo, SP, Brazil

Agenor Pereira Souza Environmental consultant, São Paulo, São Paulo, Brazil

Priscila Tavares National Institute for Space Research (INPE), São José dos Campos, SP, Brazil

Vitor Baccarin Zanetti Water and Environments Department, Civil Engineer Division, Technological Institute of Aeronautics, São José dos Campos, Brazil

Acronyms

CEMADEN	National Center for Monitoring and Early Warning of Natural Disasters
UNICAMP	University of Campinas
IMaRS	Institute for Marine Remote Sensing
INPE	National Institute for Space Research
IG-SMA/SP	Institute of Geology – Secretariat for the Environment of the State of São Paulo
FFLCH-USP	Faculty of Philosophy, Languages and Human Sciences – University of São Paulo
USP	University of São Paulo
NPH-UNISANTA	Núcleo de Pesquisas Hidrodinâmicas – Universidade Santa Cecília
UERJ	Rio de Janeiro State University
UNISANTOS	Catholic University of Santos
ITA	Technological Institute of Aeronautics
UNESP	Universidade Estadual Paulista

Part I
Introduction

Chapter 1
The METROPOLE Project – An Integrated Framework to Analyse Local Decision Making and Adaptive Capacity to Large-Scale Environmental Change: Decision Making and Adaptation to Sea Level Rise in Santos, Brazil

José A. Marengo, Frank Muller-Karger, Mark Pelling, and Catherine J. Reynolds

Abstract Assessment of the risks due to exposure and sensitivity of coastal communities to coastal flooding is essential for informed decision-making. Strategies for public understanding and awareness of the tangible effects of climate change are fundamental in developing policy options. A multidisciplinary, multinational team of natural and social scientists from the USA, the UK, and Brazil developed the METROPOLE Project to evaluate how local governments may decide between adaptation options associated with SLR projections. METROPOLE developed a participatory approach in which public actors engage fully in defining the research problem and evaluating outcomes.

Using a case study of the city of Santos, in the coast of the State of Sao Paulo in Southeastern Brazil, METROPOLE developed a method for evaluating risks jointly

J. A. Marengo (✉)
National Center for Monitoring and Early Warning of Natural Disasters (CEMADEN), São José dos Campos, São Paulo, Brazil
e-mail: jose.marengo@cemaden.gov.br

F. Muller-Karger
Institute for Marine Remote Sensing/IMaRS, College of Marine Science, University of South Florida, St. Petersburg, FL, USA
e-mail: carib@usf.edu

M. Pelling
Department of Geography, King's College London, London, UK
e-mail: mark.pelling@kcl.ac.uk

C. J. Reynolds
International Ocean Institute, College of Marine Science, University of South Florida, St. Petersburg, FL, USA
e-mail: cjreynolds@usf.edu

L. H. Nunes et al. (eds.), *Climate Change in Santos Brazil: Projections, Impacts and Adaptation Options*, https://doi.org/10.1007/978-3-319-96535-2_1

with the community, comparing 'no-action' to 'adaptation' scenarios. At the core of the analysis are estimates of economic costs of the impact of floods on urban real estate under SLR projections through 2050 and 2100. Results helped identify broad preferences and orientations in adaptation planning, which the community, including the Santos municipal government, co-developed in a joint effort with natural and social scientists.

Keywords Sea-level rise · Adaptation options · Adaptive capacity · Participatory approach · Santos · METROPOLE project

1.1 Introduction

How decision makers and the public perceive and respond to potential local impacts of large-scale change, including economic and health risks, depends on social, cultural and political context and on how scientific evidence is presented. The METROPOLE (*An Integrated Framework to Analyse Local Decision Making and Adaptive Capacity to Large-Scale Environmental Change)* project focuses on identifying the factors that facilitate a shift in knowledge, attitudes, values and decision making about local climate risks and adaptation strategies among decision-makers and stakeholders in case study of coastal communities in Brazil, United Kingdom, and the United States (Marengo et al. 2017a, b; Paterson et al. 2017). The Belmont Forum-G8 Initiative Collaborative Research through grants from the FAPESP-Sao Paulo State Research Foundation, the US National Science Foundation and UK Natural Environment Research Council and Economic and Social Research Council, supported this project. The goal was to combine expertise from diverse scientific backgrounds to develop an integrated framework to analyse local decision-making and the adaptive capacity to local environmental change driven by large-scale processes like climate.

The hypothesis is that risk knowledge is best understood as being co-produced by science and by the social, political and cultural context. The research developed at METROPOLE concurrently analysed social context factors that affect adaptation planning and policy changes (adaptive capacity) and responses by local stakeholders when presented with interactive computer-based scenario simulations in participatory planning meetings to cope with sea level rise (SLR) and coast flooding. The project used (1) state-of-the-art visualization tools developed in the US and Brazil, (2) sophisticated survey and choice evaluation tools, and (3) a risk assessment Adaptive Capacity Index developed in the UK. The visualization tools integrate scientific and economic data at the community level for local jurisdictions, and illustrate potential impacts, economic risk, adaptation options, and cost-benefits analyses projected over time. The data included changes in sea level, temperature, storm frequency, precipitation and other variables in the past decades and high resolution projections in 5–10 year increments to 2100 under the IPCC AR5 sea level scenarios.

Expected results include a new framework to evaluate the impact of integrating scientific, economic, and cultural context data on adaption planning and decision-making. This will improve the ability of scientists to interact with stakeholders by developing an understanding of social context, and garner knowledge of best practices for the Brazilian case study.

1.2 Background

Nearly 7% of all human communities have developed in areas where the elevation is less than 5 m. from historical sea level (McGranahan et al. 2007; Wong et al. 2014; Reguero et al. 2015). Most of the world's 60 million poor people living in low elevation areas reside in just 15 countries, including Brazil, the United States of America and the United Kingdom (Barbier 2015). Urban low elevation coastal zones are expanding faster than elsewhere and this trend is expected to continue into the future (Seto et al. 2011). Globally, between 660,000 and 1,200,000 km^2 of land, 93–310 million inhabitants and 3100–11,000 billion USD of built capital are currently located at elevations less than the present 100-year sea level flooding event, but by mid-century there could be an increase in global flood losses for the 136 largest coastal cities (Reguero et al. 2015). Of these, approximately 40 million are exposed to major flood risks, and these risks are expected to increase over the next 50 years.

One of the clear signals of present climate change is sea level rise. There is mounting evidence of other changes, including warmer temperatures in many localities, and changes in the intensity and frequency of extreme meteorological events, including wind, rain, storms and waves that can generate coastal flooding. Damages are compounded when tidal fluctuations and surges due to severe storms are superimposed on these estimates. A rising sea level combined with high tides and storm surges is expected to impact the human built environment along coastal zones of the world as well as coastal ecosystems such as wetlands, coral reefs, beaches, and estuaries. Higher sea level typically leads to increase in coastal erosion, high risk of flooding, and contamination of fresh water sources through saltwater intrusion (Mcleod et al. 2010). Many of these coastal ecosystems are already impacted by human uses that have weakened their resilience (Hinkel et al. 2010). Extremes in meteorological events can also lead to erosion of coastal areas, landslides, and floods. These can have direct impacts on human communities, but also important indirect impacts through changes in biodiversity and other ecosystem services (IPCC 2014).

Sea-level rise is a tangible and tractable effect of climate change that poses significant challenges to society from the next 50 to 100 years, or earlier (Hauer et al. 2016). Global mean sea level rose by 0.19 (0.17–0.21) mm $year^{-1}$ over the period 1901–2010 based on historical tide gauge records; these rates are observed globally on average, as measured using satellite data collected since 1993. Between 1993 and 2010, the average global sea level rise rate was near 3.2 (2.8–3.6) mm $year^{-1}$.

Similarly high rates likely occurred between 1920 and 1950 (IPCC 2014). A gradual increase in average sea level of 1 m often cited as a possible scenario within a 100-year timeframe (Rhein et al. 2013; Wong et al. 2014), would seriously affect some coastal populations in Brazil (ECLAC 2011), the USA, and the UK (GEI Consultants 2015, 2016).

1.3 The METROPOLE Project

To improve resilience, policymakers need to understand current adaptation processes and obstacles, and plan accordingly to be effective. These processes depend in great measure on how decision-makers and the public perceive and respond to changes and the perception of risk. In order to evaluate how local government may respond to risks associated with sea level rise projections, a group of natural and social scientists from the United States (US), United Kingdom (UK), and Brazil developed the *METROPOLE* Project.

The METROPOLE study goals were to determine to what extent stakeholder beliefs, values, and preferences regarding adaptation options and funding choices may facilitate or hinder adaptation. The METROPOLE project encompassed a three-part, integrated environmental, economic, and social analysis embedded in a municipal planning effort involving stakeholders and decision makers in Brazil, the UK and the US. The first part included the use of the COastal Adaptation to Sea Level Rise Tool (COAST) model (Catalysis Adaptation Partners 2015) to show visualizations of SLR, infrastructure impacts, costs/benefits for adaptations, and small group discussions to define stakeholder estimates for action. The second piece involved administering pre- and post-workshop surveys to participants, to identify links between risk experiences, beliefs, values and attitudes about local government priorities for possible adaptation actions and public financing, and to assess change after seeing the COAST visualizations and discussing scenarios. The third element was the Adaptive Capacity Index (ACI), an assessment of institutional and individual interactions that shape local and regional adaptive capacity. The project was conducted in: the city of Santos (state of São Paulo, Brazil), city of Selsey (West Sussex, United Kingdom), and cities in Broward County (Florida, United States). The three study sites present both similarities (coastal cities threatened by the risk of sea-level rise) and differences (the size of the population and of the local economy), which make the comparison of the attitudes and values about local climate risks and adaptation strategies among decision-makers and stakeholders particularly challenging. This paper focuses on the Brazilian component in the city of Santos.

METROPOLE used the approach of Daniels and Walker (2001) and Burch (2010)) to explore the complex issue of how communities of different cultural backgrounds respond to risk and adaptation related to climate change. The IPCC defined this as the process of adjustment to actual or expected climate and its effects, including either moderate harm, or the opportunity to exploit beneficial opportunities. For this study, the IPCC Glossary (IPCC 2012) was adopted to establish the theoretical framework for adaptation and evaluation of risks, hazards, and vulnerability. The exception is that in the context of METROPOLE, "mitigation" means risk management or reduction of risk due to a hazard, and not reduced emissions of greenhouse gases.

The METROPOLE project involves developing visualization tools that integrate scientific information and socioeconomic data at the municipal level in each country, and illustrate the potential impacts, economic risks, adaptation options and analyses of cost-benefit projected over time. The central theme of the METROPOLE project is, therefore, to show, in an integrated way, how some coastal areas under different climate regimes and human pressures would be affected by SLR caused by climate change, and if society and the government would be prepared or not to take proper and fast adaptation measures. Recent studies on climate adaptation enhance the importance of engaging or activating communities and supporting community roles in understanding climate change and adaptation needs (Ross et al. 2015). Responses from cities to improve their resilience are urgent but policymakers need to understand current adaptation to plan comprehensively and spend effectively (Georgeson et al. 2016).

On the three case studies, Santos is a coastal city in the State of Sao Paulo in southeastern Brazil. Santos and areas adjacent to the city is strategic economic center for Brazil, with a large concentration of industries located along the coastal zone. The Port of Santos, the key economic asset for the municipality of Santos, is responsible for the transport of products from the largest industrial park in Brazil and handles some 25% of Brazil's foreign trade. Founded in 1546, Santos is one of the oldest settlements of Brazil. It occupies an area of 281 km^2, of which 39.4 km^2 lies in the insular domain (São Vicente Island) and 231.6 km^2 correspond to the mainland part of the municipality (Marengo et al. 2017a). Around 99.3% of the Santos population live in the insular domain. The Port of Santos alone is responsible for the transport of products from the largest industrial park in Brazil, handling around 25% of Brazil's foreign trade (ICF-GHK 2012).

The municipality of Santos is also a portrait of the social asymmetry of the country, featuring upscale neighbourhoods especially closer to the shoreline and poor neighbourhoods concentrated at the Northwestern Zone of the island, on the hill-slopes and the wet lowlands. The irregular occupation of the hill-slopes and mangroves, the pollution generated by industries located around the area, and deforestation of the Atlantic Rainforest, which reduced the water retention capacity of the soil and increased the continental runoff, accelerated common processes in the area such as landslides, mudslides and floods, putting a large contingent of the population under constant threat. The region is affected by tropical, subtropical and mid-latitude weather systems. During summer, when convective activity is greater, the SACZ (South Atlantic Convective Zone) influences the rainfall regime, with a cloud band and rainfall remaining semi-stationary for several days. Frontal systems are common in the area, mainly during autumn-winter. Storm surges have historically affected this region. Nowadays, storm surges typically cause destruction or urban infrastructure and damages related to traffic interruption at the southeastern ending of the Santos coastline (ICF-GHK 2012).

1.4 Sea Level Rise and Adaptation Options: The Case of Santos, Sao Paulo/Brazil

For the Latin America and the Caribbean Atlantic coast, SLR between 1950 and 2008 rose ~2 mm $year^{-1}$ (Losada et al. 2013) and the Brazilian coastal areas are being affected by coastal erosion and coastal inundation (Souza et al. 2005; PBMC 2014), with the southern part of the State of São Paulo and Rio de Janeiro seeing rates of between 1.8 and 4.2 mm $year^{-1}$ since the 1950s (Alfredini et al. 2013; Harari et al. 2007). According to (Harari and Camargo 1995), the mean sea level in Santos has risen at a rate of 1.2 mm $year^{-1}$ since 1940s, with an increasing trend in the past decade. This rate is slower than the global mean sea level rise rate. This suggests that additional factors such as estuarine circulation patterns, land subsidence and/or anthropogenic interferences like dredging may be affecting local sea level rate.

Flooding events normally occur in two main areas of Santos: at the Northwestern Zone NWZ and at the Southeastern Zone SEZ. The latter is closer to the mouth of the Santos estuarine channel, along the seafront (Marengo et al. 2017a). At the NWZ, flooding is caused by the combination of heavy rainfall and high tides, so that the waters from the watersheds discharges summed to the superficial continental runoff flowing down towards the estuarine channel are blocked by the tidal waters rising into the existing drainage system. At the SEZ, flooding is caused by coastal inundation related to storm and tidal surges, in general associated to extra-tropical cyclones passage. SLR may affect the Port of Santos, the key economic asset for the municipality. Most sections are built to withstand water level increases of as much as 3 m above the current mean sea level (ICF-GHK 2012). The interaction between SLR and flood frequency may be of greatest concern.

The METROPOLE project incorporated an evaluation of risks and impacts of SLR and tested the city's Adaptive capacity. SLR risks were estimated using the COAST platform (Merrill et al. 2008; Marengo et al. 2017a). The METROPOLE project incorporated an evaluation of risks and impacts of SLR and tested the city's Adaptive capacity. SLR risks were estimated using the COAST platform. COAST estimates SLR and storm surge impacts by calculating damage from storm surge events cumulatively over time, given a changing base water level. It then calculates relative benefits of various adaptation scenarios in terms of cumulative *avoided* damages over time. The model is intended to be used by municipalities, state agencies and other groups interested in benefit-cost analysis for adaptation strategies aimed at reducing damages from SLR and storm surge.

COAST allows users to (a) calculate how much building damage may be avoided over time if such strategies are implemented; (b) confirm whether the projected benefits outweigh the costs; and (c) evaluate which strategies seem the most cost-effective. The process includes creation of data projections for coastal changes and potential economic and environment consequences of flooding scenarios resulting from a rise in sea level and extreme storm surge events. Financial ramifications are calculated using inputs from local stakeholders and confirmed by expert engineering review. Use of the tool in public process also allows stakeholders to articulate and modify potential adaptation strategies. The conceptual model developed for application of COAST in both the NWZ and SEZ affected by hydro-meteorological risks in Santos is shown in Marengo et al. (2017a) . Data required to run COAST for the two case studies (district, land area, construction area, assessed value of the land, value of construction and value of the property) was obtained from the Municipality of Santos.

To advance the understanding of connections between stakeholder beliefs, values and preferences regarding adaptation options and funding choices, as well as to improve understanding of barriers to adaptation in Santos, decision makers, citizens, representatives of public and private sectors and of NGOs of Santos were engaged with the METROPOLE process through two stakeholder workshops. These workshops were directed to engage decision makers, citizens, and representatives of the public and private sectors to develop and evaluate adaptation options for NWZ and SEZ of Santos. To create the data for the workshops, our team and municipal managers reviewed the estimated SLR/flood risks and discussed potential adaptation actions. After consulting with other staff and elected officials, the municipal managers selected several realistic and potentially useful combinations of actions to be discussed by stakeholders at Workshop 1. The workshops presented and discussed maps of future flooding projections due to sea level rise for 2050 and 2100. Workshop participants were shown the respective estimates of economic damages to real estate for the SEZ and NWZ zones of Santos. The small group discussions at these workshops focused on adaptation options for the community of Santos. As a result of these workshops, participants coming from various sectors of society living in Santos proposed some adaptation measures, that were later on incorporated in the COAST model (Marengo et al. 2017a b).

For the city of Santos, the most preferred adaptation options were fortification (66%) and accommodation (30%). For the NWZ, the fortification (50%) and accommodation (43%) actions were also preferred, while relocation was the least preferred option, with 4% in the SEZ and 7% in the NWZ (Marengo et al. 2017a). Table 1.1 shows that the adaptation measures selected by participants (i.e. fortification and accommodation) would be cost effective in both the lower scenario of sea level rise (0.36 m; for the period 2010–2100) and for the higher scenario (0.45 m by 2100). Santos adopted dredging and mangrove restoration for the northwestern sector, while for the southeastern sector reinforcement of existing walls and beach nourishment were the selected options. The community evaluated beach nourishment, dune restoration, structural enforcement of existing sea-walls, water pumping and improvement of tide control gates in existing drainage canals. In the southeastern sector of the city the damages from a 100-year flood in 2050 under low and high SLR scenarios would be US$34 and US$38 million, while in 2100 these damages could reach US$60 and US$75 million, respectively (Table 1.1).

However avoided damages for the NWZ might be greater than was modelled because of challenges encountered in obtaining accurate real estate valuation data from the Municipality of Santos. According to the Municipality of Santos, the NWZ shows buildings up to four floors, and there are a lot of commerce activities (stores,

Table 1.1 Cumulative damages without and with adaptation scenarios in Santos. Scenarios for a 100-year flood with SLR, for 2010–2100, with fortification (beach nourishment + dune restoration, structural enforcement/improvement of existing sea-walls) for SEZ

Item	Santos, Brazil	Santos, Brazil
	SLR scenario	SLR scenario
	Low	High
Cumulative damage without adaptation (billions USD)	$0.242	$0.467
Cumulative damage with adaptation (billions USD)	0	0
Avoided damages (benefit, billions USD)	$0.242	$0.467
Cost of adaptation measures (millions USD)	$10.1	$10.1
Benefit-cost ratio	23.9	28.7

Modified from Marengo et al. (2017b)

supermarkets, restaurants, offices). The SEZ zone is mostly upper-level class residential with apartment buildings, hotels and shops that can reach up to 20 floors or more on the seaside.

Commercial buildings were not included in the list of properties. This represents an underestimate of risk and of damages. COAST also did not estimate damage from winds, erosion forces, or rainwater drainage backups, which might also cause building damage, substantially threaten health and other property, and changes local ocean circulation, salinity, etc. that may in turn also affect local sea level. Natural sedimentary processes in tidal flats and the probable expansion and/or retraction of mangroves into the estuarine system, or the resilience of these ecosystems as well as impacts to public services, urban infrastructure, or business interruptions or clean-up costs after extreme weather events were not included, Data from commercial properties are not geo-referenced and could not be included in the COAST running.

1.5 Adaptive Capacity Index (ACI) Analysis Sao Paulo/Brazil

The Adaptive Capacity Index (ACI) has been developed to provide a theoretically grounded measurement tool and coupled analytical framework that can help practitioners and researchers surface the negotiated pathways through which adaptive capacity accrues and is deployed within administrative regimes (Paterson et al. 2017). Evaluating the ACI is an analytical process that engages a range of stakeholders in a comprehensive and thoughtful assessment of community resilience and ability to proactively seek change. The METROPOLE Project focused the ACI assessment on key actor organisations within the environmental risk scene. Respondents commented on contemporary capacities for the 2015 timeframe and for two others (2005 and 2010), in order to develop a trend analysis and compare current and historical shifts in adaptive capacity. Developing the ACI required conducting surveys. The METROPOLE team specialists conducted 24 interviews in Santos. The interviewees included a broad cross-section of government, civil, and private sector actors, representative institutions and organizations active in evaluating risk scenarios in Selsey, Broward County and Santos.

Table 1.2 presents the overall ACI scores for Santos evaluated over the 2005–2015 decade (Marengo et al. 2017b; Paterson et al. 2017). The results demonstrate general variability with some subcomponents showing limited variation over time, with some factors increasing and others decreasing over the decade under examination. Santos demonstrated consistent and progressive increases in adaptive capacity between 2005 and 2015. The community's response was to attribute any trends to changing management responsibilities, shifts in political ideology, and social contracts. Overall respondents from Santos returned higher results, closer to outstanding (strong formal capacity with integrated and strategic planning across sectors. A decline in the ACI in a category was attributed to a decrease in available resources, the inability to control existing resources, and to austerity measures that have been in place and enforced at a national level. A positive trend was observed in the ability to learn and to plan for the future, suggesting that collaborative action between organizations increased in the 2005–2015 timeframe. Respondents pointed to the importance of partnership funding and co-funding mechanisms now required for capital projects as a catalyst for closer working relationships between the different levels of government. It was also pointed out that the re-organization of the structures of major governmental agencies has provided greater interconnectedness between these agencies, promoting cross learning and joint planning efforts.

In Santos, the ACI analysis showed some progression toward adaptability, but this was slow or stagnant in some factors (Table 1.2). Lack of progression across the components of the adaptive capacity index from the perspective of local actors was accounted for through: (i) a lack of organisational integration and (ii) the dominance of the adaptation agenda by Civil Defence. This suppressed leadership and innova-

Table 1.2 Overall adaptive capacity index (ACI) sub-components scores for Santos-Brazil

	Santos, Brazil		
Subcomponents	2005	2010	2015
Critical self-reflection	3.03	3.14	3.41
Ability to experiment	2.69	2.69	2.94
Ability to learn	2.91	3.12	3.41
Ability to plan for the future	2.99	3.17	3.43
Command over available resources	2.37	2.59	3.01
Organizational responsibility	3.11	3.33	3.61
Organizational architecture	2.77	2.91	3.18
Levels of capital	2.99	3.19	3.40
	22.86	24.14	26.39

Modified from Marengo et al. (2017b)

tion especially between agencies. Global economic pressures were also felt by Santos. While the global economic downturn of 2008 had a limited effect on the Brazilian economy at the time, greater impacts were noted post 2014 with more constrained resources and funding opportunities reported across all sectors and agencies. Together these pressures served to hold Santos' adaptive capacity. In Santos, the private sector expressed high levels of adaptive capacity across many components with the exception of 'command over available resources'. This suggested that adaptive capacity and the development agenda in Santos was driven much more by the private sector than either the civil or government sector (Paterson et al. 2017).

Experience from previous recessions shows that economic downturn has a detrimental effect on getting resources to environment issues in the country because resources to cope with economic recession might be considered more urgent, with climate change initiatives neglected. But recent severe storm surges in the city (April, August and October 2016), which affected the study area harshly, might contribute to keeping adaptation actions on the political agenda (Marengo et al. 2017b).

1.6 Conclusions and Final Remarks

Projections of SLR impacts from the COAST model under the no-action scenarios provided an initial estimate of the possible cost of SLR and storm surge through 2100 for some key regions of Santos. Model runs with adaptation options in place indicated the possible efficiency of these means of addressing the challenges of SLR and storm surge in the city. Model results provide useful information for Santos, a strategic area of the country, and especially for the Port of Santos, the most important coastal economic resource in Brazil. However more research is be needed to better understand possible impacts the adaptation options evaluated may have on local revenue over time.

As in the other regions where COAST has been used, in Santos, evaluations about priorities and community values were particularly meaningful and useful outcomes from the study. In recognition that the municipality of Santos needs to be creative about its future, most attendees wanted to explore feasible possibilities using structural and ecological options.

Like for many other coastal areas of the world, in Santos, the threat of SLR appears to be an issue when combined with the threat of extreme rainfall events and storm surges. Model results for this important Brazilian coastal seaport city showed that SLR alone is not likely to inundate buildings, but the Santos Municipality should consider SLR when planning for protection from large flood events. The benefit-cost ratio for elevation was highly positive for the SE Zone. Although these benefits could theoretically be obtained immediately, decisions around most adaptation strategies require substantial public support and will take years or decades to develop and implement.

The ACI analysis for Santos shows that there are choices for increasing resiliency against sea level rise that have a high benefit-cost ratio. Specifically, these communities found that fortification of the coast, elevation of real estate, and flood proofing are important investments that help mitigate higher costs due to flood risks 60–100 years from now. Based on the surveys' results, there is a clear sense from each community of the risk of floods due to intense storms, and of the benefits of engaging as a community early in a process that helps people to understand risks, impacts, and costs A majority of workshop participants prioritized pursuing physical infrastructure and green infrastructure actions now or within the coming years or decades. This will have benefits that include a social and natural science basis and that address the local and regional social, political and cultural context. Assessment of the risks due to exposure and sensitivity of coastal communities to coastal flooding is essential for informing decision-making. Strategies to promote public understanding and awareness of the tangible effects of climate change are fundamental in developing policy options.

The multidisciplinary, multinational nature of METROPOLE was key in providing local governments with scientific evidence of the need for adaptation to SLR. At the core of the analysis are estimates of economic costs of the impact of floods on urban real estate under SLR for middle- and long-term twenty-first century projections. The institutional arrangement articulated in the document involves city and state government, civil society, the private sector, and universities and research institutes in Santos. It is an important step towards efficient multilevel governance processes aiming to cope with risks associated with climate change. The METROPOLE project has helped Santos demonstrate innovation in governance issues and provide valuable lessons to other coastal locations in Brazil.

References

Alfredini, P., Arasaki, E., Pezzoli, A., & Fournierm, C. P. (2013). Impact of climate change on the Santos Harbor, São Paulo State (Brazil), TransNav. *International Journal on Marine Navigation and Safety of Sea Transportation, 7*(4), 609. https://doi.org/10.12716/1001.07.04.17.

Barbier, E. (2015). Hurricane Katrina's lessons for the world. *Nature, 524*, 285–287.

Burch, S. (2010). Transforming barriers into enablers of action on climate change: Insights from three municipal case studies in British Columbia, Canada. *Global Environmental Change, 20*(2), 287–297. https://doi.org/10.1016/j.gloenvcha.2009.11.009.

Catalysis Adaptation Partners. (2015). *COastal Adaptation to Sea level Rise Tool (COAST) version 3.0.* Catalysis Adaptation Partners, LLC, South Portland, ME, licensed through GEI Consultants, Inc.

Daniels, S. E., & Walker, G. B. (2001). *Working through environmental conflict: The collaborative learning approach.* Westport: Praeger Publishers.

ECLAC. (2011). *Efectos del Cambio Climático en la Costa de América Latina y el Caribe: Dinámicas, Tendencias y Variabilidad Climática* [Effects of Climate Change on the Coast of Latin America and the Caribbean: Dynamics, Trends and Climate Variability]. LC/W.447, the Economic Commission for Latin America and the Caribbean (ECLAC), the Spanish Climate Change Office under the Ministry of Agriculture, Food and Environment of the Government of Spain, and the Environmental Hydraulics Institute of Cantabria University, ECLAC, Santiago de Chile, Chile, 263 pp.

GEI Consultants. (2015). Final report to the University of South Florida, COAST Modeling for Broward County, Florida, USA, for the Belmont Forum-GB Initiative Collaborative Research Initiative, METROPOLE: An integrated framework to analyze local decision making and adaptive capacity to large-scale environmental change. Portland.

GEI Consultants. (2016). Final report to the University of South Florida, COAST Modeling for Selsey, U.K., for the Belmont Forum-GB Initiative Collaborative Research Initiative, METROPOLE: An integrated framework to analyze local decision making and adaptive capacity to large-scale environmental change. Portland.

Georgeson, L., Maslin, M., Poessinouw, M., & Howard, S. (2016). Adaptation responses to climate change differ between global megacities. *Nature Climate Change, 6*, 584.

Harari, J., & Camargo, R (1995) *Tides and mean sea level variabilities in Santos (SP), 1944 to 1989*. Internal Report of the Oceanographic Institute of the University of Sao Paulo, Sao Paulo, Brazil, n° 36, 15 p.

Harari, J., França, C. A. S., & Camargo, R. (2007). Variabilidade de longo termo de componentes de maré e do nível médio do mar na costa brasileira. *Afro-America Gloss News Edição, 11*(1). http://www.mares.io.usp.br/aagn/aagn11/ressimgfversao2008.pdf.

Hauer, M. E., Evans, J. M., & Mishra, D. R. (2016). Millions projected to be at risk from sea-level rise in the continental United States. *Nature Climate Change*, PUBLISHED ONLINE: 14 MARCH 2016l. https://doi.org/10.1038/NCLIMATE2961

Hinkel, J. J., Nicholls, R. J., Vafeidis, A. T., Tol, R. S. J., & Avagianou, T. (2010). Assessing risk of and adaptation to sea-level rise in the European Union: An application of DIVA. *Mitigation and Adaption Strategies for Global Change, 15*, 703–719.

ICF-GHK. (2012). *Climate change adaptation planning in Latin American and Caribbean Cities*. Complete Report: Santos, Brazil, London EC1R 5BL.

IPCC. (2012). Glossary of terms. In C. B. Field, V. Barros, T. F. Stocker, D. Qin, D. J. Dokken, K. L. Ebi, M. D. Mastrandrea, K. J. Mach, G.-K. Plattner, S. K. Allen, M. Tignor, & P. M. Midgley (Eds.), *Managing the risks of extreme events and disasters to advance climate change adaptation, A Special Report of Working Groups I and II of the Intergovernmental Panel on Climate Change (IPCC)* (pp. 555–564). Cambridge/New York: Cambridge University Press.

IPCC. (2014). Summary for policymakers. In C. B. Field, V. R. Barros, D. J. Dokken, K. J. Mach, M. D. Mastrandrea, T. E. Bilir, M. Chatterjee, K. L. Ebi, Y. O. Estrada, R. C. Genova, B. Girma, E. S. Kissel, A. N. Levy, S. MacCracken, P. R. Mastrandrea, & L. L. White (Eds.), *Climate change 2014: Impacts, adaptation, and vulnerability. Part A: Global and sectoral aspects, Contribution of Working Group II to the Fifth Assessment Report of the Intergovernmental Panel on Climate Change* (pp. 1–32). Cambridge/New York: Cambridge University Press.

Losada, I. J., Reguero, B. G., Méndez, F. J., Castanedo, S., Abascal, A. J., & Mínguez, R. (2013). Long-term changes in sea-level components in Latin America and the Caribbean. *Global and Planetary Change, 104*, 34–50. https://doi.org/10.1016/j.gloplacha.2013.02.006.

Marengo, J. A., Nunes, L. H., Souza, C. R. G., Harari, J., Muller-Karger, F., Greco, R., Hosokawa, E. K., Tabuchi, E. K., Merrill, S. B., Reynolds, C. J., Pelling, M., Alves, L. M., Aragão, L. E., Chou, S. C., Moreira, F., Paterson, S., Lockman, J. T., & Gray, A. G. (2017a). A globally-deployable strategy for co-development of adaptation preferences to sea-level rise: The public participation case of Santos, Brazil. *Natural Hazards, 88*(1), 39–53.

Marengo, J. A., Muller-Karger, F., Pelling, M., Reynolds, C. J., Merrill, S. B., Nunes, L. H., Paterson, S., Gray, A. J., Lockman, J. T., Kartez, J., Moreira, F., Greco, R., Harari, J., Souza, C. R., Alves, L. M., Hosokawa, E., & Tabuchi, E. (2017b). An integrated framework to analyze local decision making and adaptation to sea level rise in coastal regions in Selsey (UK), Broward County (USA), and Santos (Brazil). *American Journal of Climate Change, 6*, 403–424.

McGranahan, G., Balk, D., & Anderson, B. (2007). The rising tide: Assessing the risks of climate change and human settlements in low elevation coastal zones. *Environment and Urbanization, 19*, 17–37.

Mcleod, E., Poulter, B., Hinkel, J., Reyes, E., & Salm, R. (2010). Sea-level rise impact models and environmental conservation: A review of models and their applications. *Ocean & Coastal Management, 53*, 507–517.

Merrill, S., Sanford, S. R., & Lapping, M. (2008). Planners and climate change action: An approach for communities. *Maine Policy Review, 17*, 142–152.

Paterson, S. K., Pelling, M., Nunes, L. H., Moreira, F. A., Guida, K., & Marengo, J. A. (2017). The scaled asymmetries of adaptive capacity: In Florida, São Paulo and England. *Geoforum, 81*, 109–119, https://doi.org/10.1016/j.geoforum.2017.02.014.

PBMC. (2014). In T. Ambrizzi & M. Araujo (Eds.), *Base Científica das Mudanças Climáticas. Contribuição do Grupo de Trabalho 1 do Painel Brasileiro de Mudanças Climáticas ao Primeiro Relatório da Avaliação Nacional sobre Mudanças Climáticas*. Rio de Janeiro: COPPE. Universidade Federal do Rio de Janeiro, 464 pp.

Reguero, R. G., Losada, I. J., Díaz-Simal, P., Méndez, F. J., & Beck, M. W. (2015). Effects of climate change on exposure to coastal flooding in Latin America and the Caribbean. *PLoS One*. https://doi.org/10.1371/journal.pone.0133409.

Rhein, M., Rintoul, S. R., Aoki, S., Campos, E., Chambers, D., Feely, R. A., Gulev, S., Johnson, G. C., Josey, S. A., Kostianoy, A., Mauritzen, C., Roemmich, D., Talley, L. D., & Wang, F. (2013). Observations: Ocean. In T. F. Stocker, D. Qin, G.-K. Plattner, M. Tignor, S. K. Allen, J. Boschung, A. Nauels, Y. Xia, V. Bex, & P. M. Midgley (Eds.), *Climate change 2013: The physical science basis, Contribution of Working Group I to the Fifth Assessment Report of the Intergovernmental Panel on Climate Change*. Cambridge: Cambridge University Press.

Ross, D., Shaw, S., Rissik, D., Cliffe, N., Chapman, A., Hounsell, V., Udy, J. N. T., & Schoeman, J. A. (2015). Participatory systems approach to understanding climate adaptation needs. *Climatic Change, 129*, 27–42.

Seto, K. C., Fragkias, M., Güneralp, B., & Reilly, M. K. (2011). A meta-analysis of global urban land expansion. *PLoS, 1*. Available at: http://journals.plos.org/plosone/article?id=10.1371/journal.pone.0023777 Consulted in Dec 2015.

Souza, C. R., Souza Filho, P. V., Esteves, S. L., Vital, H., Dillemburg, S. R., Patchineelam, S. M., & Addad, J. E. (2005). Sandy beaches and coastal erosion. In C. R. d. G. Souza et al. (Eds.), *Quaternário do Brasil* (pp. 130–152). Ribeirão Preto: Holos, Editora.

Wong, P. P., Losada, I. J., Gattuso, J. P., Hinkel, J., Khattabi, A., McInnes, K. L., Saito, Y., & Sallenger, A. (2014). Coastal systems and low-lying areas. In C. B. Field, V. R. Barros, D. J. Dokken, K. J. Mach, M. D. Mastrandrea, T. E. Bilir, M. Chatterjee, K. L. Ebi, Y. O. Estrada, R. C. Genova, B. Girma, E. S. Kissel, A. N. Levy, S. MacCracken, P. R. Mastrandrea, & L. L. White (Eds.), *Climate change 2014: Impacts, adaptation, and vulnerability. Part A: Global and sectoral aspects, Contribution of Working Group II to the Fifth Assessment Report of the Intergovernmental Panel on Climate Change* (pp. 361–409). Cambridge/New York: Cambridge University Press.

Part II
Current Environmental Patterns and Projections in the Study Area

Chapter 2
Patterns of Evolution of Frontal Systems Along the South-Southeastern Brazilian Coast

Joseph Harari and Ricardo de Camargo

Abstract In this chapter, the meteorological characteristics of the South-Southeast Brazilian coast are analysed, in particular the frontal systems and their influence on ocean circulation and sea level. The analyses are based on meteorological fields (pressure, temperature and wind) and oceanic fields (ocean currents and sea level) calculated by the CFS-v2 – Climate Forecast System (version 2), as well as on the synoptic analyses of the (Brazilian) Center for Weather Forecasting and Climatic Studies/National Institute of Space Research (CPTEC/INPE). Regarding the temporal evolution of the meteorological and oceanographic variables, it was chosen the month of September 2009, which registered several frontal systems of great intensity.

The intensity, duration, and time interval between these events have seasonal and interannual variability, and can be predicted only a few days in advance. Sea surface currents in the coastal region between 23°S and 27°S are to the South with a low intensity before the fronts (typically 0.2 m/s) and are to the North and very intense after the passage of the frontal systems (maximum above 0.4 m/s); sea surface level at the coast may vary from −0.1 m before the passage of the fronts, exceeding +0.5 m after their passage.

Keywords Atmospheric and oceanic circulations · Cold fronts · Coastal currents · Sea surface level

J. Harari (✉)
Institute of Oceanography, University of São Paulo, São Paulo, SP, Brazil
e-mail: joharari@usp.br

R. de Camargo
Institute of Astronomy, Geophysics and Atmospheric Sciences, University of Sao Paulo, São Paulo, SP, Brazil
e-mail: ricamarg@usp.br

L. H. Nunes et al. (eds.), *Climate Change in Santos Brazil: Projections, Impacts and Adaptation Options*, https://doi.org/10.1007/978-3-319-96535-2_2

2.1 Introduction: Typical Meteorological and Oceanic Conditions

In this chapter, the meteorological characteristics of the South-Southeast Brazilian coast will be analysed (Fig. 2.1), in particular the frontal systems and their influence on ocean circulation and sea level. The Municipality of Santos is inserted in the central portion of this area, at 23.95°S 46.35°W.

The analyses will be based on the meteorological fields (pressure, temperature and wind) and oceanic fields (ocean currents and sea level) calculated by the CFS-v2 – Climate Forecast System (version 2), available at https: //rda.ucar .edu/, as well

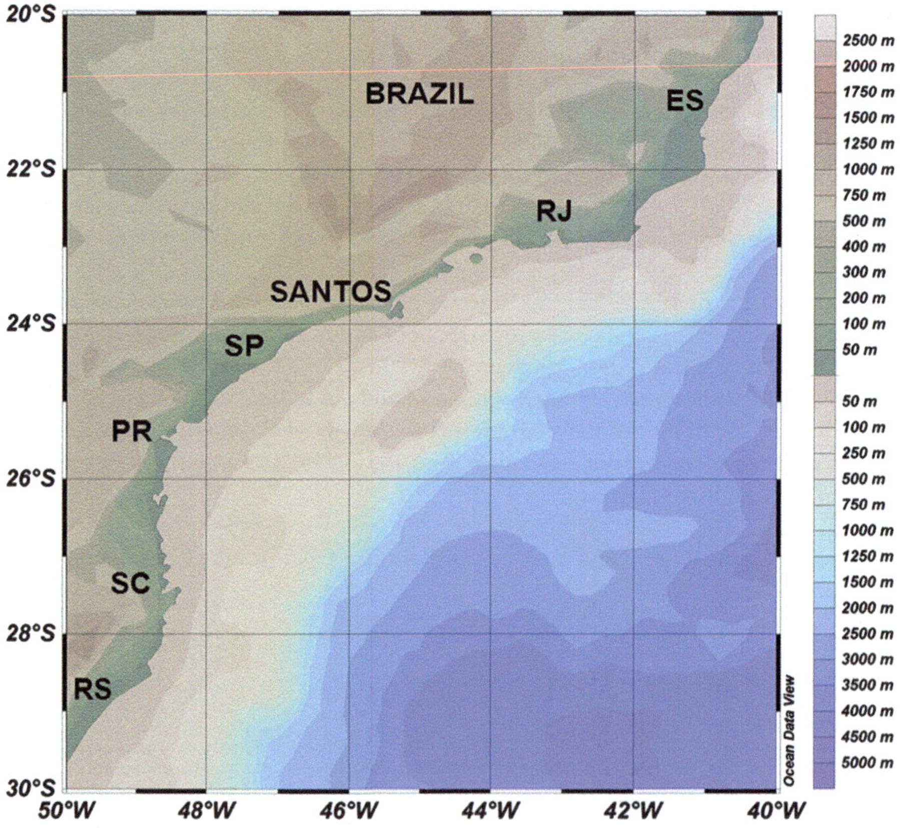

Fig. 2.1 Map of the South-Southeastern Brazilian region and Southwest Atlantic, with topography and bathymetry (in meters), involving Santos Municipality, located at 23.95°S 46.35°W; location of Rio Grande do Sul State (RS), Santa Catarina State (SC), Parana State (PR), Sao Paulo State (SP), Rio de Janeiro State (RJ) and Espirito Santo State (ES)

as on the synoptic analyses of the Center for Weather Forecasting and Climatic Studies/National Institute of Space Research (CPTEC/INPE), available at http://tempo.cptec.inpe.br/. The CFS-v2 is a third-generation product that covers the entire globe with high spatial resolution, coupling atmosphere-ocean-surface land-sea ice and has been designed to provide the best estimate of the state of these coupled domains (Saha et al. 2011). Regarding the temporal evolution of the meteorological and oceanographic variables, it was chosen the month of September 2009, which registered several frontal systems of great intensity.

In the South-Southeastern region of Brazil, atmospheric circulation on the surface of the sea depends on the South Atlantic Subtropical High Pressure Center – SASH, located around 30°S 0°W, and its interaction with the Subpolar Low Pressure Band – SPLP, located to the south of 60 °S (Seluchi and Marengo 2000). A spatial representation of the mean monthly atmospheric pressure distribution at sea level and its anomalies, in September 2009, is shown in Fig. 2.2, according to the synoptic analyses of the CPTEC/INPE.

For the month of September 2009, the mean atmospheric and oceanic conditions in the area are shown in Figs. 2.3 and 2.4, referring to the monthly average values

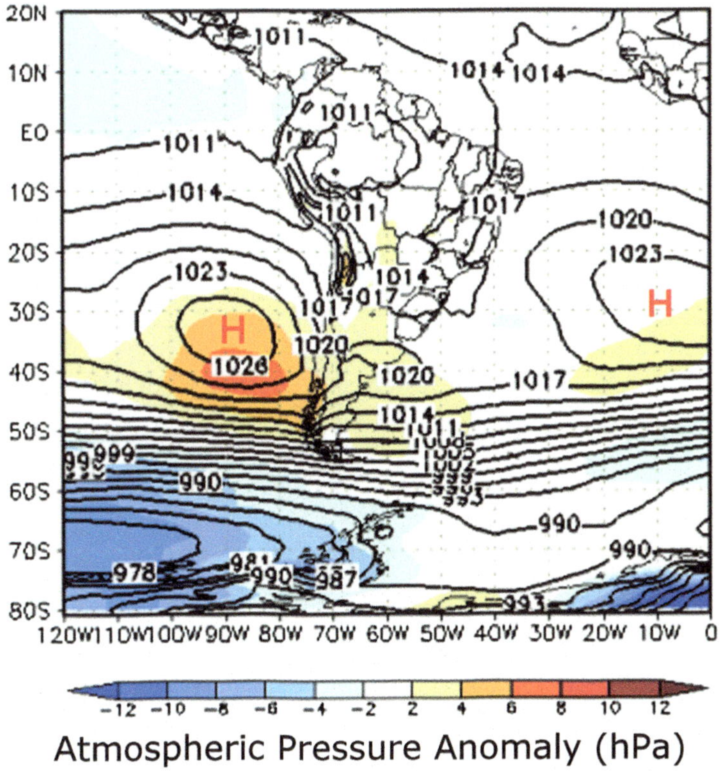

Fig. 2.2 Monthly average of atmospheric pressure at sea level in September 2009 (in hPa) and its anomalies. (Adapted from http://www.cptec.inpe.br, accessed on 10/01/2017)

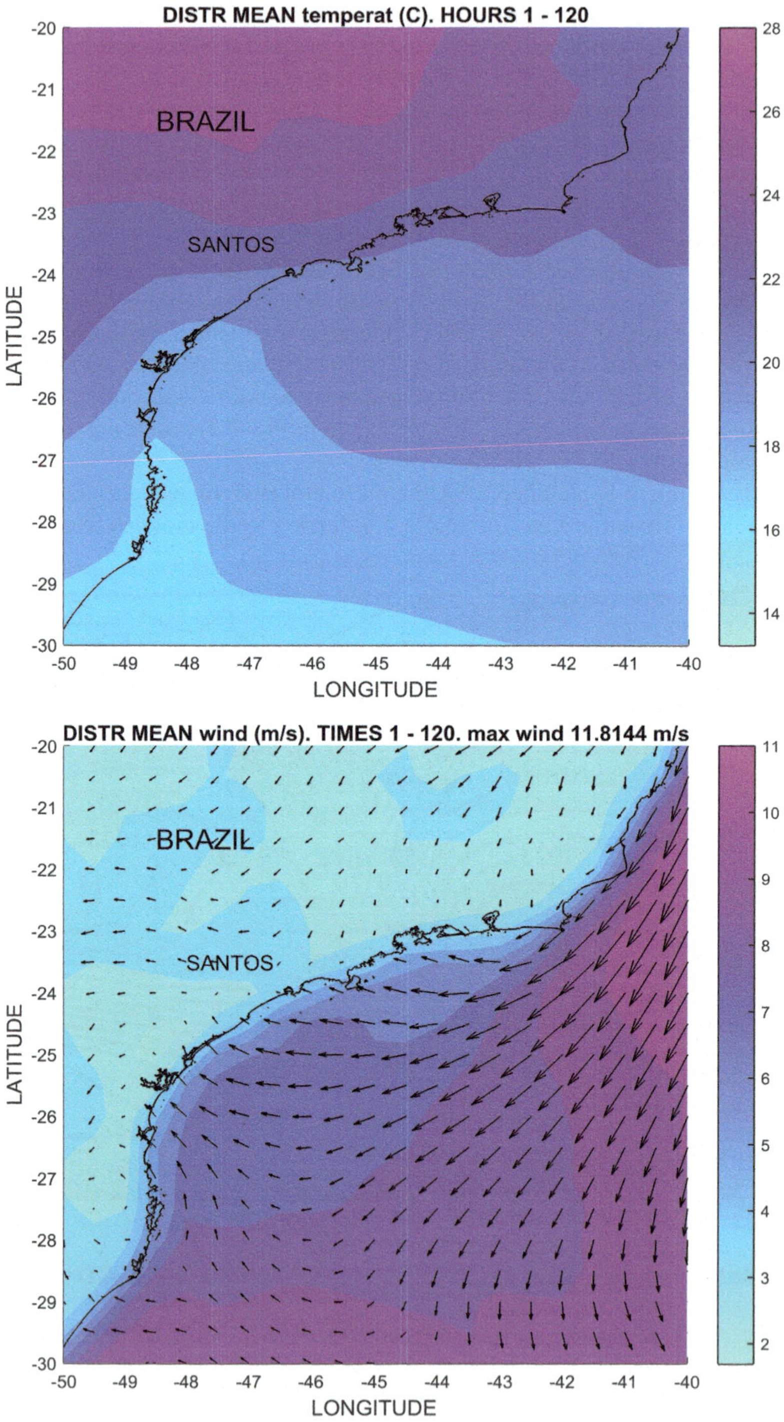

Fig. 2.3 Monthly average of air temperature (in °C) and monthly average winds (in m/s), on the surface, in September 2009, in the region 20°S – 30°S 50°W – 40°W

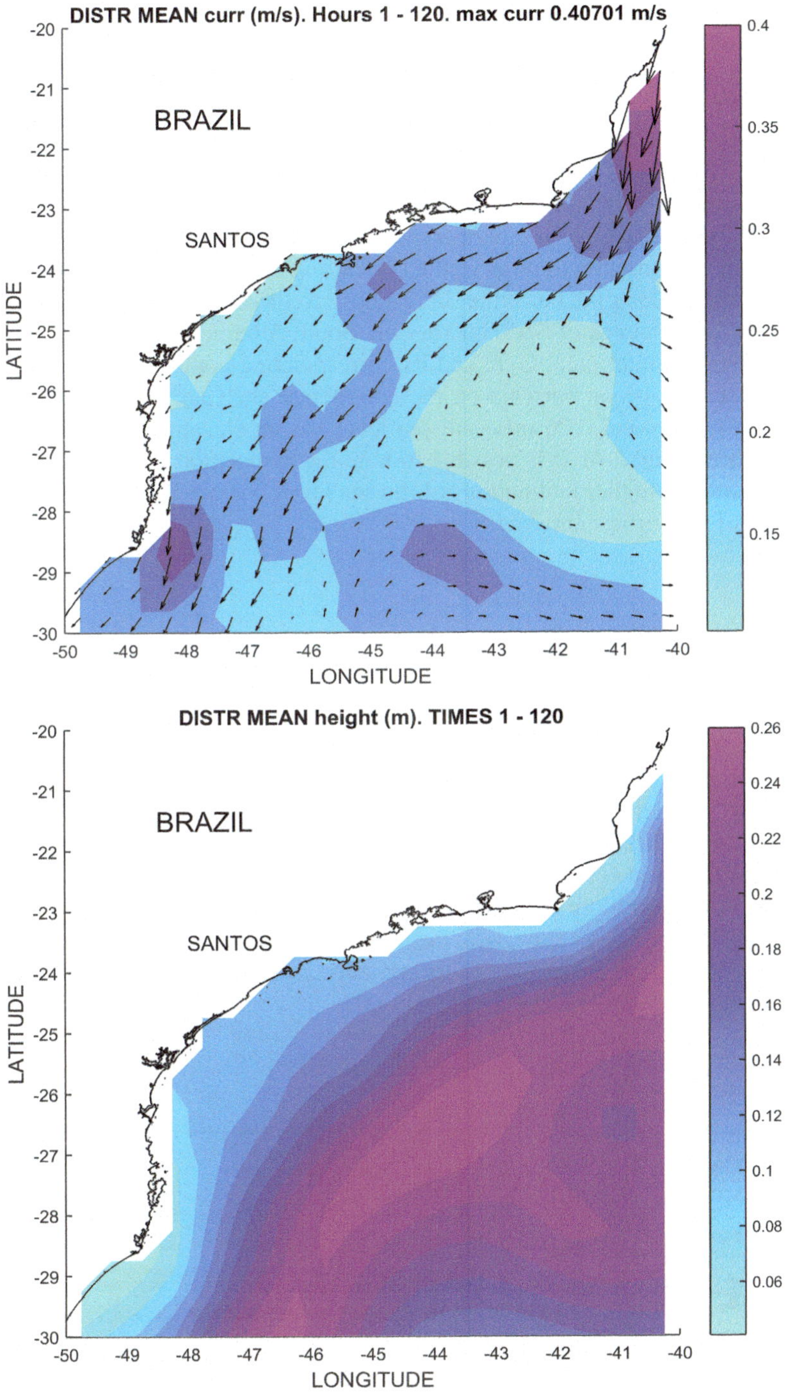

Fig. 2.4 Monthly average surface currents (in m/s) and monthly mean sea level elevation (in m), in September 2009, in the region 20°S – 30°S 50°W – 40°W

for temperature and surface winds, as well as the monthly average values of surface currents and sea level, respectively.

Under SASH influence, in the Brazilian coastal region between 23°S and 27°S, moderate surface winds of the eastern quadrant prevail, generally not exceeding 5 m/s, with high temperatures – around 20 °C – and relatively high pressures, of the order of 1020 hPa (Figs. 2.2 and 2.3). The mean temperature distribution of this month shows the influence of the frontal systems on the southern coast as well as the wind field with a positive southern component in the same region (Fig. 2.3); in the wind field, it is possible to observe the typical feature of cyclonic curvature on the edge of SASH along the coastline. In response to these typical atmospheric conditions and the influence of SASH across the South Atlantic basin, that generate the South Atlantic Subtropical Gyre and its corresponding contour current, the Brazil Current, surface ocean currents on the continental shelf are typically Southwest, with intensity around 0.20 m/s, and sea level has an elevation of +0.25 m in the offshore area (Fig. 2.4). It is possible to note the association between the stronger currents and the higher inclinations of the sea level, characterising the geostrophic standard of the surface currents.

2.2 Frontal System

In the context of Meteorology, front is a transition zone between two air masses of different densities, temperatures and humidity, and its performance in a given region imposes a great spatial contrast in the meteorological variables. This type of situation occurs when the configuration of the atmospheric motion causes an approximation between air masses of different properties, in which the basic density difference between hot air and cold air is determinant in the three-dimensional structure of the process. Thus, in a more comprehensive definition, front is the intersection of the frontal surface, or frontal zone, with the terrestrial surface (Fig. 2.5). This separation surface is called frontal surface or frontal zone, extending from the ground up to

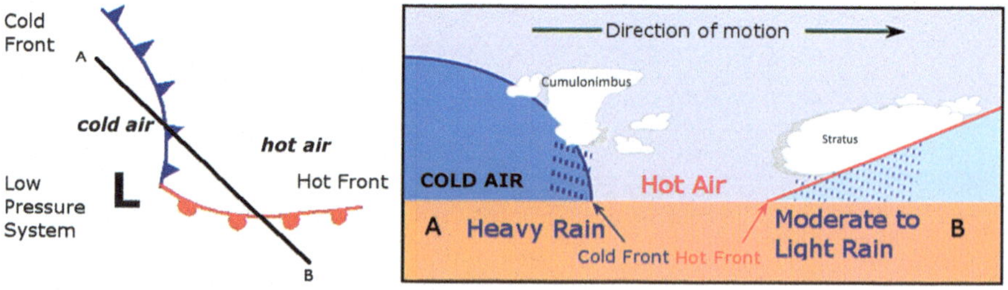

Fig. 2.5 General scheme of a cold front

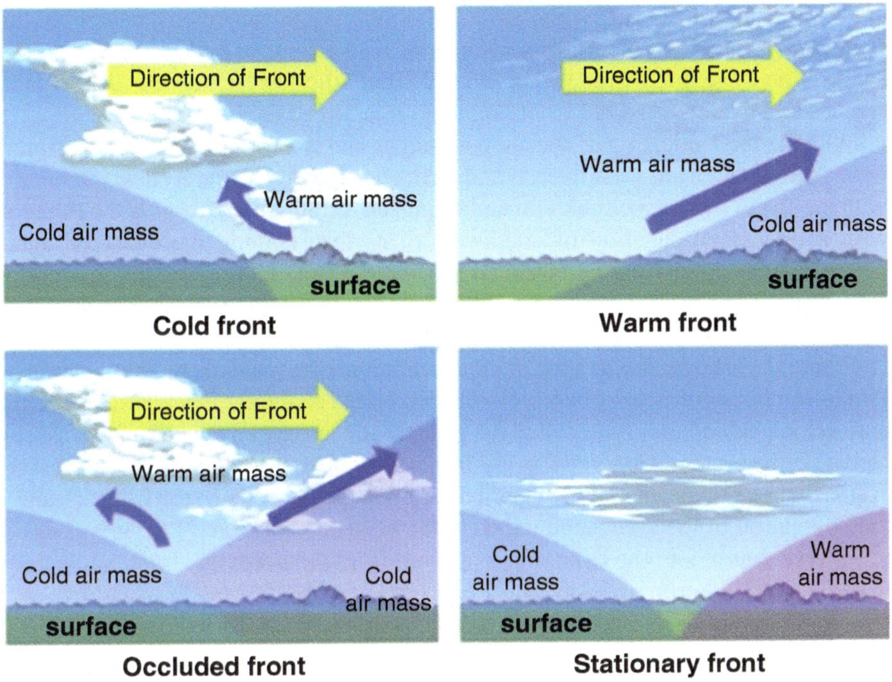

Fig. 2.6 Types of fronts: cold, hot, stationary and occluded

approximately 5000 m altitude, and its evolution is directly associated with baroclinic waves of medium latitudes (Kousky and Elias 1982; Cavalcanti et al. 2009). Therefore, the fronts act in order to reduce the horizontal temperature gradient, bringing polar air to the tropical region and tropical air to the polar region. In some cases, the frontal zone presents an intense density gradient, while in other cases this gradient is smooth. When a front advances, hot air is forced up the front surface; as the hot air rises, it cools and the water vapour turns into water, initiating the formation of clouds.

Frontogenesis is the process of forming or intensifying a front. Frontolysis is the process of destruction or weakening of a front (Satyamurty and Mattos 1989). When the temperature gradient increases, there is development and/or intensification of the system (frontogenesis); if the temperature gradient decreases, there is dissipation (frontolysis). Certain regions of the globe exhibit great frequency of frontogenesis: they are frontogenetic zones and coincide with regions of high thermal contrast.

A front can be classified by the relative movement of the masses of hot and cold air involved, and can be (Fig. 2.6): cold front, when the mass of cold air advances under the mass of hot air; warm front, when the hot air moves over the cold air; stationary front, when there is no advance of cold air nor of hot air relative to one another; and occluded front, when a cold front (the fastest moving cold sector) reaches and overcomes a hot front (and hot air is forced to rise).

With the intensification of successive troughs and crests in the Pacific, the frontal waves propagate from west to east, and change as they cross the Andes, interacting with the South America circulation and acquiring a component towards the Equator, with a typical propagation of southwest to northeast along the coast of South America, where they can reach tropical latitudes. In Brazil, the propagation of fronts causes variations in precipitation and temperature distribution in most of the country (Cavalcanti and Kousky 2009).

Figure 2.7 shows the graphical representation of a cold front propagating in South-Southeastern Brazil, based on CPTEC/INPE analyses. One can observe the anticyclone behind the front, called the post-frontal anticyclone, and the low associated with the system. Generally, every frontal system has associated a low-pressure system, whose shape and intensity will depend on the intensity of the front. Through the circulation of the associated winds to the centres of high and low, it is possible to observe that before the front, the winds are from the north quadrant, and behind the front, they are from the south quadrant

2.3 Effects of a Frontal System

In the South and Southeast regions of Brazil, the low-level winds have a Northeast and East direction, influenced by the presence of subtropical highs (Fig. 2.3). At the limit between SASH and SPLP, intense west winds and instabilities occur. These instabilities have temporal scales of days and spatial scales of thousands of kilometres, causing a pattern of atmospheric conditions typical of the evolution of cold frontal systems, which can be described as (Rodrigues et al. 2004):

1. In a pre-frontal situation, the wind typically rotates, being from Northwest, and then from Southwest and Southeast as the front passes; the winds on the day of passage and the two following days are from Southeast, bringing cold air to São Paulo.
2. The temperature increases 1 day before the passage and decrease a day later; the pressure decreases 1 day before the passage and rises the two following days.
3. After the passage of a cold front, there is usually a marked drop in temperature, increased pressure and wind gusts, when the pressure gradient is intense, and precipitation ceases.
4. In the sequence, the South winds turn again to easterly winds, with temperature and pressure increasing, returning the influence of the Subtropical High.

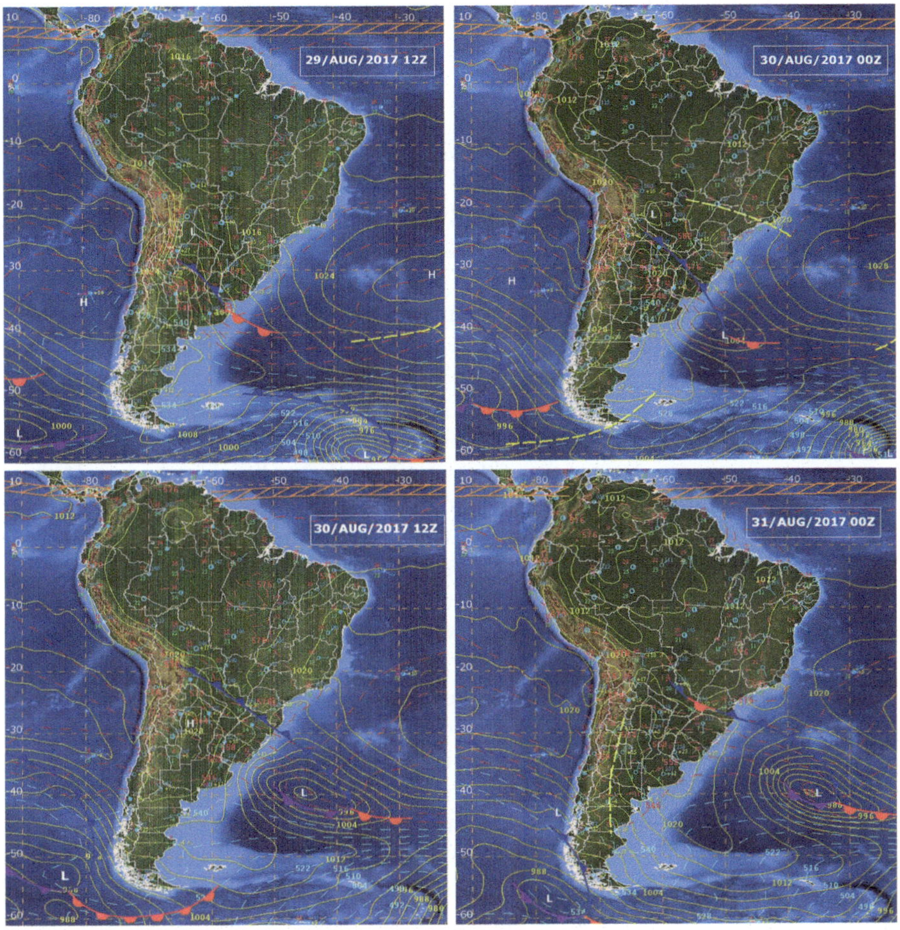

Fig. 2.7 Analyses of CPTEC/INPE from 12Z August 29 to 00Z August 31, 2017 showing the evolution of a frontal system on the south-southeast Brazilian coast. (Adapted from http://www.cptec.inpe.br, accessed on 15/09/2017)

The intensity, duration, and time interval between these events have seasonal and interannual variability, and can be predicted only a few days in advance. In general, in the south-southeast Brazilian coast between 23°S and 27°S, the easterly winds have a small intensity (typically less than 5 m/s) and long duration (5–10 days), while the North and West winds occur rapidly, with a scale of hours, and high intensity (indicating the arrivals of the frontal systems). The winds from the south are more intense (of the order of 10 m/s), and have duration of 1–3 days.

2.4 Variability of the Mean Sea Level

In first approximation, sea level can be considered as the composition of astronomical effects ("astronomical tide", whose main periodicities are diurnal and semi-diurnal) and meteorological/density influences ("meteorological tide", with longer periodicities).

The variability of the meteorological tide, or mean sea level, has an intrinsic relation with the atmospheric transients and the continental shelf geometry. In the western region of the South Atlantic, the combination of the cyclogenetic area east of the Andes and the width of the Argentine continental shelf is responsible for one of the most energetic places for the generation and propagation of meteorological tides (Camargo 2017).

In fact, the occurrence of extreme events in the oceans is directly related to an intense meteorological activity, which transfers enormous amounts of energy in the form of momentum to the surface of the sea, through the drag of the wind. In these situations, the ocean response is rapid in terms of the generation of currents and surface waves, so that the joint action of wind, current and wave can be determinant for accidents with oceanic vessels and offshore oil industry structures. In extreme cases, the intensity and persistence of severe weather conditions can be determinant for the occurrence of sea level over-elevations and flooding of low-lying coastal areas.

Meteorological events that occur on the continental shelf reach the coastal region, generating impacts on the seashore. In addition to the changes already mentioned in wave and current patterns, the response of sea level near the coast also occurs in the form of low frequency variations in relation to the astronomical tides, that is, fluctuations with periods superior to the semidiurnal and diurnal bands, whose periods are of 12 h and 24 h, respectively. These oscillations with periods of some tens of hours are called meteorological tides, and can reach significant amplitudes in extreme meteorological events, of order of a few metres, causing direct impacts in the form of coastal floods to positive sea level variations, or making maneuvers in navigation channels difficult for large vessels, in situations of negative sea level variations (Gonnert et al. 2001; Weisse 2010).

Sea level variations for a given region therefore depend on the amplitude of the astronomical tide as well as the amplitude of the meteorological tide. Even in places with small amplitudes of astronomical tide, the effects of positive meteorological tides can cause coastal flooding, or may act to prevent the continental drainage of precipitation excess, and may thus be associated with large floods.

2.5 Characterization of Extreme Sea Level Events in Santos and Their Correspondence with Synoptic Meteorological Conditions

Campos et al. (2010) identified atmospheric influence of synoptic scale on the ocean for extreme events of meteorological tide on the Brazilian southeast coast. In order to do so, they used sea level elevation data from the Port of Santos-SP, as well as wind and surface pressure fields produced by the reanalysis of the NCEP model covering the South Atlantic, for the period 1951–1990. In this work, it was possible to identify a seasonal variability as well as the evolution pattern of atmospheric systems associated with extreme events.

In the study by Campos et al. (2010) the sea level elevation data in Santos were submitted to the low-pass Lanczos filter in the time domain. This procedure allows to eliminate almost totally the influence of the astronomical forcing, since the main components have periods around 12 and 24 h. In order to analyse each of the decades, the period of 40 years of data was divided into 4, obtaining individual files of 10 years each, which were submitted to a unique methodology. The filtered series of sea level were also divided according to the seasons of the year; then only events with mean sea level elevations higher than +2 standard deviations (+37.7 cm) and below −2 standard deviations (−37.7 cm) were selected, thus separating maximum and minimum events; finally, the quantification of these cases was accomplished for the four decades. The construction of the compositions starts with the selection of filtered series values above and below +2 and −2 standard deviations, respectively. The corresponding periods were used to select the simultaneous data in the pressure and wind fields on the surface, generating seasonal files with the indexes of the extreme cases of daily meteorological tide; the compositions were then obtained by means of the individual cases; this procedure was performed for each decade.

The main conclusions of Campos et al. (2010) were the following:

1. The characteristics of events with positive elevations during autumn and negative during spring remain along the decades, in the same way as the high dispersion of the origin of winter events and the few extreme cases of summer. Thus, it can be said that the synoptic scale fluctuations associated with specific conditions have a typical behaviour, which presents little variation over the decades.
2. The explanation for the large number of negative spring events is due to the fact that the months of August and September are characteristic of the occurrence of blocking systems, typically inducing northeasterly winds along the coast associated with troughs on the continent, with consequent falling sea level. In autumn, this pattern is not frequent, being a period of strong cyclogenesis and intensification of troughs over the ocean, with Southwest winds and rising sea level, justifying the few cases of negative elevation. The fact that there is a higher occurrence

of positive cases in the autumn compared to the winter is due to the shorter period and higher frequency of the events, that is, because in winter the troughs are more intense and permanent, an extreme event can last a week, while, in the fall, cases tend to be less intense and of shorter duration, but more frequent.

3. In cases of positive elevations, the dependence of the extreme cases with the intensification, the size and the positioning of the troughs over the ocean should be highlighted. In winter and autumn – a period with a high occurrence of positive sea-level rises, it is possible to observe the intense Southwest winds, approximately parallel to the coast, covering extensive regions and close to the continent, determined by the troughs on the ocean in conjunction with high pressure on the continent. In summer and spring, there is a weakening of the troughs, with loss of vorticity and decrease of wind intensities, with less frequent flooding.
4. The great intensification of the anticyclone on the continent in the winter, increasing the pressure gradients, responsible for intense Southwest winds, should be highlighted, proving to be the period of greatest incidence of extreme cases.
5. Analysing the seasonality of negative events, winter and spring are the periods of maximum intensity of the high pressures on the ocean, as a consequence of the increase in the number and the permanence of meteorological blockings. However, in winter the centre of high pressure is, on average, more northerly and with greater intensity, whereas in the spring it is situated to the south and is less intense. In the autumn and especially in the summer, there are usually no major fetches with northeasterly winds to the point of generating intense cases of negative meteorological tide.
6. Following the temporal evolution of the synoptic conditions in the extreme periods of positive sea-level elevations, a well-defined trough is observed 2 days before the event of maximum mean sea level detected in Santos, with intense gradients and winds and axis slightly more to the west and movement towards the south. In this period, on the coast of São Paulo and Rio de Janeiro, the winds are from Northeast, with negative meteorological tide; meanwhile, the coast of Argentina, Uruguay and the Brazilian State of Rio Grande do Sul receive the influence of the southwest wind, generating positive meteorological tide, which propagates in Kelvin wave form, passing through Santos in the following days. In the day before the event of maximum elevation of the mean sea level in Santos, it is observed the displacement of the troughs to the East, accompanying the typical Western circulation of this latitude, with gain of intensity. It is possible to notice the cold fronts spreading to the northeast, invading all the South of Brazil, which are on Paraná and São Paulo states 2 days before the event of maximum elevation of the mean sea level in Santos, passing through Rio de Janeiro state the previous day and on the North of Espírito Santo state on the day of maximum elevation in Santos. That is, on the day of the occurrence of extreme tide, the front will have passed for Santos 1 or 2 days before.

7. The most favourable condition for the occurrence of an event greater than 3 standard deviations lies in the persistence of the systems in the days prior to the event itself, making clear the role of the intensification of the wave pattern coming from the south over time. Therefore, it is possible to affirm that extreme phenomena involving sea level do not tend to occur with local forcings near Santos, but depend on the time evolution and persistence of the southwest wind fetch along the entire South-Southeast Brazilian coast.

2.6 Numerical Modelling of Meteorological Tides in the Western Region of the South Atlantic

Camargo (2017) modelled the hydrodynamics of the western portion of the South Atlantic, from the north-central part of Argentina to the southeastern region of Brazil, an extensive continental shelf between latitudes 50°S and 20°S and longitudes 070°W and 040°W, in the period from 1948 to 2010. Some of the main features are the variation of the amplitudes of the astronomical tides along the coast and the presence of the Rio de la Plata, in addition to varying the width of the continental shelf. For regions directly exposed to the deep ocean, the synoptic meteorological influence may impose oscillations with characteristics of wave propagation along the coast.

Considering the meteorological tides in the Western South Atlantic, Camargo (2017) objectively analysed the general behaviour of the phenomena: their typical areas of generation, the associated atmospheric disturbances and the propagation patterns along the coast. The conditions for formation and propagation of meteorological tides are quite special, depending on the width of the Argentine platform and the cyclogenetic conditions throughout the year. The modelling results demonstrate that the main meteorological tide generation area in the Western South Atlantic is located north of 45°S (Fig. 2.8). In relation to the propagation of sea level disturbances, Camargo (2017) concludes that: (i) the highest frequency of occurrence and the largest amplitudes of meteorological tides occur in the stations located to the south; (ii) some events are generated north of 40°S, and affect only the northern part of the study region, but their amplitudes are not usually the largest; and (iii) there are events that influence the entire South Atlantic western region, whether overrising or lowering the mean sea level.

In conclusion, both from observations and hydrodynamic numerical model results, it is notable that the events in central and northern Argentina are much larger than in the southern sites, which means that the main area of mean sea level oscillations in the South Western Atlantic is located north of the 45°S parallel.

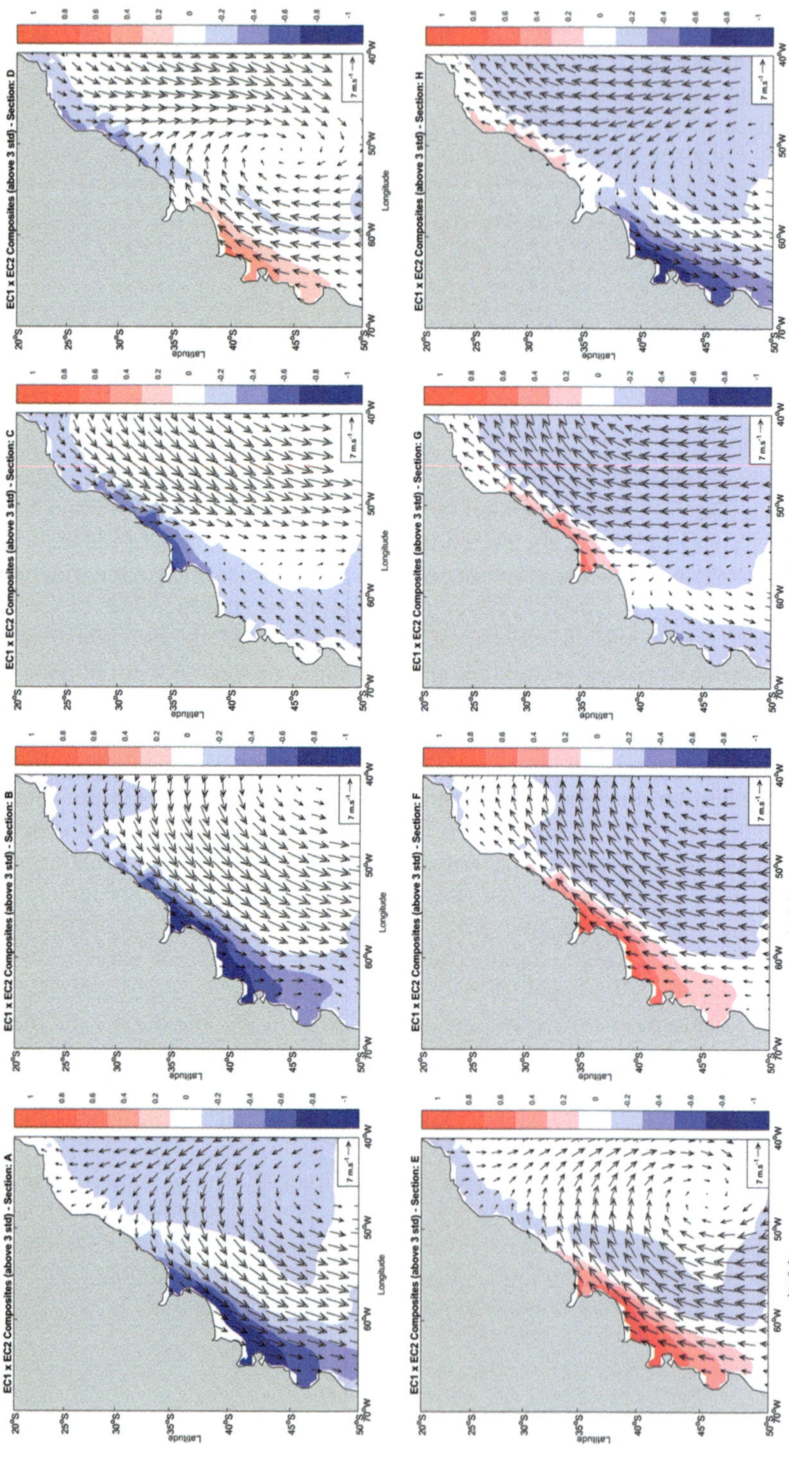

Fig. 2.8 Typical evolution of wind field anomalies and corresponding sea level anomalies in the scale of 1–10 days. (Camargo 2017)

2.7 Case Study: Effects of the Fronts of September 2009

At the end of this chapter, the meteorological conditions of September 2009 will be analysed, based on the results of the CFS-v2 model (Saha et al. 2011), especially the effects of cold fronts. This period was selected due to the occurrence of a large number of frontal systems of high intensity; the last cold front observed this month in Santos caused huge flooding of the coastal area, due to the great rise of the mean sea level.

Time series of meteorological variables were extracted from CFS-v2 model for September 2009, at the closest grid point of Santos Bay (PS), at 46.35°W 24.10°S (Fig. 2.9); these series are shown in Fig. 2.10 and the angular histogram of the winds is shown in Fig. 2.11.

In September 2009, six frontal systems propagated along the coast of South America; the first three systems did not affect much the coastal circulation in the Baixada Santista, since they dissipated before arriving at this region. The three subsequent systems had greater influence on the coastal circulation of the State of São

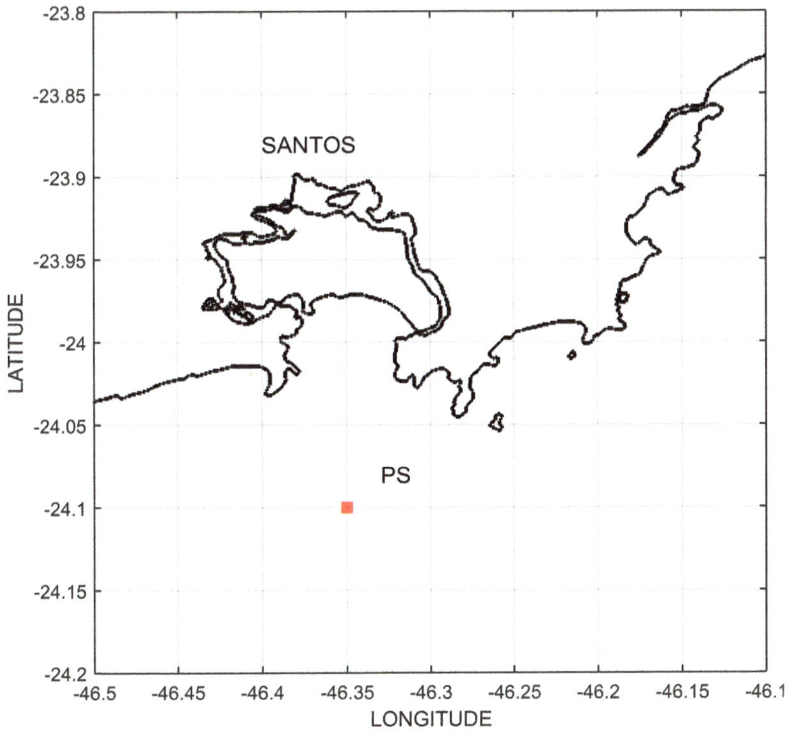

Fig. 2.9 Selected point for presentation of time series generated by the model CFSr-v2 (Point PS), off the Municipality of Santos, at 46.35°W 24.10°S

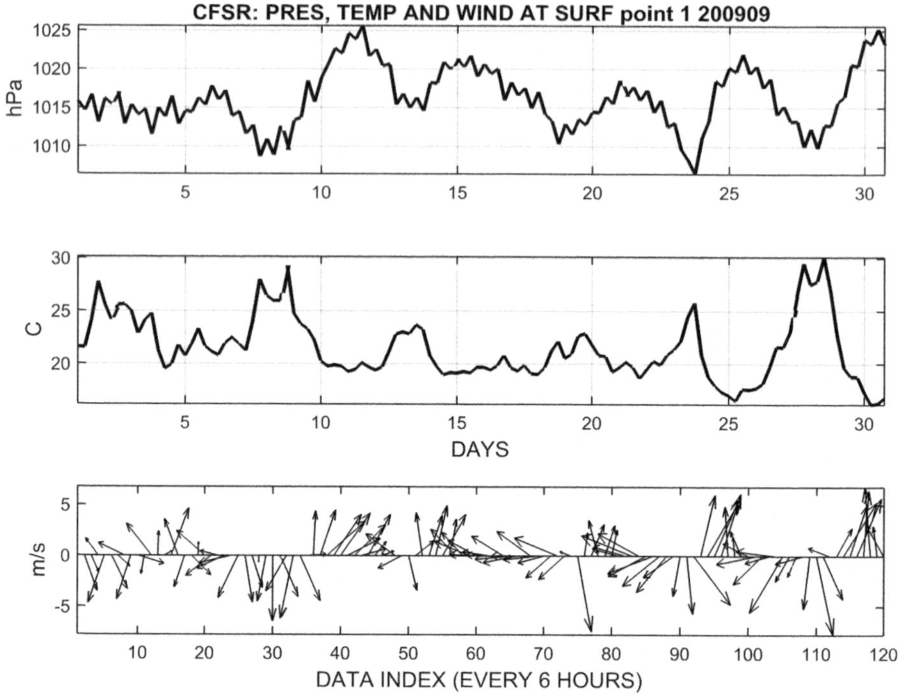

Fig. 2.10 Time series of atmospheric pressure at sea level, surface temperature and wind vector, at point PS (46.35°W 24.10°S)

Paulo, with passage in the Santos region on the 20th, 24th and 28th of the month. In fact, the 1st days of September presented good weather conditions and very high temperatures, altered only by the first frontal system, of relatively weak intensity; however, from day 10 on, several periods with winds from the South were observed, having greater intensity and duration from day 20 on (Fig. 2.10). The corresponding time series of sea level and surface currents are given in Fig. 2.12, with angular histogram of these currents in Fig. 2.13.

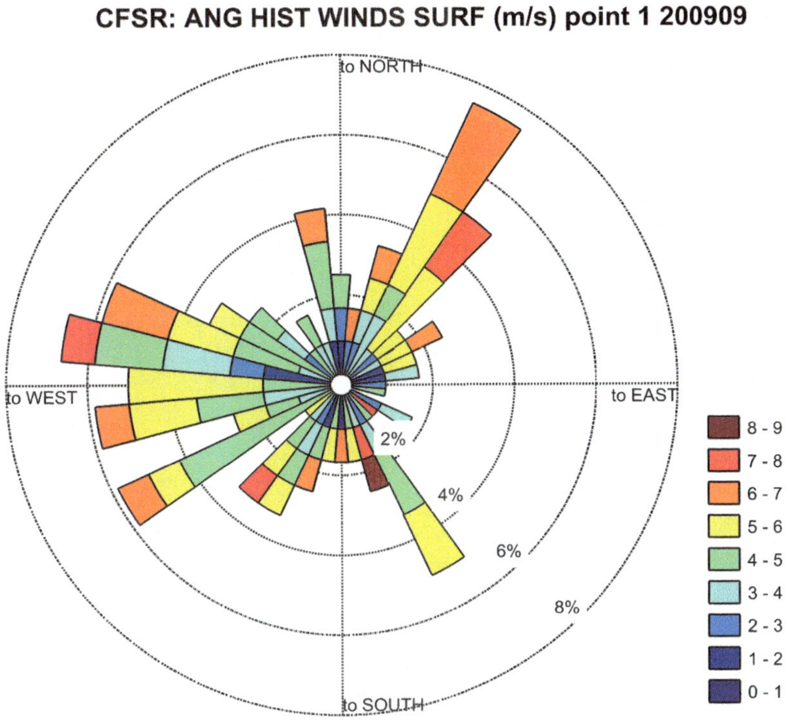

Fig. 2.11 Angular histogram of the surface winds, in September 2009, at point PS (46.35°W 24.10°S)

The effects of the six cold fronts on the meteorological and oceanographic parameters in Santos are listed in Tables 2.1 and 2.2, based on Figs. 2.10 and 2.12.

Finally, maps of the distributions of the meteorological and oceanographic variables are presented at the times of interest, 06 h on September 28 and 8 h on September 29, before and after the passage of the last cold front in Santos in September 2009, which was the most intense cold front of the analysed month:

1. Atmospheric pressure at sea level (Fig. 2.14), indicating the advance of the low pressure centre preceding the arrival of the cold front in the region of Santos.

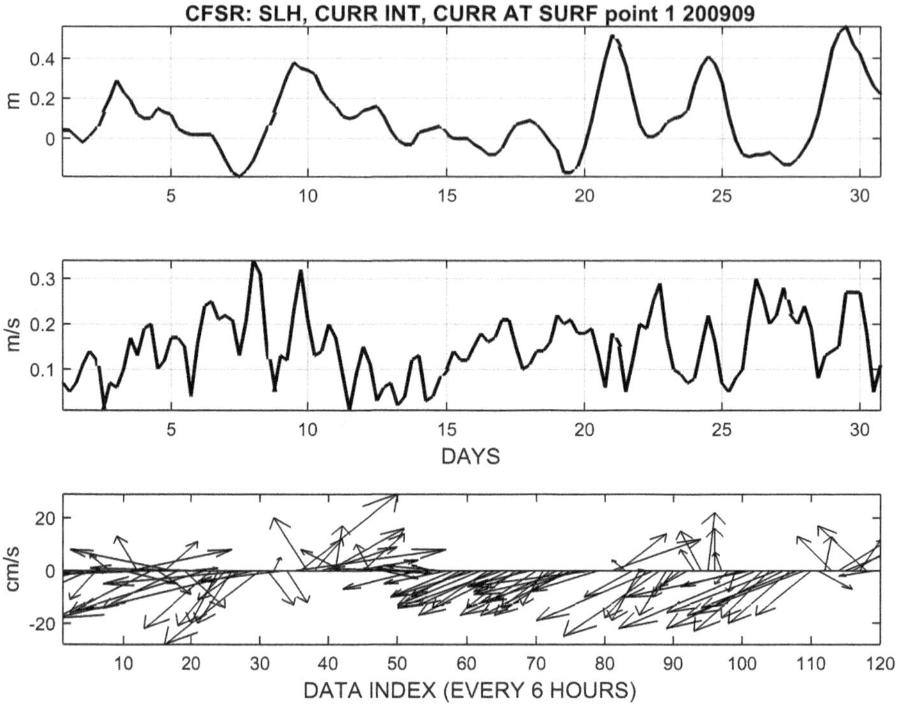

Fig. 2.12 Time series of sea level elevation, current intensity at the surface and current surface vector at PS (46.35°W 24.10°S)

2. Sea surface winds (Fig. 2.15), with North winds before the passage of the front and South winds after the passage, respectively, both with maximum intensities above 15 m/s.
3. Sea surface currents (Fig. 2.16); in the coastal region between 23°S and 27°S, the currents are to the South with a low intensity before the front (typically 0.2 m/s) and are to the North and very intense after the passage of the frontal system (maximum of 0.45 m/s).
4. Sea surface level (Fig. 2.17); in the coastal region between 23°S and 27°S, the sea level is −0.1 m before the passage of the front in Santos, and exceeds +0.5 m after the passage.

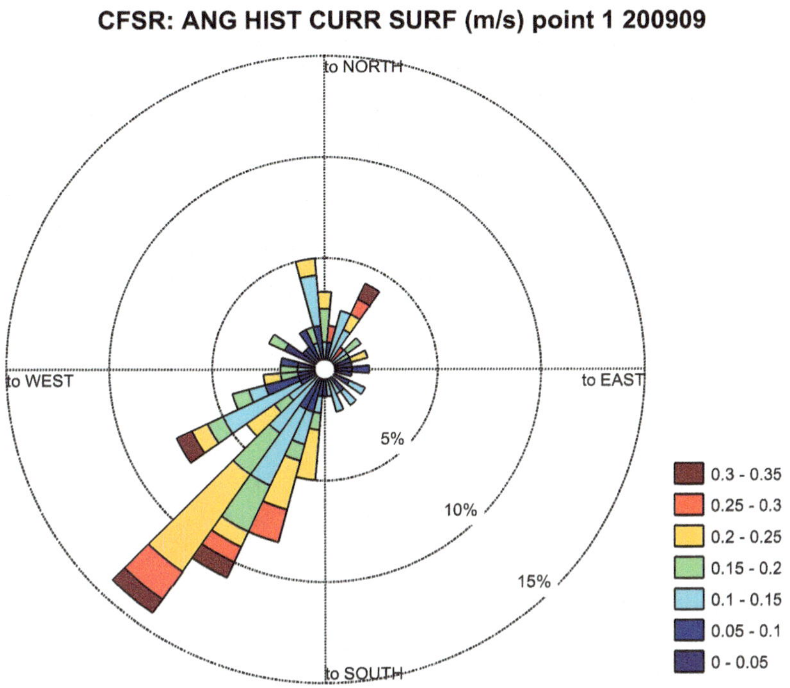

Fig. 2.13 Angular histogram of the surface currents, in September 2009, at point PS (46.35°W 24.10°S)

Table 2.1 Extreme meteorological values at PS point (46.35°W 24.10°S), associated to cold fronts in the South-Southeastern Brazilian region and their times of occurrence: atmospheric pressure at minimum sea level, minimum surface temperature, maximum intensity of the wind at the surface and maximum meridional component of the South wind

Front	Day–hour Minimum pressure (hPa)	Day–hour Minimum temperature (°C)	Day–hour Wind max. intens. (m/s)	Day–hour Max. South comp. (m/s)
1	02 – 18 h	04 – 06 h	04 – 12 h	04 – 12 h
	1012.78	19.51	5.36	4.67
2	07 – 18 h	11 – 00 h	10 – 06 h	10 – 00 h
	1008.61	19.25	6.49	4.47
3	13 – 18 h	15 – 00 h	14 – 12 h	14 – 06 h
	1014.53	19.14	5.27	4.94
4	18 – 18 h	21 – 18 h	21 – 12 h	20 – 06 h
	1010.30	18.64	7.70	3.52
5	23 – 18 h	25 – 06 h	24 – 06 h	24 – 18 h
	1006.46	16.56	7.52	6.13
6	28 – 06 h	30 – 06 h	29 – 06 h	30 – 06 h
	1009.78	16.23	7.22	6.79

Table 2.2 Extreme oceanographic values at the point PS (46.35°W 24.10°S), associated to cold fronts in the South-Southeastern Brazilian region and their times of occurrence: maximum sea level, maximum intensity of surface current and maximum south component of current

Front	Day – hour Maximum sea level (m)	Day – hour Max. current int. (m/s)	Day – hour Max. south curr. comp. (m/s)
1	03 – 12 h	03 – 12 h	03 – 18 h
	0.25	0.17	0.13
2	10 – 06 h	09 – 18 h	09 – 18 h
	0.37	0.32	0.29
3	12 – 12 h	12 – 00 h	12 – 00 h
	0.16	0.15	0.14
4	21 – 12 h	22 – 12 h	21 – 06 h
	0.48	0.25	0.14
5	24 – 18 h	24 – 12 h	24 – 12 h
	0.44	0.22	0.22
6	29 – 12 h	30 – 00 h	30 – 00 h
	0.56	0.27	0.27

2.8 Conclusions

The study shows the great complexity that characterizes the phenomenon of the evolution of cold fronts in the South-Southeast region of Brazil. However, scientific knowledge acquired since the 1980s, based on intensive measurement programs, including satellite observations, as well as sophisticated numerical models, including ocean-to-atmosphere coupling, allows highly accurate simulations and predictions.

In fact, the mapping of the fields of air pressure, temperature and humidity, winds, sea currents and sea surface elevation – regularly made available by monitoring centres and operational forecasting – has made it possible to protect people and coastal resources for cases of extreme meteo-oceanographic events.

In this way, future challenges to humanity related to global warming – such as the intensification of hurricanes and extreme meteorological events, and the rise in mean sea level – can be addressed from robust and effective scientific knowledge.

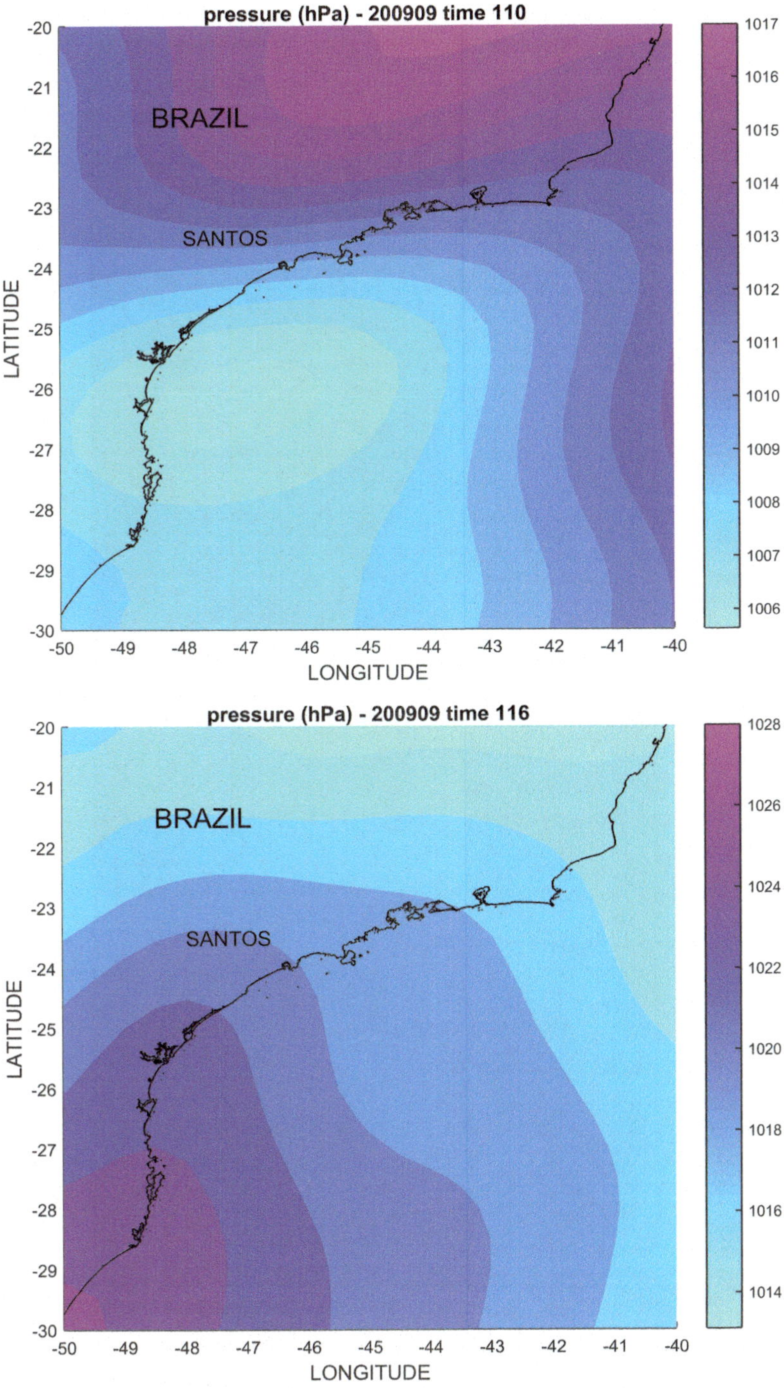

Fig. 2.14 Distribution of atmospheric pressure at sea level (hPa) in the Southwest Atlantic, on 28 September 06 h and 29 September18 h

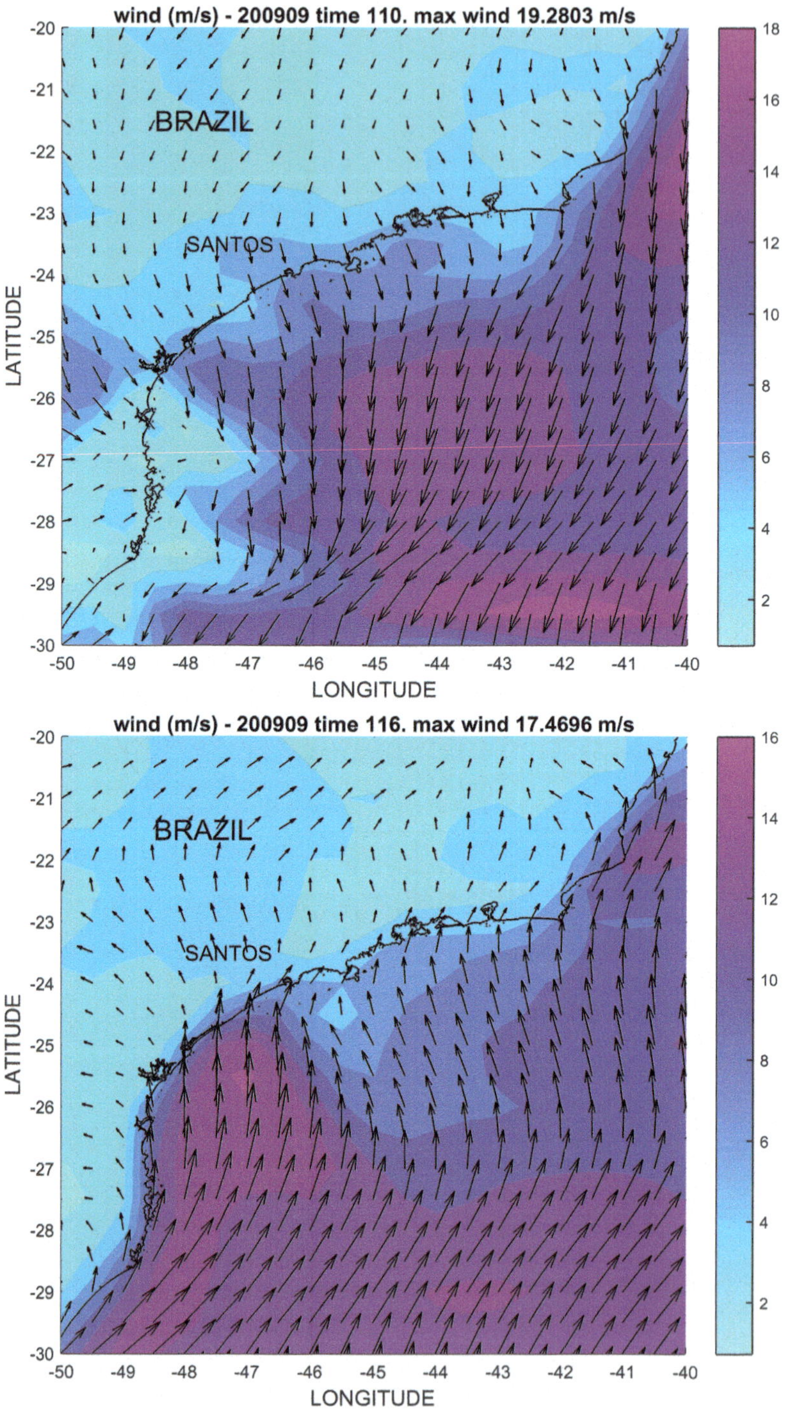

Fig. 2.15 Distribution of surface winds at the surface (m/s) in the Southwest Atlantic, on 28 September 06 h and 29 September 18 h

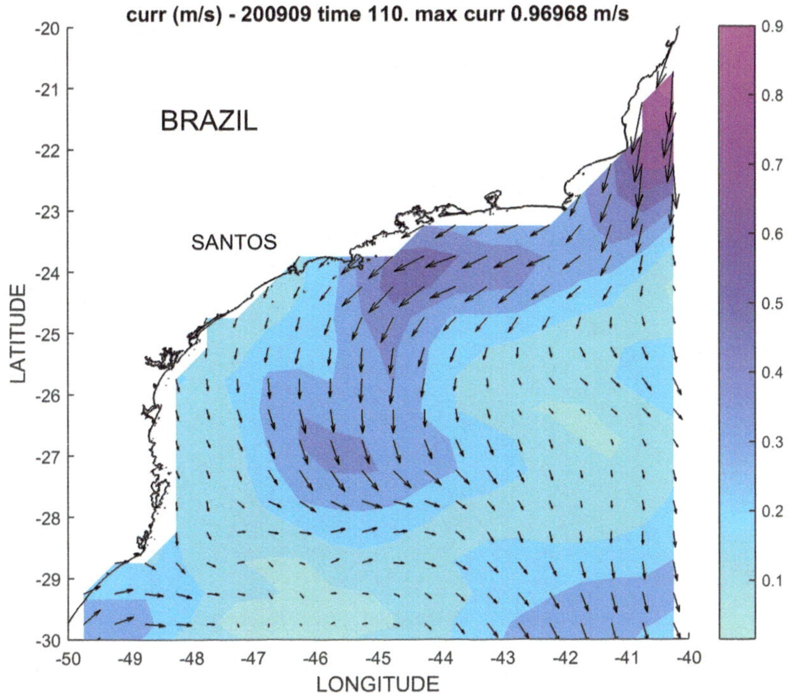

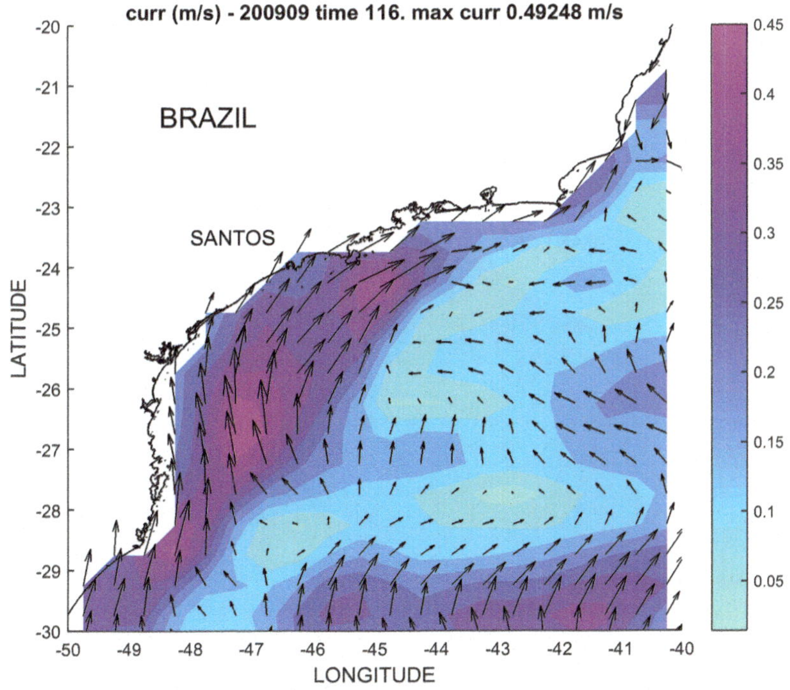

Fig. 2.16 Distribution of surface currents at the surface (m/s) in the Southwest Atlantic, on 28 September

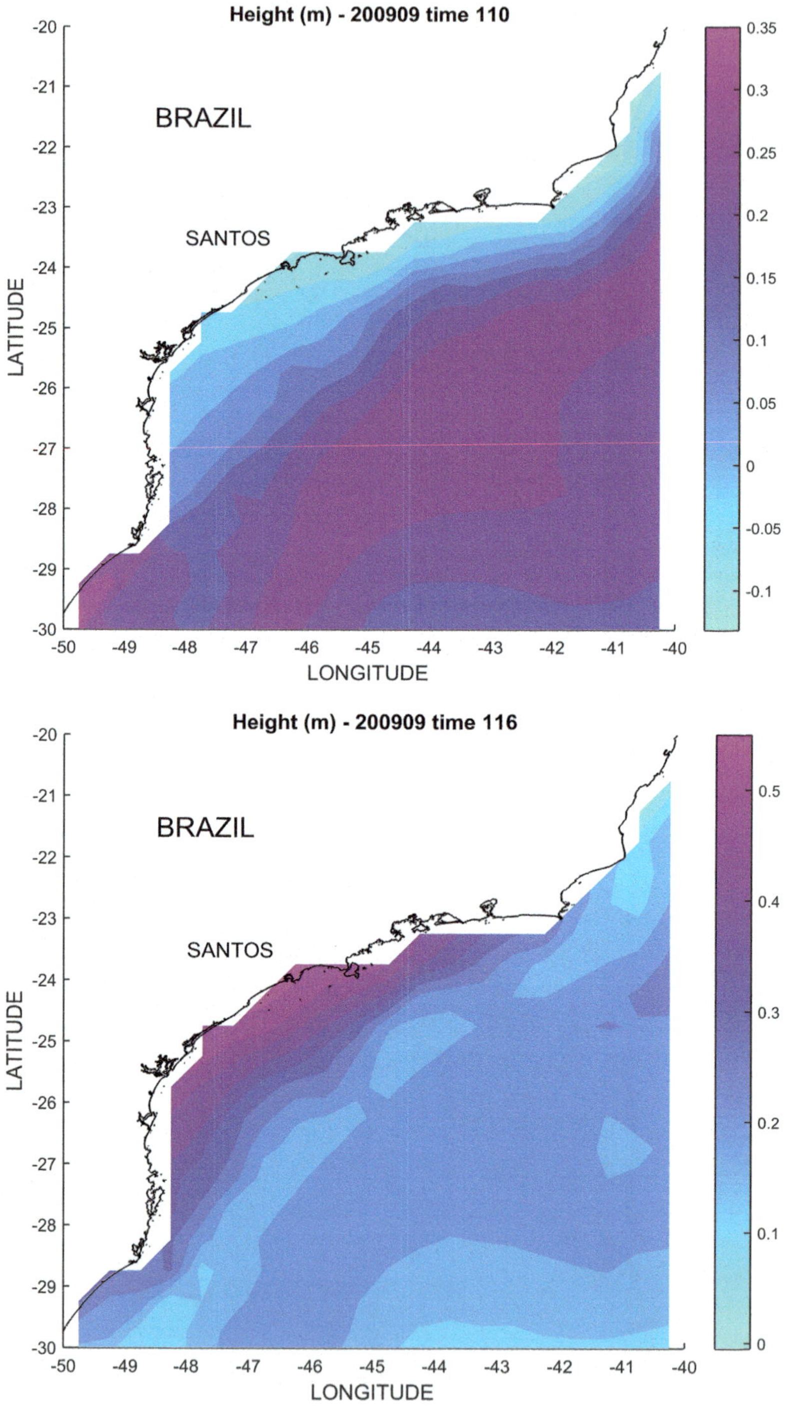

Fig. 2.17 Distribution of sea surface elevation (m) in the Southwest Atlantic, on 28 September 06 h and 29 September 18 h

References

Camargo, R. (2017). *Oscilações do nível do mar no Oeste do Atlântico Sul geradas por sistemas meteorológicos de escala sinótica: uma abordagem numérica para a descrição de suas características*. Associate Professor Dissertation, Instituto de Astronomia, Geofísica e Ciências Atmosféricas da Universidade de São Paulo, 85 p.

Campos, R. M., Camargo, R., & Harari, J. (2010). Caracterização de eventos extremos do nível do mar em Santos e sua correspondência com as reanálises do modelo do NCEP no Sudoeste do Atlântico Sul. *Revista Brasileira de Meteorologia, 25*(2), 175–184. https://doi.org/10.1590/S0102-77862010000200003.

Cavalcanti, I. F. A., & Kousky, V. E. (2009). *Frentes Frias sobre o Brasil. Capítulo 9 Tempo e Clima no Brasil*. São Paulo: Editora Oficina de Textos ISBN 978-85-86238-92-5.

Cavalcanti, I. F. A., et al. (2009). *Tempo e Clima no Brasil*. São Paulo: Editora Oficina de Textos ISBN 978-85-86238-92-5.

Gonnert, G., et al. (2001). *Global storm surges*. [S.l.]: [s.n.], 2001. 623 p. ISBN 3-8042-1054-6.

Kousky, V. E., & Elias, M. (1982). *Meteorologia Sinótica: Parte I*. São José dos Campos: INPE.

Rodrigues, M. L. G., Franco, D., & Sugahara, S. (2004). Climatologia de frentes frias no litoral de Santa Catarina. *Revista Brasileira de Geofísica, 22*(2) São Paulo. https://doi.org/10.1590/S0102-261X2004000200004

Saha, S., et al. (2011). *Updated monthly. NCEP Climate Forecast System Version 2 (CFSv2) selected hourly time-series products*. Research Data Archive at the National Center for Atmospheric Research, Computational and Information Systems Laboratory. https://doi.org/10.5065/D6N877VB. Accessed 06 Sept 2017.

Satyamurty, P., & Mattos, L. F. (1989). Climatological lower tropospheric frontogenesis in the midlatitudes due to horizontal deformation and divergence. *MWR, 117*(6), 1355–1364.

Seluchi, M. E., & Marengo, J. A. (2000). Tropical-midlatitude exchange of air masses during summer and winter in South America: Climatic aspects and examples of intense events. *International Journal of Climatology, 20*(10), 1167–1190.

Weisse, R. (2010). *Marine climate and climate change: storms, wind waves and storm surges*. [S.l.]: [s.n.]. 219 p.

Chapter 3
Patterns of Extreme Precipitation in Santos

Lucí Hidalgo Nunes, Lincoln Muniz Alves, Eduardo Kimoto Hosokawa, and José Antonio Marengo

Abstract Analysis of extreme precipitation events in different time scales over Santos is presented using the rain gauge network for the period 1980–2015. Located in the humid tropics, the area presents high annual total rainfall, particularly concentrated in spring and summer. However, the rainiest month—January—was also the one with the most homogeneous distribution. More than 40% of the days presented precipitation, which ranged from 0.1 to 332.9 mm. The precipitation variability was quite high, a fact observed in all scales evaluated, which reveals that the mechanisms that generate precipitation in the area are distinct throughout the year. The high precipitation variability is particularly important in the case of daily distribution: for instance, 25% of the rainy days responded for 71% of the total rainfall. Such concentration is a key element in the outbreak of numerous hazards, such as floods and landslides, that cause extensive property damage and loss of life, being the worst situations registered when extreme precipitation episodes coincide with storm surges.

Keywords Rainfall · Extreme events · Santos · Concentration index

L. H. Nunes (✉)
Santos, São Paulo, Brazil
e-mail: luci@ige.unicamp.br

L. M. Alves
National Institute for Space Research (INPE), São José dos Campos, SP, Brazil
e-mail: Lincoln.alves@inpe.br

E. K. Hosokawa
Municipal Government of Santos, Secretariat of Urban Development, Santos, SP, Brazil
e-mail: eduardohosokawa@santos.sp.gov.br

J. A. Marengo
National Center for Monitoring and Early Warning of Natural Disasters (CEMADEN), São José dos Campos, São Paulo, Brazil
e-mail: jose.marengo@cemaden.gov.br

L. H. Nunes et al. (eds.), *Climate Change in Santos Brazil: Projections, Impacts and Adaptation Options*, https://doi.org/10.1007/978-3-319-96535-2_3

3.1 Introduction

According to the IPCC Glossary of Terms, a climate extreme (extreme weather or climate event) is "*a value of a weather or climate variable above (or below) a threshold value near the upper (or lower) ends of the range of observed values of the variable. For simplicity, both extreme weather events and extreme climate events are referred to collectively as* 'climate extremes'" (IPCC 2012a, p. 557). Extremes reflect unusual and distinct arrangements of variables and processes that occur randomly and abruptly in a given place. Because adaptations to the frequency, magnitude or timing of extreme episodes are based on rather imperfect or inadequate knowledge (WMO 2003), extreme occurrences can severely impact a population.

Extreme precipitation, in particular, reflects the combination of numerous variables and processes that occur at different time-space scales (Berndtsson and Niemczynowicz 1988; IPCC 2012b). It is a spontaneous occurrence within the natural system and a contributor to landscape evolution. Furthermore, it can be ephemeral and spatially circumscribed, or it can last longer and cover larger areas. In the present case study, extreme precipitation is associated with local factors, and its duration and spatial extent are the result of mesoscale controls.

Extreme events are an intrinsic part of precipitation variability. However, in places of great socio-environment complexity and population vulnerability, these events can have major consequences for society and the natural environment and dramatically impact the economy and public health. Such has been the case for Santos and its neighbouring towns.

Santos and surroundings are situated on Brazil's eastern Atlantic coast, encroaches upon wetlands, watersheds, forests and mangroves. The morphology of the area consists of coastal lowlands and a seashore range: the Serra do Mar mountain range that runs parallel to the coast. Slope angle and hillslope form are aspects relevant to defining critical conditions for mass movements, which are common in the region. Floods and high tides are also frequent in the area; they strongly impact the city of Santos (see Chaps. 4, 6, 7 and 8) and are associated with numerous diseases (Chap. 10).

The coast of São Paulo state, where Santos is located, has a tropical climate regime, with high rainfall and temperatures throughout the year. The constant convection process carries heat to higher altitudes, which in turn generates clouds of great vertical development and copious precipitation, contributing significantly to the outbreak of floods. Tropical, subtropical and midlatitude controls operate in the area, and during summer the region is highly influenced by the South Atlantic Convergence Zone (SACZ), which combines humidity from both Amazonia and the Atlantic. The SACZ features a band of cloud cover and rainfall that remains semi-stationary for several consecutive days; this can lead to slope landslides and flooding (See Chap. 8).

Along the coast, the rainfall totals are high but variable. The maximum volumes are recorded at Represa Itatinga in Bertioga, a town close to Santos, where the annual average rainfall is 4750 mm, though it has reached approximately 7000 mm in some years. Daily amounts greater than 400 mm have been recorded in some sectors along the coast (Nunes 2008).

Despite the steep slopes, soils are relatively deep due to the humid tropical conditions and present high to medium values of hydraulic conductivity. This soil condition facilitates infiltration and interflow along soils and contributes to mass movements. However, intense precipitation episodes can rapidly saturate the soils and prevent infiltration, leading to flooding and landslides.

Social and economic pressures contribute to the destruction of mountain forests and to inappropriate occupational patterns, exposing people to the risk of landslides and floods triggered by heavy rain. In the past, extreme precipitation episodes displaced large quantities of weathering material, with major consequences: in 1928 and 1956, landslides caused 80 and 64 fatalities, respectively (Pichler 1957, in Nogueira 2002). Next to Santos, the city of Cubatão has an important petrochemical, chemical and steel industrial complex that could be jeopardized by an episode of extreme rain and cause an unprecedented tragedy affecting a vast region. According to Nunes (2008), Cubatão has experienced very severe rainfall events in the past: 1021.0 mm in 9 days (February 1929) and 712.0 mm in only 2 days (February 1934). The role of accumulated precipitation as a trigger of landslides and other hazardous occurrences in Santos is explored in Chap. 8.

These aspects highlight the importance of assessing current extreme rainfall patterns in this strategic but fragile region. It is important to remember that the summer holiday season—when the population of Santos can increase by up to four times—coincides with the rainiest period of the year, exposing an increasing number of people to hazardous events triggered by extreme precipitation. Therefore, this study evaluates the current patterns of extreme rainfall in Santos at different time scales, to provide information of great importance for regional and local planning in view of a sustainable and safer use of space.

Despite increasing concern that extreme precipitation episodes may be changing in both intensity and frequency due to climate change (IPCC 2007), the goal of this study was to analyse current patterns of variability and not changes in the frequency of extreme events. Event frequency is recognized as a signal of ongoing climate change (NAST 2001), and its detection would require either the analyses of a longer time series than the one evaluated here, or model projections of extremes.

3.2 Material and Methods

To determine the temporal variability of rainfall in the municipality of Santos, we used the Saboó pluviometric station (SABESP), located at 23° 55′ 12″ S and 46° 20′ 40″ W, and 32 m altitude. For the analysed period (36 years, from 1980 to 2015), we conducted analyses at the annual, seasonal, monthly and daily time scales to observe the variability in rainfall, with emphasis on the extreme positive volumes, which have a considerable impact on Santos. Located near both the port and the commercial centre, the Saboó station reports on conditions that affect activities important to the municipality. It is one of the control stations used by the Municipal Civil Defence to monitor rainfall that causes disruption in the municipality. This same pluviometric station was used in the analyses described in Chap. 8.

In recent decades, various indicators have been developed and adopted to characterise extreme climatic events involving excess or scarcity of rain. The quantile technique, originally proposed by Pinkayan (1966) and widely used by Xavier et al. (2002), was used in this work to identify the distribution patterns of rainfall in Santos, as well as the more and less humid years—bearing in mind that water shortage, to an extent that would qualify as a proper dry season, does not occur in the area at any time of year. We also examined whether the patterns were systemic and, in particular, whether they presented some regularity. As noted by Xavier et al. (2007), the quantiles respond well to the asymmetries—an intrinsic characteristic of rain—in which mean and median rarely coincide.

The quantile technique is based exclusively on rainfall, and its main advantage is its simplicity. However, an understanding of the climatic characteristics of the region is necessary for a better interpretation of the results. In this technique, a quantile of order p (defined as $0<p<1$) is a numeric value the splits the distribution into two parts, with probabilities p to the left of this theoretic quantile, and 1-p to the right, such that the values above and below that established by a given quantile total 100.0% (Xavier et al. 2007).

The quantile technique has been used by various authors: Karl and Knight (1998) observed that values greater than the 90th percentile accounted for 40% of the daily extreme events in the United States. Studying daily rainfall in the state of São Paulo, Brazil, with data from 287 stations, Liebmann et al. (2001) defined an extreme event as that in which the daily rainfall was equal to or exceeded 3%, 4% or 5% of the annual average at each station. The authors observed that most of the extreme episodes were recorded from October to March, and that the relationship between the number of days with extreme events and the annual total is weak. Considering the number of days whose rainfall totals were greater that the 90th, 95th and 97.5th percentiles for 196 stations in India, for the years 1910–2000, Roy and Balling (2004) found an increase in the extreme precipitation events in the country. Examining seasonal precipitation extremes for the southern Andes during the period 1963–2003, Robledo and Penalba (2008) classified the daily volumes based on the 50th, 75th and 95th percentiles, concluding that the occurrences were most concentrated in the winter.

The quantiles used in this work refer to the orders Q(0.15), Q(0.35), Q(0.65), Q(0.85) and Q(0.95), whose purpose is to delimit the levels (or ranges) of severity of the extremes. Thus, $X_1 , X_2 , \ldots , X_n$ is a series of rainfall data accumulated during a certain period of the year, for a given location, over n years of observation: once the quantiles are calculated, the values of year i correspond to a particular class of event, as listed in Table 3.1. Note that the seasons "dry" and "very dry" refer to the relative totals of the other classes, since, as previously indicated, the study area does not experience a typical dry season. The classes were chosen empirically to reflect the local conditions (humid tropical climate); hence the greater number of classes of rainy extremes, attesting to the permanently humid conditions of the region.

Table 3.1 Intervals and categories of the quantile technique used in this study

Interval	Category
$X_i \leq Q(0.15)$	Very Dry (VD)
$Q(0.15) < X_i \leq Q(0.35)$	Dry (D)
$Q(0.35) < X_i \leq Q(0.65)$	Normal (N)
$Q(0.65) < X_i \leq Q(0.85)$	Rainy (R)
$Q(0.85) < X_i \leq Q(0.95)$	Very Rainy (VR)
$X_i \geq Q(0.95)$	Extremely Rainy (ER)

Based on seasonal averages, it is thus possible to identify the years in which excess or scarcity of rain occurred, as well as the degree of severity, in the municipality of Santos during the period 1980–2015. We calculated the number of events per season, considering the months of January, February and March (JFM) as summer; April, May and June (AMJ) as autumn; July, August and September (JAS) as winter and October, November and December (OND) as spring. To define the seasons, we took into account the main atmospheric processes: for example, the SACZ, which is an important mechanism for promoting precipitation in the area, is more common in March than in December. Thus, we defined summer as occurring from January to March, and definition of the other seasons followed from there.

When defining the extreme occurrences, the quantile technique reflects the rainfall patterns normally associated with disaster events. Therefore, the identification of rainfall events at different scales of intensity may guide mitigation and adaptation actions for specific and strategic municipal sectors.

Complementing the analysis of extreme events, we evaluated parameters such as standard deviation, coefficient of variation, and number of rainy days, as well as the degree of concentration of daily rainfall. To evaluate the concentration of rainfall volumes over the historical series, and thus the (ir)regularity of rainfall throughout the period, we used the Concentration Index (CI) (Martín-Vide 2004). The CI defines the contribution of daily rainfall volumes, particularly the extremes, for the entire series and categorises them into 1-mm classes.

3.3 Results

Occurrences of climatic extremes, such as the severe drought in the southeast of Brazil in 2014–2015, entail significant economic impacts (Marengo et al. 2015; Nobre et al. 2016) and have been a constant subject of debate and scientific study in recent decades (IPCC 2014). Within this scenario, we must not only consider the climatological characteristics, but also analyse the behaviour of the variability of precipitation, seeking to identify changes in the rainfall regime that may represent alterations of long-term trends.

Figure 3.1 shows a well-defined annual cycle and a variable rainfall regime caused by the interaction between the atmosphere, the ocean and the regional

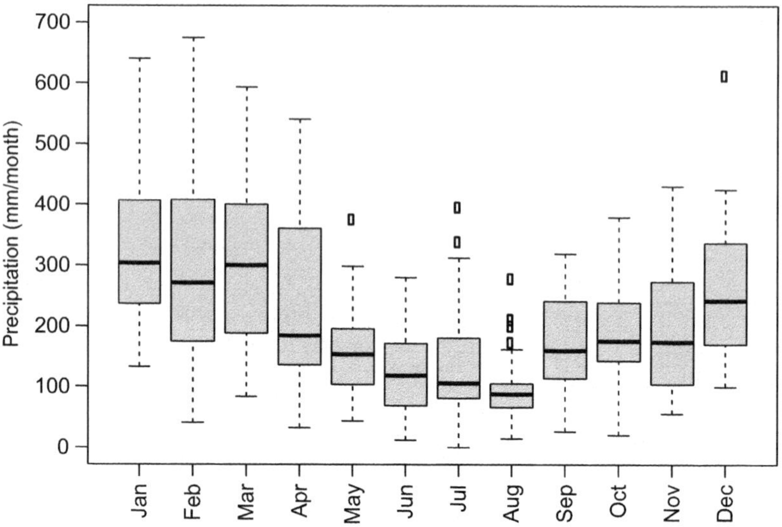

Fig. 3.1 Boxplot of monthly rainfall for Santos/Saboó from 1980 to 2015. Values expressed as (mm/month). The vertical line inside the box shows the median value. The box represents the 25th and 75th percentile (interquartile range) values

Table 3.2 Average monthly mean, median, maximum and minimum values and the difference between them, and the standard deviation and coefficient of variation for the series (1980–2015)

	Avg	Median	Max	Min	Max/Min	StdDev	CV	Avg rainy days
Jan	327.7	117.6	640.6	132.6	4.8	127.7	39.0	17
Feb	293.3	176.4	674.9	40.6	16.6	184.8	63.0	14
Mar	308.1	148.5	770.5	83.3	9.2	165.0	53.6	15
Apr	237.1	132.4	540.4	32.1	16.8	140.3	59.2	12
May	157.9	71.9	376.4	43.2	8.7	79.7	50.5	11
Jun	129.9	70.3	279.8	11.7	23.9	74.2	57.1	8
Jul	133.0	90.7	396.4	0.0	396.4	613.1	461.1	10
Aug	97.3	54.9	278.8	14.4	19.4	61.8	63.5	10
Sep	177.4	77.0	318.7	26.2	12.2	81.1	46.4	13
Oct	195.7	94.1	378.7	20.1	18.8	99.2	50.7	15
Nov	196.0	111.2	430.4	55.6	7.7	115.6	59.0	13
Dec	257.4	113.8	613.9	99.9	6.1	126.1	49.0	15

physiography. Relative to the extreme monthly values (size of the lower and upper whiskers), the wide variability throughout the year is confirmed. However, it is in the dry period (JAS) that we observe a predominance of outliers. On average, January is the rainiest month; however, February was the month in which the most extreme values were concentrated, with urban impacts as a direct consequence.

Table 3.2 presents information on the monthly and daily levels, confirming that despite the greater volumes of rainfall in the area, there was no rain in the month of

July 2008, which widened the difference between the maximum and minimum totals for the month. Conversely, although the highest totals were recorded in January, the month was the most consistent, as demonstrated by both the coefficient of variation and the ratio between the maximum and the minimum monthly rainfall volumes in the analysed period. The uniformity observed in January emphasises that the mechanisms promoting precipitation in the area, which differ throughout the year, are more consistent in the summer. Some months presented very high maximum volumes, particularly March 1996, the rainiest month in the time series, with 770.5 mm.

Table 3.2 also shows the number of days that had precipitation (starting at 0.1 mm), independent of the amount. January, the rainiest month, had rainfall on more than half of its days; February had rain on half of its days; and December, March and October, on nearly half of their days. June presented the fewest rainy days on average. In general, the average of rainy days per year is high: 154 days, which is typical for a humid tropical regime. The correlation between the monthly rainfall average and the rainy days per month was 0.86, revealing a strong relationship between these two elements. The correlation between the annual averages and the number of rainy days per year was 0.38, showing that a very rainy year does not necessarily consist of many rainy days.

Table 3.3 shows that 36.5% of the total annual rainfall is concentrated in the months of January–March, representing an average of 929.1 mm. The high totals for these months are associated with the convection of the South Atlantic Convergence Zone (SACZ) and cold fronts that come from the south, which, in this case, culminate in a sequence of days with recorded precipitation. In addition to these mechanisms, the summer convection, favoured by the high temperatures and constant humidity, generates strong, localized and brief rains, usually in late afternoon. In the winter (JAS), precipitation diminishes and is almost totally dependent on cold fronts; hence, in years with less incursion of extratropical masses, the reduction in rainfall is accentuated, as discussed by Monteiro (1973). During the least humid season (relative, given that the area does not undergo the water deficit typical of a proper dry season), the average rainfall is 405.1 mm, which corresponds to 16.3% of the annual average of 2508.2 mm. Table 3.3 also shows that the two most rainy seasons (summer and spring) contribute on average more than 60.0% of the annual total, and the intraseasonal variability is high, particularly in the least rainy period (autumn and winter), as the coefficient of variation values indicate.

Thus, the mechanisms that promote precipitation in the rainiest period (spring and summer) are most persistent. Furthermore, the spring and summer are the two seasons with the most days of recorded rainfall, though even in the autumn and winter, the number of rainy days is high. Considering the very wet and extremely wet events, as defined by the quantile technique (Table 3.1), the spring presented the highest number of extremes occurrences, while summer presented the fewest (see Fig. 3.3). The daily maximum occurred in summer: 332.9 mm of rain fell in only 1 day (16 January 1992), the value exceeding the monthly average of the rainiest month in the series. That 1-day rainfall corresponded to 13.3% of the series' annual average.

Table 3.3 Seasonal totals, percentage of rainfall in relation to annual average, seasonal variability (CV), average number of rainy days, number of extreme and very extreme wet events, and the seasonal daily maximum

Summer	Total (mm)	929.1
	%	36.5
	CV	122.7
	Avg Rainy Days	15
	Very wet events	3
	Wettest day (mm)	332.9 (16 Jan 1992)
Autumn	Total (mm)	524.9
	%	21.0
	CV	239.6
	Avg Rainy Days	11
	Very wet events	4
	Wettest day (mm)	188.9 (5 Apr 2010)
Winter	Total (mm)	405.1
	%	16.3
	CV	162.7
	Avg Rainy Days	11
	Very wet events	4
	Wettest day	148.9 (9 Sep 1996)
Spring	Total (mm)	649.2
	%	26.1
	CV	132.7
	Avg Rainy Days	14
	Very wet events	6
	Wettest day	194.0 (16 Dec 2000)

Despite the uniformity of the precipitation in the summer months, which presented high volumes in all years of the study period, the city is routinely confronted with the problems caused by intense rains, such as landslides and floods or associated phenomena such as strong storm surges, which reveals the lack of planning appropriate to the characteristics of the area, an issue explored in Chap. 8. Nevertheless, the totals for the rainiest period in the municipality of Santos diminished over the 36 analysed years (Fig. 3.2), despite the marked interannual variability of the summer months.

Figure 3.3 presents the frequency distribution of the seasonal extreme events, based on the quantile technique. In general, the rainfall totals for the municipality of Santos largely fit into the category of normal or common for each season of the year. However, although the number of dry events is relatively higher than that of wet events, the most important aspect related to urban impacts may be the peak of wet events that show up in the transition seasons of autumn (AMJ) and spring (OND). We also note few moderately wet events in winter (JAS). While there are relatively few events in the most extreme class (extremely wet), the daily totals in this category for the spring and the summer are the highest, generating much disruption within the city—particularly in the summer as it coincides with school holidays—and can affect port operations central to the local, regional and national economies.

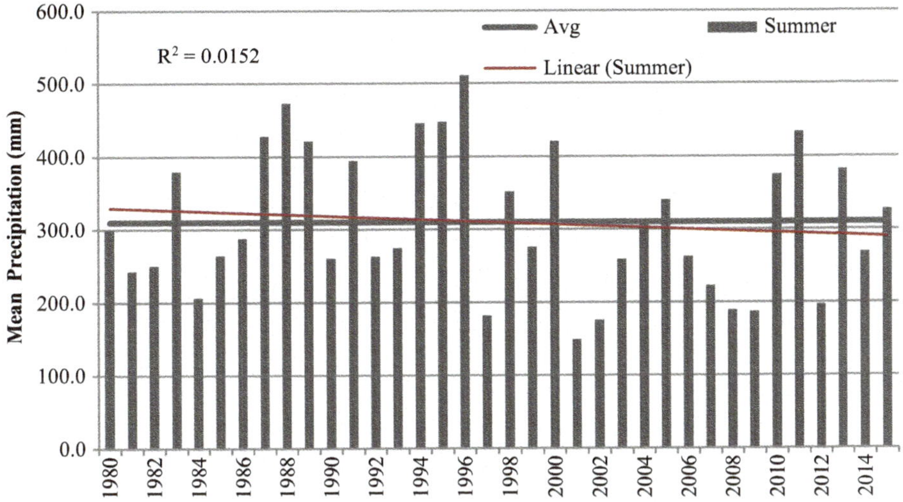

Fig. 3.2 Precipitation variability in Santos/Saboó during summer (JFM)

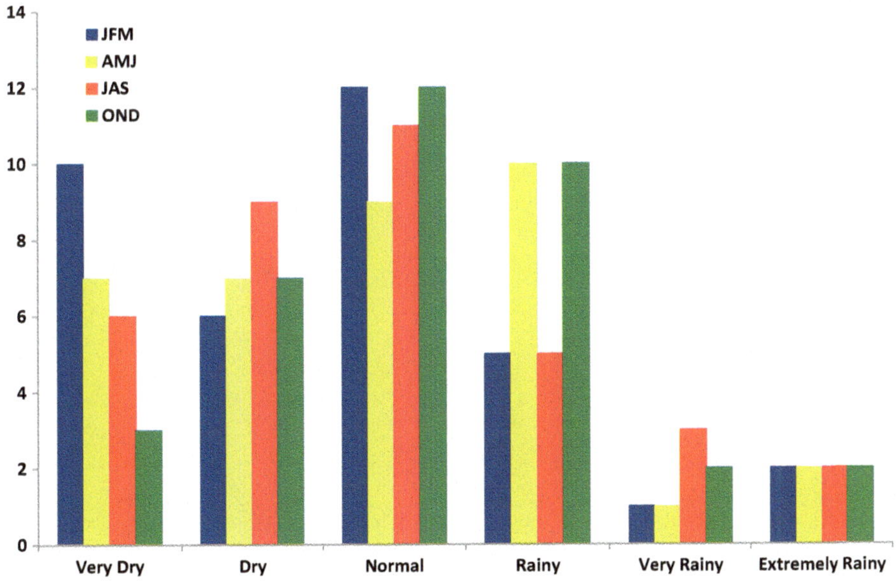

Fig. 3.3 Seasonal frequency distribution of extreme events for Santos/Saboó. Colours represent different seasons. Category: very dry, dry, normal, rainy, very rainy and extremely rainy conditions. (See Table 3.1)

Finally, the analysis of the Concentration Index (CI) provided information regarding the daily distribution of precipitation in the analysed period. The daily rainfall totals for all days in the series (1980–2015) were categorised into 1-mm classes, as follows: 0.1–0.9, 1.0–1.9, 2.0–2.9 and so on. The last class, 332.0–332.9, contained the highest recorded value in the series: 332.9 mm (Table 3.3).

Over the 36-year period, precipitation occurred on 5610 days, representing 43% of all days in the period. Not every class contained a record of rainfall; for example, the classes between that of the second highest (201.0–201.9) and the highest daily rainfall class (332.0–332.9) had frequencies of 0 (as did some other classes). Because the daily rainfall frequencies are higher for the lowest rainfall classes and vice versa, regardless of the climatic regime, the relationship between the sum of rainy days and the accumulated rainfall is exponentially negative (Martín-Vide 2004). Figure 3.4 presents this information for the Saboó station, considering the daily rainfall classes, between 1980 and 2015.

In the case of a series with perfectly regular daily rainfall distribution, the ideal extreme CI values would be 0 for the same daily quantities and 1 for a single day of rain. Thus, the degree of irregularity of daily rainfall can be evaluated through comparison of its values in relation to the ideal case, i.e. 100% regular distribution. The farther the series' distribution is from the ideal curve, the more irregular the distribution of daily rainfall.

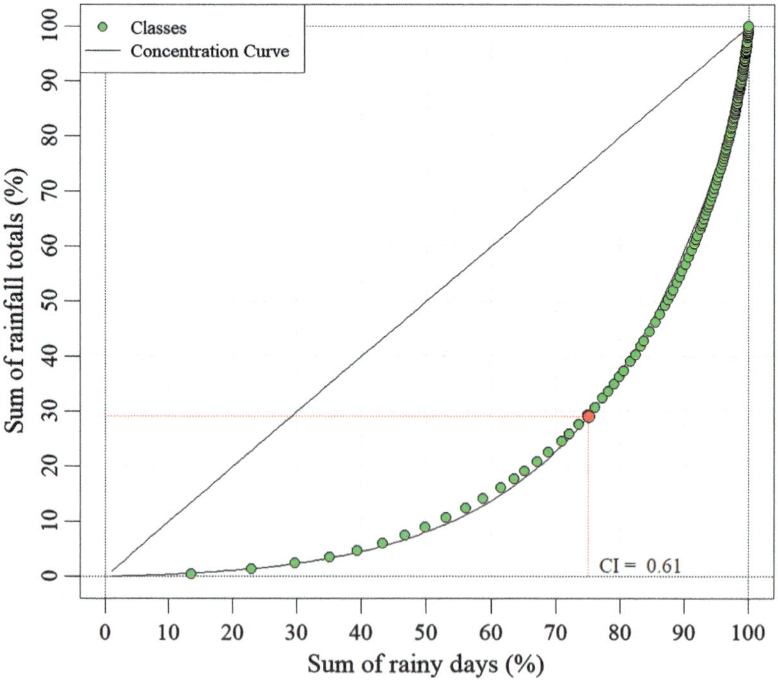

Fig. 3.4 Concentration curve for Saboó in relation to the perfect daily rainfall distribution

Figure 3.4 shows the degree of irregularity in the daily rainfall distribution at the Saboó station relative to the hypothetical ideal.

In addition to the visual confirmation, the CI value calculated for Saboó station (for more information, see Martín-Vide 2004) was 0.61, which, according to Martín-Vide (2004), indicates the poor distribution of the rains. Demonstrating this characteristic, the rainfall accumulated over 25% of the rainiest days was equal to 71% of the total rainfall, which emphasises the irregularity of the rainfall distribution.

3.4 Final Remarks

The analysis of extremes is an important element of characterising precipitation, because although they are rare by definition, they have the potential to trigger alterations in the physical environment and strongly impact human activities.

The assessment of negative (below normal) and positive (above normal) extreme events in a series of rainfall records for the central area of Santos describes the conditions experienced by a large portion of the population that resides in or travels through the municipality.

The extremes were defined by the quantile technique and categorized into five classes: two classes for the drier episodes and three for the wetter episodes. Common or normal events were grouped into a sixth class.

At the annual scale, the area presents considerable inter- and intra-annual variability, with volumes concentrated in the spring-summer period. The autumn months, but especially the winter (April–September), are more variable—the occurrence of precipitation during this period is strongly influenced by the presence of cold fronts, which vary from year to year. The summer is more uniform, despite the presence of more mechanisms that generate rain. At the monthly scale, January is the rainiest as well as the most consistent, proving that the controls that promote precipitation in this period are more regular.

Other parameters, such as the coefficient of variation, were calculated to enrich the analysis of variability. The definition of the Concentration Index (CI) is particularly important for showing the poor distribution of daily rainfall in Santos, with most of the rainy days presenting low volumes and with few events contributing significantly to the total rainfall. As addressed in Chap. 8, the main disturbances that affect the municipality are related to above-average and accumulated rainfall, demonstrating the importance of its definition for the area.

The relatively short analysed period (36 years), the type of analysis used, and the fact that the study was based only on a pluviometric series does not allow conclusions about the possible alteration of pluviometric variability in the region, nor about any pattern that may be representative of climate change. However, this study indicates that, given the local climatic regime (wet tropical), extremely high volumes are possible, such as 770.5 mm in a single month or even 332.9 mm in a single day. In the case of high tide (see Chap. 6) and a sequence of rainy days (Chap. 8), all orders of disruption and damage can strike not only the municipality but also the entire region, which could impact the nation were the operation of the port of Santos to be affected.

Acknowledgements To the Civil Defence of Santos, for ceding the data from the Saboó station.

References

Berndtsson, R., & Niemczynowicz, J. (1988). Spatial and temporal scales in rainfall analysis-some aspects and future perspectives. *Journal of Hydrology, 100*, 293–313. https://doi.org/10.1016/0022-1694(88)90189-8.

IPCC Intergovernmental Panel on Climate Change (2007) Climate change 2007: Synthesis report. https://www.ipcc.ch/publications_and_data/publications_ipcc_fourth_assessment_report_synthesis_report.htm. Accessed 17 April 2017.

IPCC Intergovernmental Panel on Climate Change. (2012a). Glossary of terms. In: *Managing the risks of extreme events and disasters to advance climate change adaptation*. https://www.ipcc.ch/pdf/special-reports/srex/SREX-Annex_Glossary.pdf. Accessed 17 Apr 2017.

IPCC Intergovernmental Panel on Climate Change. (2012b). *Managing the risks of extreme events and disasters to advance climate change adaptation, Special Report of the IPCC*. Cambridge: IPCC.

IPCC Intergovernmental Panel on Climate Change. (2014). *Climate change 2013 – the physical science basis*. Cambridge: IPCC.

Karl, T. R., & Knight, R. W. (1998). Secular trends of precipitation amount, frequency, and intensity in the United States. *BAMS, 79*, 231–241. https://doi.org/10.1175/1520-0477(1998)079<0231:STOPAF>2.0.CO;2.

Liebmann B, Jones C, Carvalho, LMV de (2001) Interannual variability of daily extreme precipitation events in the State of São Paulo, Brazil. JCLI 14: 208–218. https://doi.org/10.1175/1520-0442(2001)014<0208:IVODEP>2.0.CO;2.

Marengo, J. A., Nobre, C., Seluchi, M., et al. (2015). A seca e a crise hídrica de 2014-2015 em São Paulo. *Revista USP, 106*, 31–45. https://doi.org/10.11606/issn.2316-9036.v0i106p31-44.

Martín-Vide, J. (2004). Spatial Distribution of a daily precipitation concentration index in Peninsular Spain. *International Journal of Climatology, 24*, 959–971. https://doi.org/10.1002/joc.1030.

Monteiro, C. A. d. F. (1973). *A dinâmica climática e as chuvas no Estado de São Paulo: estudo geográfico sob a forma de atlas*. São Paulo: Instituto de Geografia/USP.

NAST National Assessment Synthesis Team. (2001). *Climate change impacts on the United States: the potential consequences of climate variability and change, Report for the US Global Change Research Program*. Cambridge: NAST National Assessment Synthesis Team.

Nobre, C. A., Marengo, J. A., Seluchi, M. E., et al. (2016). Some characteristics and impacts of the drought and water crisis in Southeastern Brazil during 2014 and 2015. *JWARP, 8*, 252–262. http://file.scirp.org/Html/13-9402744_63776.htm.

Nogueira, F. R. (2002). *Gerenciamento de riscos ambientais associados a escorregamentos: contribuição às políticas públicas municipais para áreas de habitação subnormal*. Thesis, Universidade Estadual Paulista.

Nunes, L. H. (2008). *Critérios para a definição de eventos extremos – estudo de caso para o litoral central paulista*. Paper presented at the 8th Brazilian Symposium of Geographical Climatology, University of Uberlândia, Alto Caparaó, MG 24–29 August 2008.

Pichler, E. (1957). Aspectos geológicos dos escorregamentos de Santos. *Boletim da Sociedade Brasileira de Geologia, 6*(2), 68–77.

Pinkayan, S. (1966). *Conditional probabilities of ocurrence of wet and dry years over a large continental area*. Boulder: Colorado State University. http://hdl.handle.net/10217/61293 . Accessed 13 Sept 2017.

Robledo, F. A., & Penalba, O. C. (2008). *Study of the frequency of extreme daily precipitation on the south of Andes Mountain Range. Temporal variability in the period 1961–2003 and relation with Antarctic Oscillation*. Paper presented at the 20th. Conference on Climate Variability and Trend, New Orleans Ernest N. Morial Convention, 20–24 January 2008.

Roy, S. S., & Balling, R. C., Jr. (2004). Trends in daily extreme precipitation indices in India. *International Journal of Climatology, 24*, 457–466.

WMO World Meteorological Organization. (2003). *Technical note agrometeorology related to extreme events*. WMO 943. https://library.wmo.int/opac/doc_num.php?explnum_id=3442. Accessed 23 May 2017.

Xavier, T. d. M. S., Silva, J. F., & Rebello, E. R. G. (2002). *A técnica dos quantis e suas aplicações em meteorologia, climatologia e hidrologia em regiões brasileiras*. Brasília: Thesaurus Editora de Brasília, Brasília.

Xavier, T. M. S., Xavier, A. F. S., & Alves, J. M. B. (2007). *Quantis e eventos extremos. Aplicações em Ciências da Terra e Ambientais*. Fortaleza: RDS.

Chapter 4
Projections of Climate Change in the Coastal Area of Santos

Sin-Chan Chou, José A. Marengo, Adan J. Silva, André A. Lyra, Priscila Tavares, Celia Regina de Gouveia Souza, Joseph Harari, Lucí H. Nunes, Roberto Greco, Eduardo K. Hosokawa, Luiz E. O. C. Aragão, and Lincoln M. Alves

Abstract The objective of this work is to assess the projections of climate change in the city of Santos. The assessment is based on the downscaling of two global climate model simulations using the Eta Regional Climate Model at 20-km and 5-km resolutions, under RCP4.5 and RCP8.5 scenarios for the period between 1961 and 2100. The higher horizontal resolution simulations reproduce in more detail the surface characteristics, such as the topography, vegetation cover, and coastline, and

---

S.-C. Chou (✉) · A. J. Silva · A. A. Lyra · P. Tavares · L. E. O. C. Aragão · L. M. Alves
National Institute for Space Research (INPE), São José dos Campos, SP, Brazil
e-mail: chou.chan@inpe.br; laragao@dsr.inpe.br; Lincoln.alves@inpe.br

J. A. Marengo
National Center for Monitoring and Early Warning of Natural Disasters (CEMADEN), São José dos Campos, São Paulo, Brazil
e-mail: jose.marengo@cemaden.gov.br

C. R. G. Souza
Institute of Geology – Secretariat for the Environment of the State of São Paulo (IG-SMA/SP), São Paulo, SP, Brazil

Post-Graduate Programme on Physical Geography – Faculty of Philosophy, Languages and Human Sciences, University of São Paulo (FFLCH-USP), São Paulo, SP, Brazil
e-mail: celia@sp.gov.br

J. Harari
Institute of Oceanography, University of São Paulo, São Paulo, SP, Brazil
e-mail: joharari@usp.br

L. H. Nunes
Santos, São Paulo, Brazil
e-mail: luci@ige.unicamp.br

R. Greco
Institute of Geosciences, University of Campinas (Unicamp), Campinas, SP, Brazil
e-mail: greco@ige.unicamp.br

E. K. Hosokawa
Municipal Government of Santos, Secretariat of Urban Development, Santos, SP, Brazil
e-mail: eduardohosokawa@santos.sp.gov.br

L. H. Nunes et al. (eds.), *Climate Change in Santos Brazil: Projections, Impacts and Adaptation Options*, https://doi.org/10.1007/978-3-319-96535-2_4

capture the extreme climate events. Evaluation of the model simulations of the present climate show reasonable agreement with observed climatology. Frequency distributions of precipitation and temperature values show that the 5-km run approaches the observed precipitation better than the 20-km resolution run. The assessment of climate change projections indicates that warming in the region reaches about 2 °C until the end of the twenty-first century, and that precipitation reduces in the entire region. Trends of climatic extreme indices show increase of hot days, warm nights, and in the length of consecutive dry days with the increase of the atmospheric greenhouse gas concentrations. Projections of the minimum surface pressure off the coast of Southeast Brazil show weakening tendency under RCP8.5 scenario.

Keywords Santos · Climate projections · Dynamical downscaling · Extreme-climate indices · Storms

4.1 Introduction

Events of storm surges have become more frequent mainly since the end of the nineties, as registered by long-term observations along the coasts of the States of São Paulo and Rio de Janeiro. Storm surges have caused coastal erosion along these coastlines, some economic losses related to destruction of urban structures and disruptions in ports and tourism activities (see Chap. 6). These surges are related to quasi-stationary or slow-moving extratropical cyclones off the coasts of South and Southeast Brazil.

In the city of Santos, one of the most important coastal municipalities in Brazil, long-term observations have detected the increase of sea level (see Chap. 5). This work assesses the trend of extratropical cyclonic storms affecting the coast of Southeast Brazil in the past and in the future climate, under the emission scenarios RCP4.5 and RCP8.5. The assessment is based on the detection of extratropical cyclones reproduced by the downscaling of global climate model simulations using the Eta Regional Climate Model at 20-km and 5-km resolutions. The downscaling reproduces, at higher resolution, the simulations from global climate models the HadGEM2-ES and MIROC5, under RCP4.5 and RCP8.5 scenarios.

This chapter assesses the trend of extratropical cyclonic storms affecting the coast of Southeast Brazil in the past and in the future climate, under the emission scenarios RCP4.5 and RCP8.5. The assessment is based on the detection of extratropical cyclones reproduced by the downscaling of global climate model simulations using the Eta Regional Climate Model at 20-km and 5-km resolutions. The downscaling reproduces, at higher resolution, the simulations from global climate models the HadGEM2-ES and MIROC5, under RCP4.5 and RCP8.5 scenarios.

4.2 Downscaling of Global Climate

In order to integrate for decades and centuries, global climate models adopt resolutions that may vary around 1° and 3° of latitude and/or longitude. At these resolutions, the representation of topography, vegetation cover, coastline etc., is inevitably coarse. To assess the impact, vulnerability and adaptation to climate change, higher spatial resolution is required as the related problems have local scale features.

The Eta Regional Climate Model is developed by INPE (Chou et al. 2012; Marengo et al. 2012) to provide climate change projections over South America at higher resolution. The model dynamics uses finite volume and step-mountain coordinate. The recent version has upgrades as described in Mesinger et al. (2012) and has been evaluated for long-term simulations (Pesquero et al. 2010; Chou et al. 2012, 2014a, b).

The model is setup with 20-km resolution in the horizontal and 38 layers in the vertical. The downscaling runs are driven by the global climate models MIROC5 and HadGEM2-ES, under the RCP4.5 and RCP8.5 emission scenarios, for the period between 1961 and 2100. The model domain covers the entire South America and Central America. These Eta model simulations are evaluated in Chou et al. (2014a) and the projections assessed in Chou et al. (2014b). Assessment of the global climate models can be found in Flato et al. (2013).

A second nesting is produced by downscaling the runs to 5-km resolution. Due to the higher resolution, the domain is reduced to cover part of Southeast Brazil, where the city of Santos is located. This higher resolution version uses the non-hydrostatic version of the Eta model and its evaluation can be found in Lyra et al. (2017).

The Eta model continuous simulations are carried out without the need to apply nudging to the lateral boundaries or to internal domain. In addition, the model topography does not need smoothing for the long-term runs.

Model evaluation carried out by Yin et al. (2013) of the simulations of 11 CMIP5 global climate models showed that the HadGEM2-ES model had the best performance of the surface conditions and the atmospheric circulation. HadGEM2-ES was one of the models, among 19, that presented the highest spatial correlation between simulations and rainfall observations (Gulizia and Camilloni 2015) for the South American region, south of the equator. Downscaling of the projections of the global climate models, HadGEM2-ES and MIROC5, were carried out by the Eta model for the entire South America in the spatial resolution of 20 km (Chou et al. 2014a). These projections showed more intense temperature increase in the central part of the country, expanding throughout the country until the end of the twenty-first century. On the other hand, the projections showed reduction in rainfall in the region that extends from the southern part of the Amazon up to the Southeast region.

In the Fifth Assessment Report (AR5) from the Intergovernmental Panel on Climate Change (IPCC) (2013), the greenhouse gas concentration scenarios are based on the Representative Concentration Pathways (RCP), which are expressed in terms of radiative forcing at the end of the twenty-first century. The scenarios used in this work are RCP8.5 and RCP4.5 (van Vuuren et al. 2011), which raised the radiative forcing pathway respectively to 8.5 Wm^{-2} and 4.5 Wm^{-2} in 2100.

4.2.1 *Validation of the Downscaling*

Prior to the assessment of the climate change projected by the downscaling of the Eta regional model for the Santos area, the present climate simulations should be evaluated in order to provide some indication of the reliability of the projections.

4.2.1.1 Simulated Climatology

The mean temperature of the simulations varies from about 18 °C during winter up to about 26 °C in summer (Fig. 4.1). The southern part of the area, toward the coast, is always warmer than inland areas. This horizontal temperature gradient is due to the presence of Serra do Mar mountain range whose orientation follows approximately the coastline. These temperatures are slightly cooler than the reported observed climatology.

Mean precipitation is also more intense toward the coast, where values reach about 8 and 10 mm/day during summer and reduce to about 5 mm/day during winter (figures not shown). The precipitation gradient across the coast, reducing inward, is similar to the pattern observed by Barbosa (2008). Precipitation simulations generally show underestimate during rainy season and overestimate during winter. This feature is inherited by the global climate model HadGEM2-ES. Runs driven by Era-Interim as the lateral boundaries do not show this shift in the annual cycle of precipitation.

4.2.1.2 Frequency Distribution

The mean climatology is very similar between the 20-km and the 5-km resolutions Eta model runs. An important distinction between the runs from these two resolutions is revealed by the frequency distribution, which shows that the higher spatial resolution simulation can reach more extreme values and can approach the observed distribution.

The frequency distribution (Fig. 4.2) of maximum temperatures shows good agreement between the 5-km simulations and the observations, particularly in the extreme values of about 38 °C. The simulated mean annual cycle of minimum temperature is warmer than the observations. The coldest temperatures are generally observed in July, but model driven by HadGEM2-ES generally delays the coldest temperatures by about a month. The frequency distribution of the minimum temperatures of the two model resolutions is very similar, but the 5-km version runs produce somewhat warmer minimum temperatures.

The increase of resolution to 5 km has been beneficial to the simulation, in particular for the maximum temperatures and extreme precipitation.

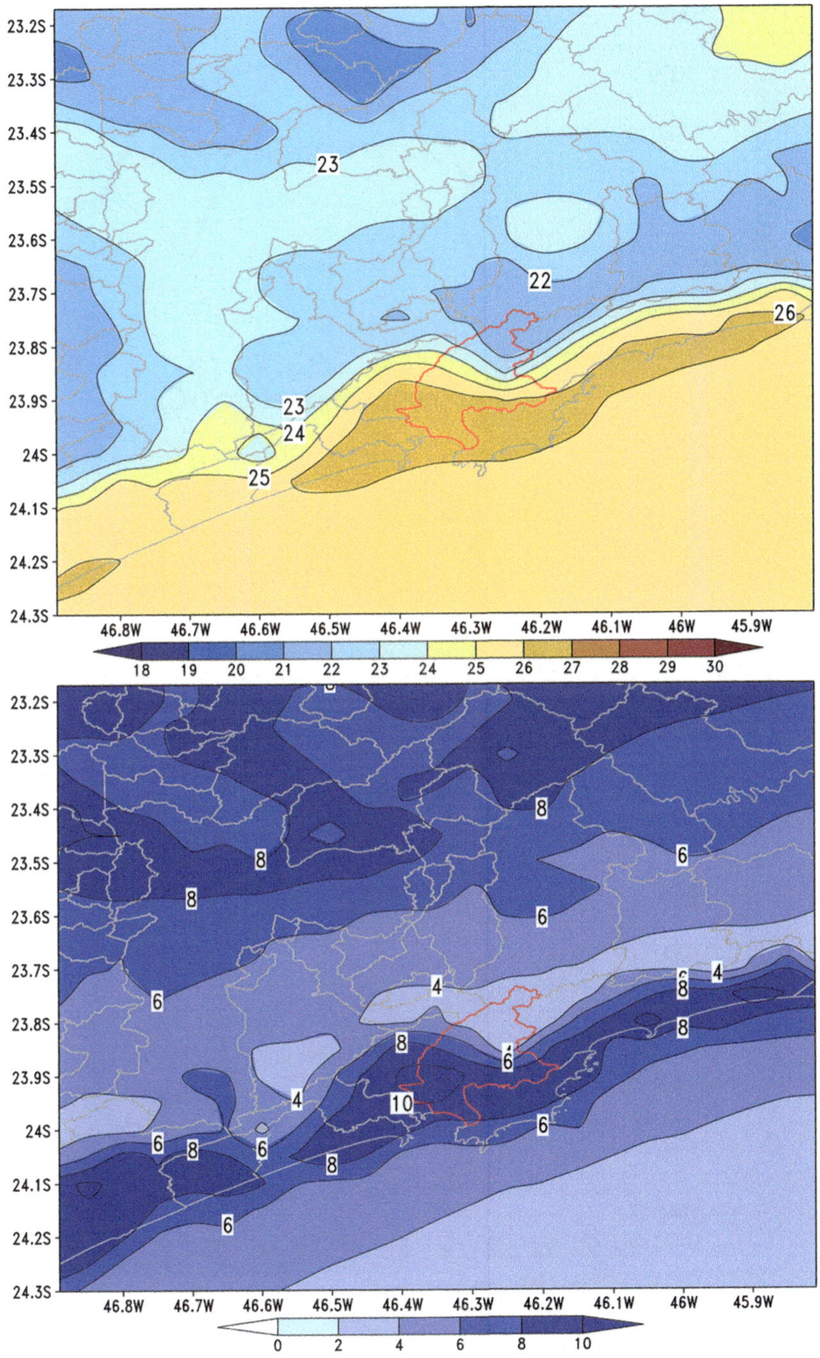

Fig. 4.1 Downscaling simulations of the Eta model 5-km resolution for Santos area for the 30-year period, from 1961 to 1990. The borderline of the city of Santos is contoured in red. The mean December–January–February temperature (°C) is shown in the top row and the mean December–January–February precipitation (mm/day) is shown in the bottom row, for the period 1961–1990 (left column)

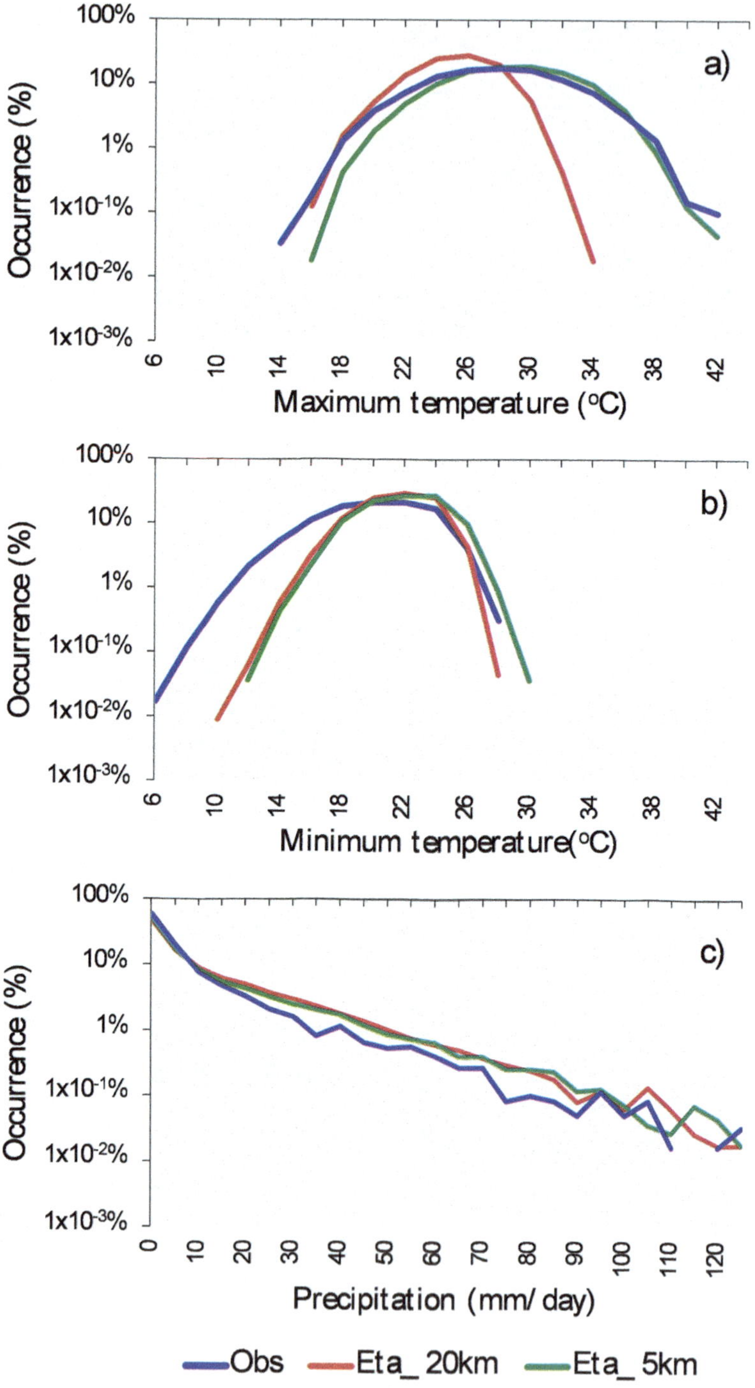

Fig. 4.2 Frequency distributions of (**a**) maximum and (**b**) minimum temperatures (degree Celsius), and (**c**) precipitation (mm/day). The blue line refers to the observations, the red dotted lines refer to Eta at 20-km resolution and green dotted lines refer to the Eta-5 km resolution runs. The y-axis are plotted in log scale in order to show the low frequency occurrences, although those values may not be frequent, they are related to the extreme events

4.2.2 Projections of Climate Change

4.2.2.1 Mean Temperature and Precipitation

Projections of air temperature at 2 m indicate a more intense warming in the continent (Fig. 4.3). This increase in temperature change intensifies as the CO_2 concentration of the scenario increases, reaching about 4 °C at the end of the twenty-first century. Projections show stronger warming during summer (DJF).

Projections of precipitation change in Santos in summer (DJF) indicate reduction in the first period, from 2011 to 2040. Whether driven by HadGEM or MIROC5, or under the RCP4.5 or RCP8.5 scenarios, all four downscaling options, indicate rainfall reduction, with more intensity in the Eta projections nested into HadGEM2-ES in the RCP8.5 scenario. However, the trend changes in the second period, from 2041 to 2070, especially in the Eta runs nested into MIROC5, which becomes neutral. At the end of the century, there is disagreement between the downscaling runs from HadGEM and those from MIROC5. While the Eta nested into HadGEM runs reduces precipitation, the Eta into MIROC5 increases in the region. Cavalcanti et al. (2017) projected monthly precipitation changes for the end of the twenty-first century for the Baixada Santista Metropolitan Region, which includes the city of Santos, from results of the Eta Regional model at 40-km horizontal resolution and 38 vertical levels. The authors also found decrease of precipitation from middle winter to early spring months, indicating drier conditions in the dry season. Figure 4.3b shows the strong reduction of precipitation toward the end of the century, indicated by the Eta run driven by HadGEM2-ES under RCP8.5 scenario.

4.2.2.2 The Frequency Distribution

The frequency distribution of precipitation is plotted on a logarithmic scale to clearly illustrate the extreme strong precipitation rates, which generally exhibit small frequency. This distribution takes into account the daily precipitation throughout the entire year, not only in the summer or winter months. The temperature distributions are displaced toward warmer values in the future time slices at the end of the century in both RCP4.5 and RCP8.5 scenarios (Fig. 4.4a). While the 2-m temperature distribution curves in the first and second future time slices are close to each other in both RCPs, in the third future time slice, the temperature distribution in RCP8.5 is clearly detached and increases fast. The precipitation distribution curves (Fig. 4.4b) are relatively close to one another in the two RCPs and in the three future time slices. However, in Santos, the RCP8.5 scenario does not show clear extreme values at the end of the century. These extreme values of precipitation are projected to occur in the middle of the century in both the RCP4.5 and RCP8.5 scenarios.

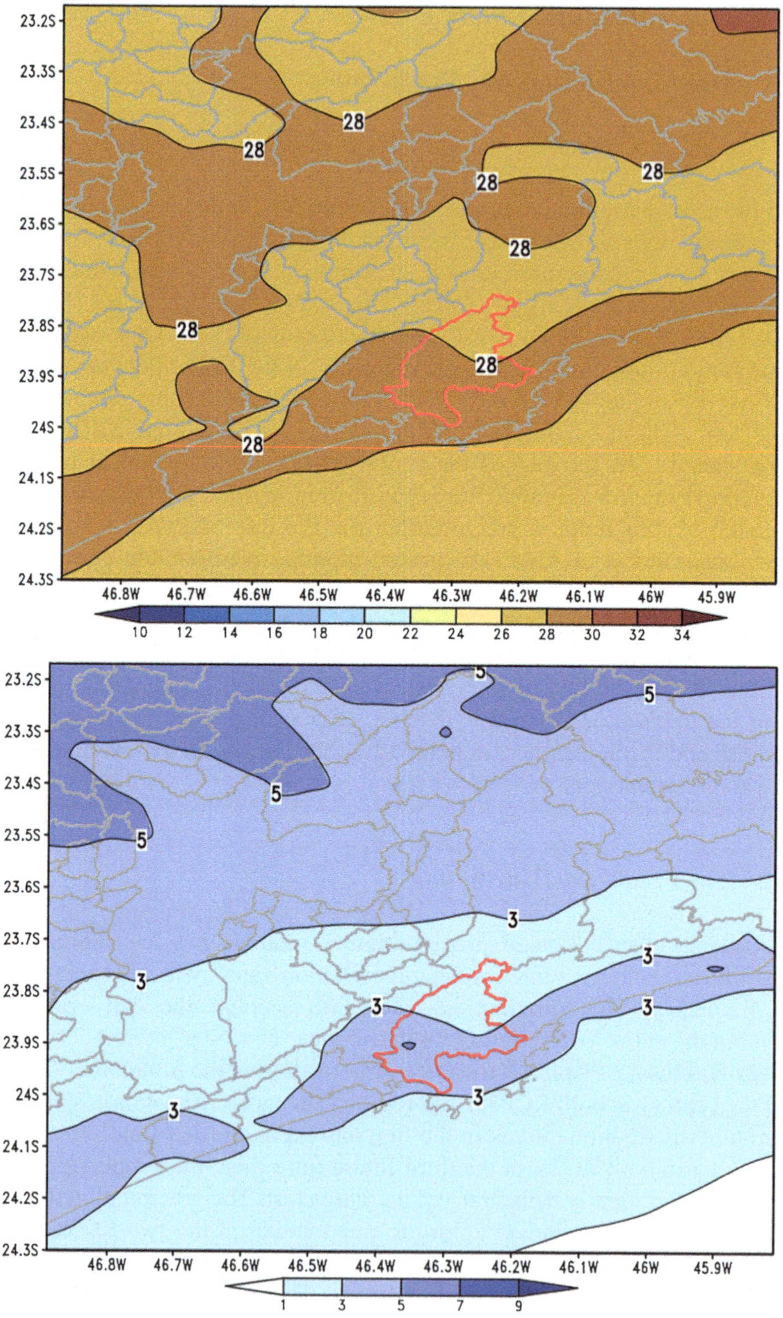

Fig. 4.3 As in Fig. 4.1, but for the future period 2071–2100

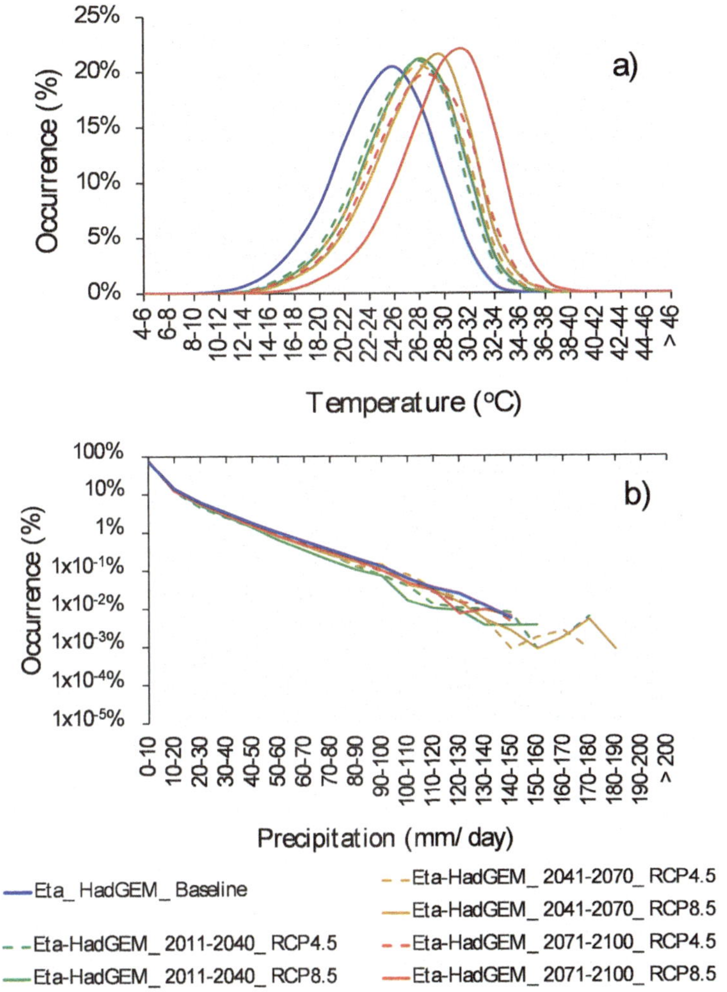

Fig. 4.4 Frequency distributions of Eta-5 km resolution runs (**a**) temperatures (°C), and (**b**) precipitation (mm/day), for the baseline and the three future timeslices. The blue line refers to the period 2011–2040, the green lines refer for the period 2041–2070, and red lines refer to the end of the century, 2071–2100, from the. Only precipitation frequency of occurrence is plotted in log scale

4.2.2.3 Trends of the Climatic Extreme Indices

Some extreme indices according to Alexander et al. (2006) are calculated. These extreme indices are also shown in Marengo et al. (2017).

The projections of temperature increase for the city of Santos show values that reach between 2 and 4.5 °C at the end of the twenty-first century, with an increase in the amplitude of temperature variability (Fig. 4.5). The occurrence of hot nights

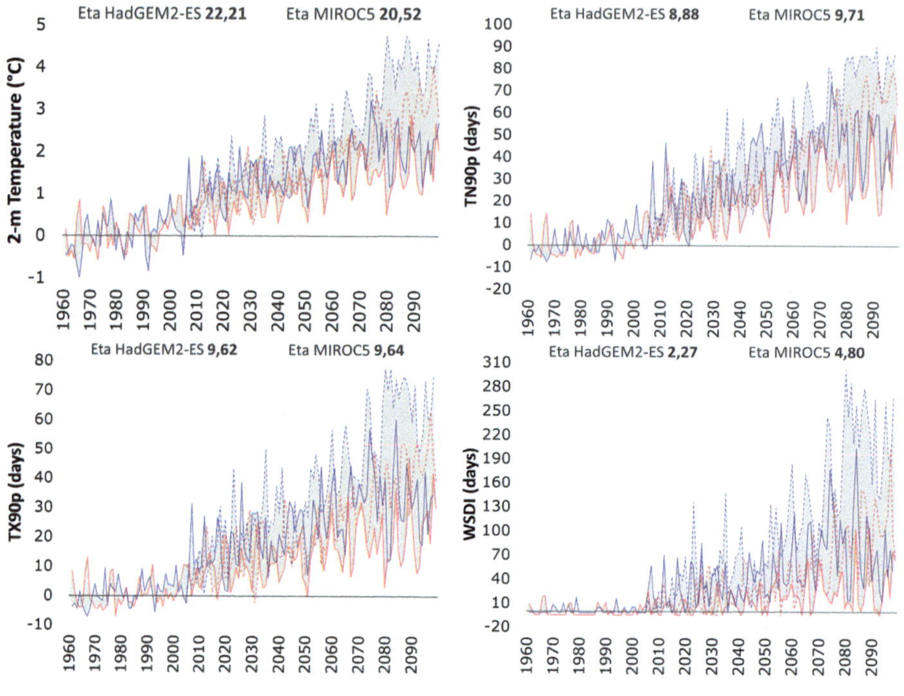

Fig. 4.5 Temperature trends of changes of climatic extremes in the city of Santos: (**a**) the annual mean 2-m temperature (°C), (**b**) TN90p, the 90th percentile of the annual minimum temperatures, (**c**) TX90p, the 90th percentile of the annual maximum temperature, and (**d**) the length of heat waves (days). The downscaling runs driven by HadGEM2-ES are in blue and driven by MIROC5 are in red. The solid lines refer to the RCP4.5 and the dashed lines refer to the RCP8.5. The grey shaded area denotes the amplitude of the related uncertainties. The values on top of the graphs are the mean values of the historical period

(TN90p) increase in the frequency by about 40–80% of the days of the year at the end of the century. In addition, the heat waves index show increased duration trend toward the end of the century.

The projections of total annual rainfall indicate a greater number of events of negative anomaly in comparison with the present period in Santos (Fig. 4.6). There are several events of above normal rainfall, but negative anomaly rain events predominate. The projections of the consecutive days with no-rain index (CDD) indicate an increase in the duration of the dry days, which may have caused a reduction in total annual rainfall over the years as the concentration of greenhouse gases increases. Consecutive wet days (CWD) index, in general, shows the tendency for shorter lengths of wet days. The RX5day index indicates the maximum accumulated rainfall in 5 days for each year. Projections of this index show increased variability in the future climates. The RX5day values exhibit larger fluctuations in future periods than in the baseline climate.

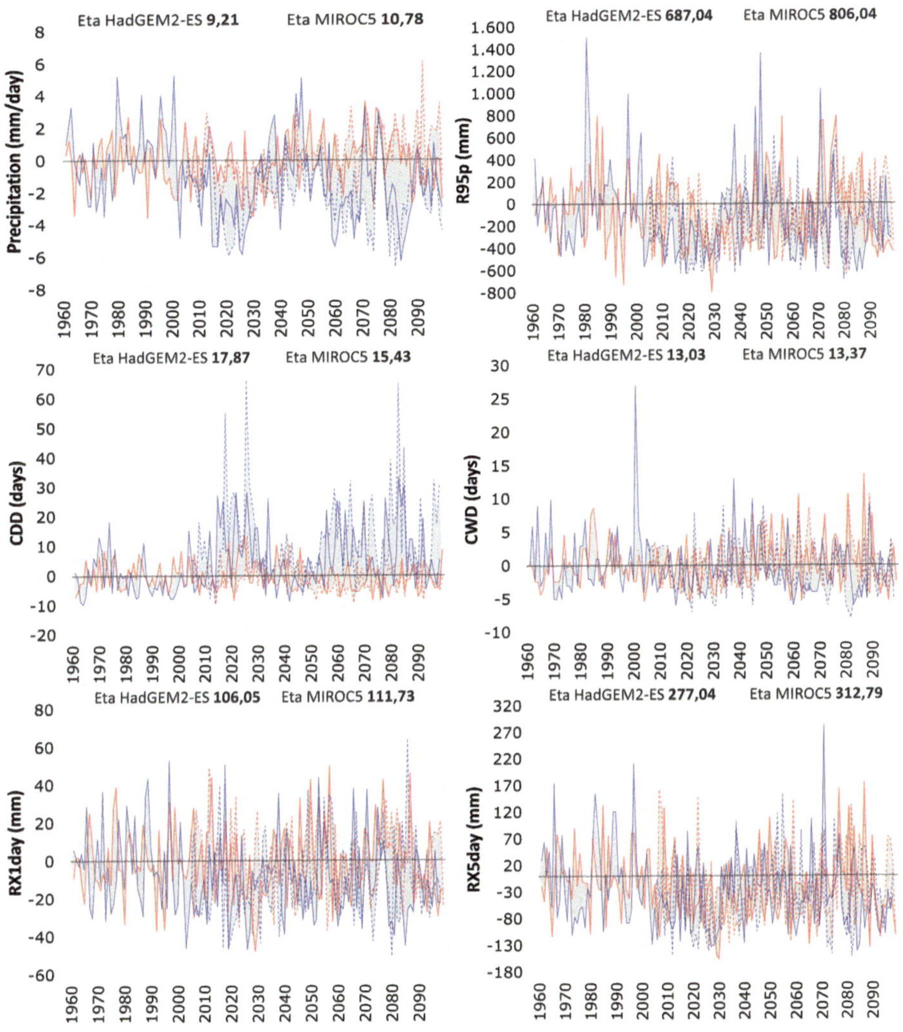

Fig. 4.6 Trends of changes of climatic extremes in the city of Santos: (**a**) the annual mean precipitation; (**b**) R95p, the 95th percentile of the annual precipitation; (**c**) CDD, the consecutive wet days, (**d**) CWD, the consecutive wet days, (**e**) RX1DAY, the annual maximum daily precipitation; and (**f**) RX5DAY, the annual maximum 5-day accumulated precipitation. These downscaling runs driven by HadGEM2-ES are in blue and driven by MIROC5 are in red. The solid lines refer to the RCP4.5 and the dashed lines refer to the RCP8.5. The grey shaded area denotes the amplitude of the related uncertainties. The values on top of the graphs are the mean values of the historical period (Source: PBMC 2016)

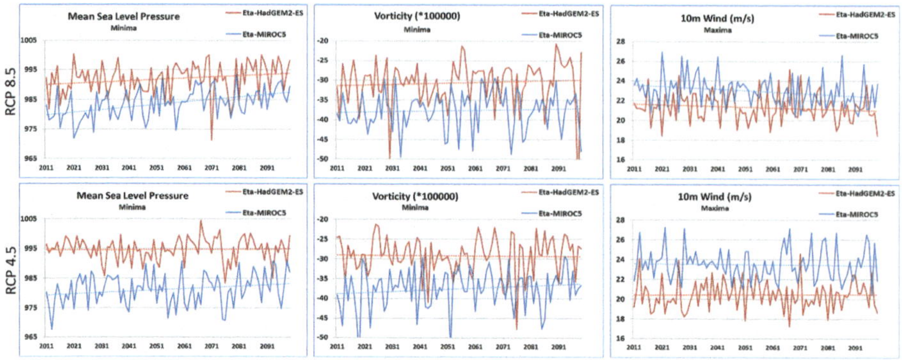

Fig. 4.7 Trends in the annual minimum surface pressure (hPa), 10-m winds (ms-1) and the relative vorticity (× 10–5 s-1) of the 10-m winds are taken off the coast of South and Southeast Brazil from the downscaling runs. The top row refers to the RCP8.5 and the bottom row refers to the RCP4.5 scenario runs

4.2.2.4 Trends in Storms

The storm surges that frequently hit the coast of Santos are related to the passage of extratropical cyclones over southern Atlantic Ocean.

The minimum surface pressure is obtained from an area off the coast of South and Southeast region. This area exhibits a local maximum density of cyclogenesis during summer, autumn, and spring (Reboita et al. 2010). Once the minimum pressure is located, the 10-m winds and the relative vorticity of the 10-m winds are obtained. The time series of these variables are plotted in Fig. 4.7. Under RCP4.5 scenario, the surface pressure in the Eta driven by HadGEM2-ES shows no trend throughout the twenty-first century, whereas the Eta run driven by MIROC5 shows weakening of the minim surface pressures. However, under RCP8.5 scenario, all runs show the positive trends in the minimum surface pressure in the Eta model output. The corresponding relative vorticity of the 10-m winds and the 10-m winds themselves show positive and negative trend, respectively.

4.3 Conclusions and Some Considerations

Projections of climate change in the region around the city of Santos are assessed based on the downscaling of the Eta Regional Climate Model at 20-km and 5-km, driven by the global climate models, HadGEM2-ES and MIROC5, under the RCP4.5 and RCP8.5 greenhouse gas emission scenarios. The downscaling runs were carried out for the period between 1961 and 2100. The mean climatological temperature and precipitation features are better described by the 5-km runs as the topography and the coastline are better represented in higher spatial resolution. In addition, the frequency distribution is also better reproduced by the higher resolution runs as the extreme values approach closer to the observations.

The projections indicate a moderate warming, but a strong reduction in total annual precipitation toward the end of the century. Indices of extreme temperatures show clear warming trends with increasing variability. On the other hand, for precipitation, although there is no clear trend of the extremes of precipitation, their variability increases through the century. The coastal winds, with a slight increasing strength, agree reasonably with the observations during the baseline period. However, during the future periods, the storms related to extratropical cyclones show a slight decrease in the strength of the 10-m winds and a weakening of the low-pressure values.

These results are only reached due to the high-resolution downscaling runs provided by the regional Eta model at 20-km and 5-km resolutions. The warming and the total annual precipitation reduction trends are robust signals in all the four projections at 20-km resolution. The character of the storms that hit the area is not clear in the feature as the signs of the trends vary from the baseline to the future climates. Care should be taken as these downscaling reflect the downscaling of the climate of two global models among the about 46 global climate models provided by the CMIP5. The downscaling of various different global climate models is desirable but causes a large computational workload. Still, the information provided by these projections can support follow-up studies on adaptation measures in different socioeconomic sectors that have activities in the City of Santos.

Acknowledgments This work was partially funded by CNPq 308035/2013-5, CNPq 306757/2017-6, FAPESP 2012/51876-0, FAPESP 2014/21048-4, FAPESP 2014/00192-0, and FAPESP 2017/06627-6.

References

Alexander, L. V., Zhang, X., Peterson, T. C., Caesar, J., Gleason, B., et al. (2006). Global observed changes in daily climate extremes of temperature and precipitation. *Journal of Geophysical Research, 111*, D05109. https://doi.org/10.1029/2005JD006290.

Barbosa, J. P. M. (2008). *Avaliação de técnicas empíricas e estatísticas de identificação de extremos de precipitação para o litoral paulista e entorno*. MSc dissertation. Universidade Estadual de Campinas, Campinas.

Cavalcanti, I. F. A., Nunes, L. H., Marengo, J. A., Gomes, J. L., Silveira, V. P., & Castellano, M. S. (2017). Projections of precipitation changes in two vulnerable regions of São Paulo State, Brazil. *American Journal of Climate Change, 6*, 268–293. https://doi.org/10.4236/ajcc.2017.62014.

Chou, S. C., Marengo, J. A., Lyra, A. A., Sueiro, G., Pesquero, J. F., Alves, L. M., Kay, G., Betts, R., Chagas, D. J., Gomes, J. L., & Bustamante, J. F. (2012). Downscaling of South America present climate driven by 4-member HadCM3 runs. *Climate Dynamics, 38*, 635–653. https://doi.org/10.1007/s00382-011-1002-8.

Chou, S. C., Lyra, A., Mourão, C., Dereczynski, C., Pilotto, I., Gomes, J., Bustamante, J., Tavares, P., Silva, A., Rodrigues, D., Campos, D., Chagas, D., Sueiro, G., Siqueira, G., & Marengo, J. (2014a). Assessment of climate change over South America under RCP 4.5 and 8.5 down-

scaling scenarios. *American Journal of Climate Change, 3*, 512–527. https://doi.org/10.4236/ajcc.2014.35043.
Chou, S. C., Lyra, A., Mourão, C., Dereczynski, C., Pilotto, I., Gomes, J., Bustamante, J., Tavares, P., Silva, A., Rodrigues, D., Campos, D., Chagas, D., Sueiro, G., Siqueira, G., Nobre, P., & Marengo, J. (2014b). Evaluation of the Eta simulations nested in three global climate models. *American Journal of Climate Change, 3*, 438–454. https://doi.org/10.4236/ajcc.2014.35039.
Flato, G., Marotzke, J., Abiodun, B., Braconnot, P., Chou, S. C., Collins, W., Cox, P., Driouech, F., Emori, S., Eyring, V., Forest, C., Gleckler, P., Guilyardi, E., Jakob, C., Kattsov, V., Reason, C., & Rummukainen, M. (2013). Evaluation of climate models. In T. F. Stocker, D. Qin, G.-K. Plattner, M. Tignor, S. K. Allen, J. Boschung, A. Nauels, Y. Xia, V. Bex, & P. M. Midgley (Eds.), *Climate change 2013: The physical science basis. Contribution of working group I to the fifth assessment report of the intergovernmental panel on climate change* (pp. 741–866). Cambridge, UK: Cambridge University Press. https://doi.org/10.1017/CBO9781107415324.020. ISBN: 978-1-107-66182-0.
Gulizia, C., & Camilloni, I. (2015). Comparative analysis of the ability of a set of CMIP3 and CMIP5 global climate models to represent precipitation in South America. *International Journal of Climatology, 35*(4), 583–595.
IPCC. (2013). Climate change 2013: The physical science basis. Contribution of working group I to the fifth assessment report of the intergovernmental panel on climate change. Stocker, T. F., D. Qin, G.-K. Plattner, M. Tignor, S. K. Allen, J. Boschung, A. Nauels, Y. Xia, V. Bex and P. M. Midgley (Eds.), Cambridge: Cambridge University Press, 1535 pp.
Lyra, A., Tavares, P., Chou, S. C., Sueiro, G., Dereczynski, C. P., Sondermann, M., Silva, A., Marengo, J., & Giarolla, A. (2017). Climate change projections over three metropolitan regions in Southeast Brazil using the non-hydrostatic Eta regional climate model at 5-km resolution. *Theoretical and Applied Climatology*. https://doi.org/10.1007/s00704-017-2067-z.
Marengo, J. A., Chou, S. C., Kay, G., Alves, L. M., Pesquero, J. F., Soares, W. R., Santos, D. C., Lyra, A. A., Sueiro, G., Betts, R., Chagas, D. J., Gomes, J. L., Bustamante, J. F., & Tavares, P. (2012). Development of regional future climate change scenarios in South America using the Eta CPTEC/HadCM3 climate change projections: Climatology and regional analyses for the Amazon, São Francisco, and the Paraná River Basins. *Climate Dynamics, 38*(9–10), 1829–1848.
Marengo, J. A., Scarano, F. R., Klein, A. F., Souza, C. R. G., & Chou, S. C. (2017). Impacto, vulnerabilidade e adaptação das cidades costeiras brasileiras às mudanças climáticas: Relatório Especial do Painel Brasileiro de Mudanças Climáticas. In J. A. Marengo & F. R. Scarano (Eds.), *Painel Brasileiro de Mudanças Climáticas*. Rio de Janeiro: PBMC, COPPE – UFRJ. 184 p. ISBN: 978-85-285-0345-6.
Mesinger, F., Chou, S. C., Gomes, J. L., Jovic, D., Bastos, P., Bustamante, J. F., Lazic, L., Lyra, A. A., Morelli, S., Ristic, I., & Veljovic, K. (2012). An upgraded version of the Eta model. *Meteorology and Atmospheric Physics, 116*(3), 63–79. https://doi.org/10.1007/s00703-012-0182-z.
PBMC. (2016). *Impacto, vulnerabilidade e adaptação das cidades costeiras brasileiras às mudanças climáticas: Relatório Especial do Painel Brasileiro de Mudanças Climáticas*. In: J. A. Marengo & F. R. Scarano (Eds.). PBMC, COPPE – UFRJ. Rio de Janeiro. 184 p. ISBN: 978-85-285-0345-6.
Pesquero, J. F., Chou, S. C., Nobre, C. A., & Marengo, J. A. (2010). Climate downscaling over South America for 1961–1970 using the Eta model. *Theoretical and Applied Climatology, 99*(1–2), 75–93.
Reboita, M. S., da Rocha, R. P., Ambrizzi, T., & Sugahara, S. (2010). South Atlantic Ocean cyclogenesis climatology simulated by regional climate model (RegCM3). *Climate Dynamics, 35*, 1331–1347. https://doi.org/10.1007/s00382-009-0668-7.

van Vuuren, D. P., Edmonds, J., Kainuma, M., Riahi, K., Thomson, A., Hibbard, K., Hurtt, G. C., Kram, T., Krey, V., Lamarque, J.-F., Masui, T., Meinshausen, M., Nakicenovic, N., Smith, S. J., & Rose, S. K. (2011). The representative concentration pathways: An overview. *Climatic Change, 109*(1–2), 5–31.

Yin, L., Fu, R., Shevliakova, E., & Dickinson, R. E. (2013). How well can CMIP5 simulate precipitation and its controlling processes over tropical South America? *Climate Dynamics, 41*(11–12), 3127–3143.

Chapter 5
Projection and Uncertainties of Sea Level Trends in Baixada Santista

Joseph Harari, Ricardo de Camargo, Celia Regina de Gouveia Souza, and Lucí Hidalgo Nunes

Abstract The main factors that produce variations in sea level are: wind, through friction on the surface of the sea, in the form of surface waves and currents; atmospheric pressure; astronomical effects, associated with the relative positions of Earth, Sun and Moon. Other complementary factors also affect the sea level: eustasy, an effect dependent on the volume of water in the global ocean; steric effects, resulting from the variation of the volume of the water; halosteric effects, associated with salinity variations; isostasy, varying the average level of the sea due to changes in the position of the seabed and the topography of the coasts; local variations of the Earth's gravity field.

The mean sea level in Santos has great variability, so that a large number of complete years should be considered for assessing the real sea level trend, because several long period periodicities are present in the records; the computed trends show great variations, and some decades have a definite decreasing trend of mean sea level; yet, earlier decades tend to have lower trends than recent ones, so the rise in mean sea level can be considered to accelerate over time.

J. Harari (✉)
Institute of Oceanography, University of São Paulo, São Paulo, SP, Brazil
e-mail: joharari@usp.br

R. de Camargo
Institute of Astronomy, Geophysics and Atmospheric Sciences, University of Sao Paulo, São Paulo, SP, Brazil
e-mail: ricamarg@usp.br

C. R. G. Souza
Institute of Geology – Secretariat for the Environment of the State of São Paulo (IG-SMA/SP), São Paulo, SP, Brazil

Post-Graduate Programme on Physical Geography – Faculty of Philosophy, Languages and Human Sciences, University of São Paulo (FFLCH-USP), São Paulo, SP, Brazil
e-mail: celia@sp.gov.br

L. H. Nunes
Santos, São Paulo, Brazil
e-mail: luci@ige.unicamp.br

L. H. Nunes et al. (eds.), *Climate Change in Santos Brazil: Projections, Impacts and Adaptation Options*, https://doi.org/10.1007/978-3-319-96535-2_5

Keywords Sea level variabilities and trends · Factors affecting sea level · Tidal measurements and analysis · Uncertainties of mean sea level trends analysis

5.1 Introduction

The surface level in the oceans and coastal areas has a large number of spatial and temporal variabilities, ranging from a few centimetres to thousands of kilometres, from a few seconds to decades. In some specific places, there are very adequate sea-level time series that allow the characterization of their long-term variability and trends. However, in most of the world ocean, in situ data are scarce and such characterisation is not possible, or is only possible through interpolations and extrapolations in space and time.

El Niño/Southern Oscillation (ENSO) is the most important ocean-atmosphere coupled phenomenon to cause global climate variability on interannual time scales. An ENSO event causes significant changes in the distribution of surface temperature of the Pacific Ocean, with duration of 15–18 months, with profound effects on the climate. During a "normal" year, i.e. without the El Niño phenomenon, the trade winds blow from east across the tropical Pacific Ocean, causing an excess of water in the western Pacific, so the surface of the sea is about 0.5 m higher on the coasts of Indonesia than in South America. This pattern provokes the upwelling of deeper, colder, nutrient-laden waters on the western coast of South America that feed the marine ecosystem, promoting immense fish populations off the coast of Chile and Peru. However, in an El Niño event – which occurs irregularly at intervals of 2–7 years, with a mean of 3–4 years – the winds blow with less force throughout the centre of the Pacific Ocean, resulting in a decrease of the surface gradient and the deep water upwelling, causing a warmer accumulation of water on the west coast of South America, and consequently resulting a decrease in primary productivity and fish populations.

ENSO can be monitored by the ENSO Multivariate Index (MEI), which uses six observed variables on the tropical Pacific (Wolter and Timlin 2011): sea level pressure, zonal and meridional components of surface wind, sea surface temperature, surface air temperature and total sky cloudiness – see MEI values in Fig. 5.1: positive MEI values, in red, represent the occurrence of the El Niño phenomenon, while negative values, in blue, represent opposite conditions, of La Niña. In the last decade, the phenomenon of El Niño was observed in 2006–2007 (with low intensity), 2009–2010 (with moderate intensity) and 2015–2016 (with strong intensity); on the other hand, La Niña conditions were observed in 2008 (with moderate intensity) and in 2012–2014 (with strong intensity).

In addition to the variability, in the last decades, mainly since 1950, the oceans started to present trends, mainly due to the influence of anthropic activities around the world (Cazenave and Remy 2011). Among these trends, the most important are related to ocean warming and sea level rising.

As for the elevation of the mean sea level, measurements of satellite altimetry since 1993 indicate an overall average elevation of 3.28 cm/decade (Fig. 5.2), but

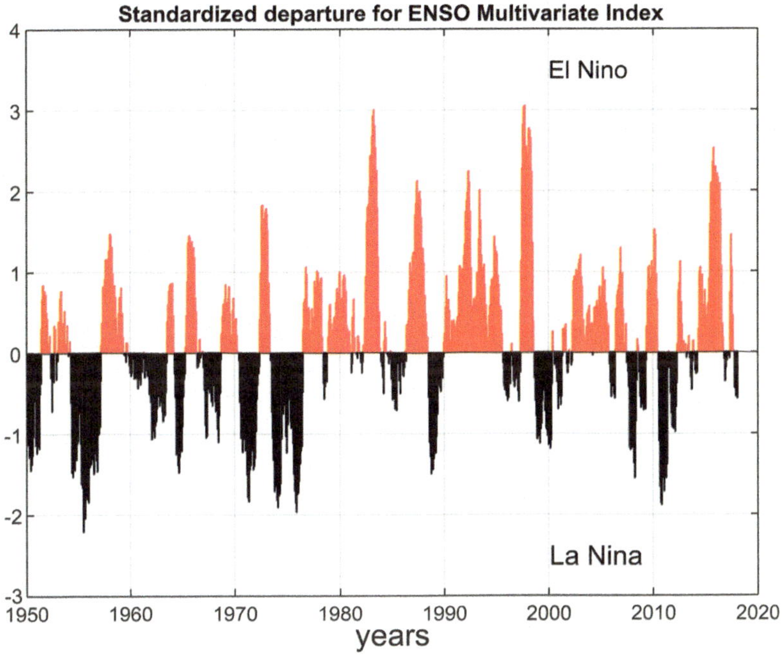

Fig. 5.1 ENSO Multivariate Index (MEI), for the years 1948–2016. (Wolter and Timlin 2011)

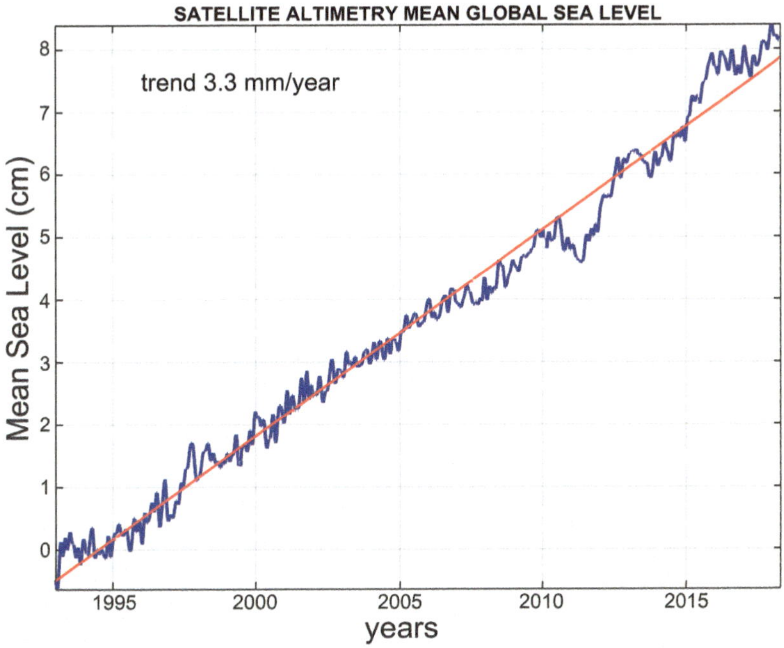

Fig. 5.2 Variation of the global mean level of the oceans, based on satellite altimetry, since 1993. (Adapted from AVISO 2016)

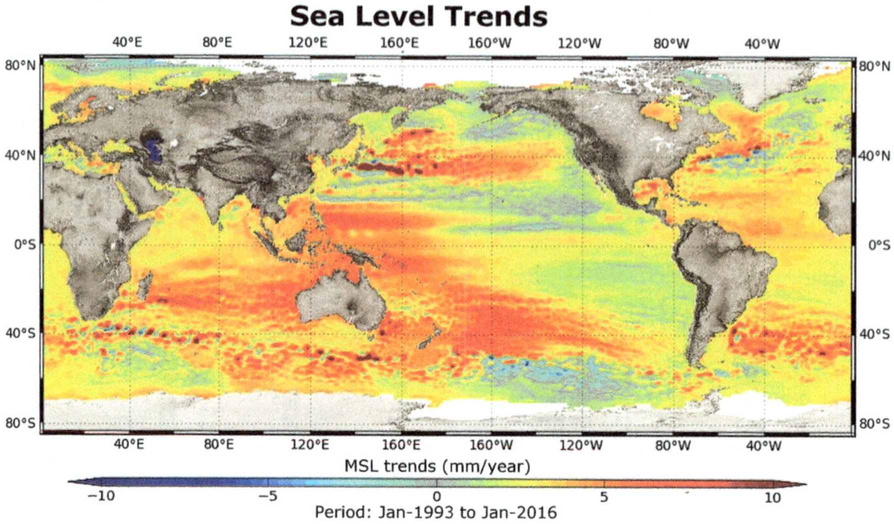

Fig. 5.3 Spatial distribution of mean sea level trends, as measured by satellite altimetry, since 1993. (Adapted from AVISO 2016)

with huge spatial variations of the rates (Fig. 5.3), practically between −10 and +10 cm/decade (Llovel et al. 2010).

On the other hand, there are aspects of sea level that are virtually invariant over time; as an example, between the extremes of the Panama Canal there is a 20 cm difference between the lower (Atlantic) and the higher (Pacific) levels.

What causes sea level variations? How are these variations measured and analysed? Which are the most important periodicities? Are there long-term trends in sea level? Specifically in the region of Santos (Brazil), how does the level of the sea behave? These questions will be answered in the course of this chapter.

5.2 Factors That Produce Sea Level Variations

The main factors that produce variations in sea level are (Stewart 2005; Tomczak and Godfrey 1994):

1. Wind, through friction on the surface of the sea, transfers mechanical energy, which manifests itself in the form of surface waves and currents, these occurring from the surface to depths of some tens of meters.

   1.1 The waves constitute high frequency oscillations of sea level, with periods varying from tenths of seconds (capillary waves) to several minutes (wind waves) or even some days (storm surges). Capillary waves have amplitudes and wavelengths of a few centimetres, while wind waves have amplitudes that can reach 2–3 m, with wavelengths of a few hundred meters; the storm surges correspond to typical sea level elevations of 1–3 m, with wavelengths of hundreds to thousands of kilometres. Wind waves can propagate over very large distances, thousands of kilometres away, and then acquire regular features, being called swell. Waves can also be generated by earthquakes and tsunamis, which are known as tsunamis, whose periods typically are longer than 5 min; these waves have small amplitude in the deep ocean, but can reach spectacular effects on the coastline of 10 m or more.

1.2 The currents produced by wind are more intense at the surface and decay linearly with depth. Convergence or divergence of these currents cause rising or lowering of the sea level, respectively. Coastal currents generated by the wind may also cause variation of the surface level by accumulating waters against the coast, or removing waters from the coast. Under typical conditions, as a result of transport by the wind-generated currents, the sea level varies about 1 m, in periods of 12–24 h, for distances in the range of hundreds to thousands of kilometres. In the case of very strong winds, operating at great distances and/or for very long periods, the associated sea level variations correspond to storm surges. Although wind currents are the most important, currents generated by other effects (such as density) also produce sea level increase or decrease, depending on convergence/divergence in the ocean or accumulation/removal on the coast.

2. Atmospheric pressure produces variations in sea level; high pressures tend to lower the sea level and low pressures tend to raise the level. Although the hydrostatic pattern is not necessarily always prevailing, it indicates the order of magnitude of the phenomenon, so that the increase of 1 hPa produces decrease of about 0.01 m at sea level (and vice versa). Another effect associated to atmospheric pressure is related to its spatial variations. When pressure differences are large between close positions at sea (high gradients), as in tropical cyclones, intense currents are generated, the configuration of which may result in sea level variation of ±1.5 m.
3. Astronomical effects, associated with the relative positions of the Earth, the Sun and the Moon, produce sea level oscillations with well-defined periodicities, in diurnal scale (periods around 24 h), semi-diurnal (around 12 h), ter-diurnal (8 h) and long periods (monthly, semi-annual and annual). The most important aspect of tides is the exact repetition of tidal waves for any ocean location (tidal components). On the other hand, the propagation of deep ocean tide waves to shelf and coastal regions produces in the shelf co-oscillating systems, which eventually cause the amplification of tidal waves, and also produce small period waves (with typical periods of 6 h, 3 h, 2 h, etc. …). The astronomical tides have amplitudes ranging from virtually zero to 10 m, with wavelengths of thousands of kilometres. In each point of the ocean, the spring tides are those that have greater daily amplitudes, since they occur in the full moon or new moon, with additive effects of the sun and the moon. On the other hand, the neap tides are those that have the smallest daily amplitudes, since they occur in the first and third quarter

of the moon, when the attractive effects of sun and moon tend to cancel. Finally, tidal propagation in semi-closed bays and water bodies produces oscillations with natural frequencies of oscillation, called "seiches", with typical periods of 10–20 min, and amplitudes of few centimetres.

There are other complementary effects that also affect the level of the sea (Carton et al. 2005; Antonov et al. 2002):

1. Factors dependent on eustasy, i.e. on the volume of water in the global ocean, including the variation of the mass of water present in the ocean by catching ice masses or their melting. This effect is linked to the glaciations, having periods of recurrence of the order of tens or even hundreds of thousands of years; however, scientists consider that this effect has come to have great importance in a much smaller time scale, in the order of decades, due to the global warming induced by the climatic variations observed mainly since 1950.
   (a) Steric effects, resulting from the variation of the volume of the water, due to the thermal expansion or contraction, with annual frequency according to the seasons, but with a much more important and less known component, linked to the global variation of the temperature of the oceans, a determinant factor in the global climate change.
2. Halosteric effects, associated with salinity variations, affecting water density as a result of melting of ice, increase or decrease of river discharge and variations in precipitation. A decrease in water density corresponds to an increase in its level, in order to balance the hydrostatic pressures with the neighbouring regions. Again, the influence of global climate change may cause the effects associated with salinity variations to become very important in sea level variations.
3. Factors related to isostasy, in particular to glacioisostasy, varying the average level of the sea due to changes in the position of the seabed and the topography of the coasts. Another effect that can be amplified by global climatic variations.
4. Local and regional variations of the Earth's gravity field. The existence of gravimetric anomalies caused by the existence of seamounts, or proximity to the coast with large topographic accidents, or zones of greater density in the crust or underlying mantle, which can reach several mGal over short distances, cause a corresponding rise in sea level for each positive deviation. Short-period gravimetric anomalies (usually linked to the hydrological cycle in the continent) cause the rise and fall of the sea level in the coast.

Finally, secondary effects can also be considered in sea level variations, such as large rivers discharge or heating of coastal waters used for cooling nuclear power plants.

This chapter will address variations in the sea surface level by focusing on periodicities greater than a few hours and wavelengths greater than thousands of kilometres. That excludes the effects of high frequency and small time scale waves, whose periods are between tenths of seconds and several minutes, the so-called gravity waves, i.e. the waves whose main restoring effect is the force of terrestrial gravity.

An interesting aspect to consider is that in the real ocean all the above-mentioned effects occur continuously and superimposed. For example, effects of very strong winds, occurring during spring tides, with strong atmospheric pressure gradients, tend to cause exaggerated sea-level rises, which may be associated with large floods of low coastal areas.

5.3 Forms of Measurement of the Sea Level

Tide-gauges are devices that record sea level (IOC 2015); on the coast, can be floating-buoy devices, pressure recorders or radar; in the open ocean, pressure gauges are used, and altimeter measurements from satellites or airplanes.

Figure 5.4 shows the general scheme of a floating-buoy tide gauge. A stilling well is a vertical tube with a hole at its base through which sea water can flow. The level inside will be, in principle, the same as that of the open sea outside, but energetic wave motion will be damped inside due to the hole acting as a 'mechanical filter'. In the well is a float, which rises and falls with the water level, and is attached via a wire over pulleys to a chart recorder driven by an accurate clock. The rise and

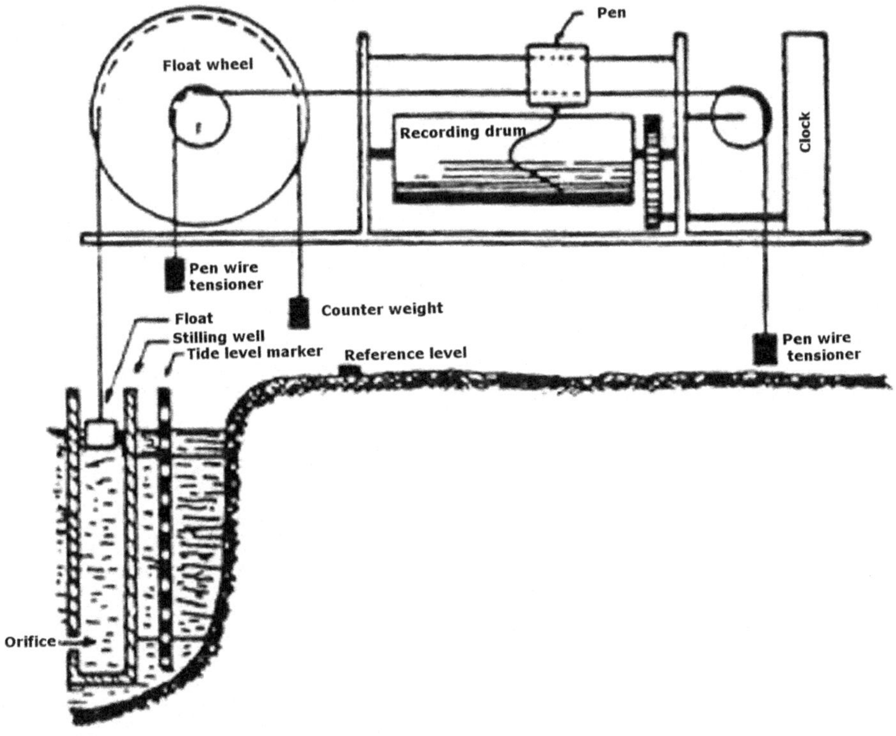

Fig. 5.4 Scheme of a floating-buoy tide gauge. (Adapted from Franco 1977)

fall of the water level is thereby recorded as a line traced by a pen on paper charts. This type of gauge is of historical importance as they were used for almost two centuries (although with modern improvements such as replacing the paper charts with modern electronic data loggers) and so data from them make up the data sets of sea level change that are nowadays archived and used for studies into long-term climate change. In continuous tidal records, commonly appears short-period oscillation, as for example seiches (Fig. 5.5) and storm surges may also be present.

Pressure gauges are usually pressure sensors placed at the bottom of the sea that measure the pressure (or height) of the water column and record it internally at certain time intervals (atmospheric pressure must be subtracted in these measurements). In deep locations, these devices are anchored for long periods so, in their recovery, they are released from the anchors by means of acoustic communication, with attached buoys bringing them back to the surface, where they are then collected by the ships. Pressure markers are also used on the coastline along with barographs, which facilitates the subtraction of atmospheric pressure.

Since the 1990s, radar tide gauges are also used: they operate out of the water and send electromagnetic pulses down, which return reflected from the surface of the sea. The time interval between the emission of the pulses and the reception of the echoes indicate the distance of the transmitter to the surface of the sea, from which the information about the sea surface height is extracted (Fig. 5.6).

The sea surface level can also be measured by altimeters (active sensors) placed on satellites or airplanes; these sensors send electromagnetic signals and receive them reflected: the time interval for receiving the echo, and its intensity and shape, indicate the level of the sea surface (Fig. 5.7), the wind intensity and the significant height of the surface waves.

Radar altimeters on board satellites permanently transmit high-frequency signals to the Earth (more than 1700 pulses per second) and receive echoes from the surface of the sea. The signals emitted and reflected are analysed so very accurate information about the ocean may be obtained, namely (Fu et al. 1994; Souza 2009):

1. Accurate measurements of the electromagnetic signal travel time between the satellite and the surface of the sea, multiplied by the speed of light in the medium

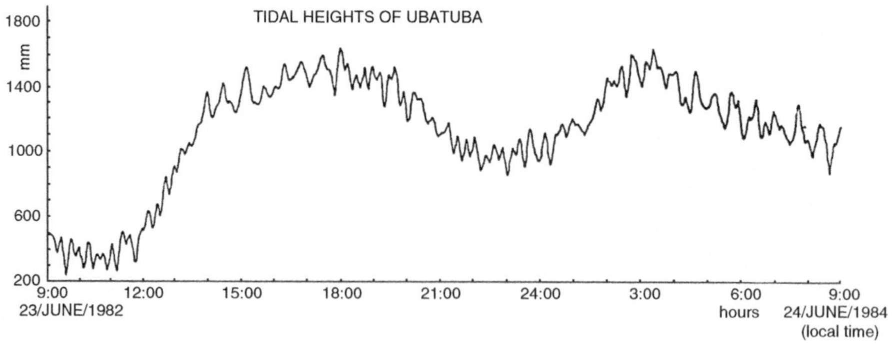

Fig. 5.5 Tide gauge record from Ubatuba (Brazil), with seiches superimposed to the tides

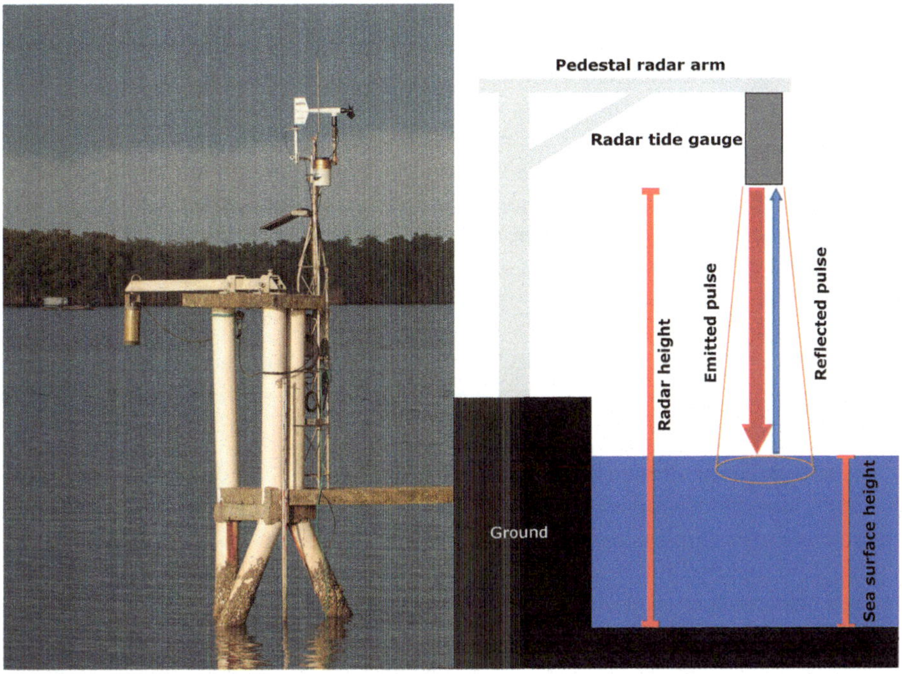

Fig. 5.6 Image of radar tide gauge (left) and its scheme (right), with emitted (E) and reflected (R) electromagnetic pulses, which give the distance of the radar to the sea surface; when subtracting this distance from the radar height to the bottom, the sea surface height is estimated. (Photo by Eng. Francisco Luiz Vicentini Neto/scheme by Eng. Tiago Cortez). The image also shows a meteorological station with a solar panel

Fig. 5.7 Scheme of sea level measurement through satellite altimetry

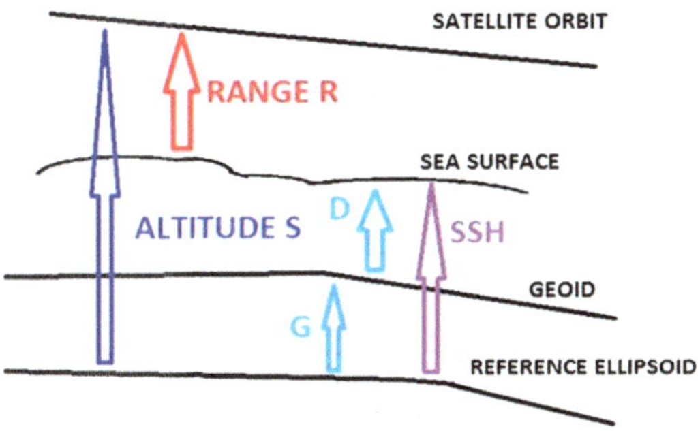

in which the electromagnetic waves travel, determine the height of the satellite relative to the surface ("ALTIMETRIC RANGE = R"). Satellite positioning systems, such as the Doris system, allow to determine the position of the satellite in relation to an arbitrary reference surface (the reference ellipsoid) with high accuracy ("SATELLITE ALTITUDE = S"). Therefore, the "SEA SURFACE

HEIGHT = SSH" is determined as: SSH = S – R (Fig. 5.7). The reference ellipsoid is an approximation of the surface of the Earth, a sphere flattened at the poles, so that the altitude of the satellite above this ellipsoid is obtained with an accuracy of 3 cm.

2. The radar altimeter receives the reflected altimetric signal (or echo), with intensity that varies over time. Where the surface of the sea is flat, the amplitude of the reflected altimetric signal increases markedly from the moment the leading edge of the radar signal hits the surface. However, with rough seas, the altimetric signal reaches the crest of a wave at sea, and then a series of other ridges, which cause the amplitude of the reflected altimetric signal to increase more gradually.
3. The echo power, given by the backscatter coefficient (sigma0), is used to characterize the surface wind intensity (WSPD), but there is no information about its direction.

For the modern altimetric satellites (TOPEX/POSEIDON, JASON 1, JASON 2, …, etc.) have been adopted altitude of 1330 km, orbit inclined 66° to the terrestrial Equator (so that the measurements are restricted between 66° latitude North and South) and a period of repetition of 10 days, i.e. the satellite passes over the same geographic position every 10 days, uniformly sampling the surface of the Earth. Updated information on altimetric satellites in operation (and already deactivated) is regularly provided by AVISO (2016), including access to measurements (www.aviso.altimetry.fr/en/my-aviso.html).

The sea surface height that would exist without any disturbance (due to winds, currents, tides, etc.) is called geoid G, being due only to variations in gravity generated by variations of the mass distribution within the Earth (Pavlis et al. 2012). Therefore, the geoid corresponds to an equipotential surface of terrestrial gravity, and has spatial variation with amplitude of the order of 100 m.

The height of the SSH sea surface can therefore be considered as the sum:

$$\text{SSH} = \text{GEOID}(\text{G}) + \text{DYNAMIC TOPOGRAPHY}(\text{D}),$$

$$\text{SSH} = \text{G} + \text{D}.$$

The dynamic topography D comprises a permanent stationary component (due to the average circulation associated with permanent winds, rotation of the Earth, etc.) and a highly variable component (due to winds, atmospheric pressure, solar radiation, etc.); the amplitude of the dynamic topography is of the order of 1 m, with a maximum of 10 m.

To derive the dynamic topography D, the easiest way would be to subtract the geoid G from the SSH sea surface. In practice, as the geoid is still not sufficiently known, the mean sea level is subtracted (Schaeffer et al. 2012).

However, before using altimetric data, it is necessary to make geophysical corrections (in addition to corrections due to possible instrumental errors). As an altimetric signal travels through the atmosphere, it undergoes diffraction, due to the effects of the ionosphere, dry atmosphere, wet atmosphere (water vapour) and precipitation. Sensors in satellite (from other instruments or using multiple frequencies of altimetry) and independent information (generally from global atmospheric models) allow to correct these effects, so that the final value R is estimated to be within 2 cm precision (Ablain et al. 2009). In addition to these corrections, the altimetric signal must have two more corrections (Tran et al. 2010): the "electromagnetic error", due to the fact that the concave shape of surface wave troughs tend to concentrate and better reflect the altimetric pulse while the ridges tend to disperse it; and the "sea state error" due to the fact that, for wind waves, troughs tend to have larger areas than pointy ridges.

In many respects, the orbit of an altimetry satellite is a compromise between spatial and temporal resolutions (Souza 2009): a satellite revisiting the same spot often covers fewer points than a satellite with a longer orbital cycle. On the other hand, the joint operation of several high – precision altimetric satellites allows a significant increase in the spatial – temporal resolution of observations; Topex/Poseidon – ERS and Jason – Envisat are good examples of altimetric satellites operating together. Topex/Poseidon and Jason-1 have a 10-day repetition period, so they spend about the same points very often, but their land tracks are about 315 km away in the equator – more than the average extent of an oceanic eddy. On the other hand, ERS-2 and Envisat only revisit the same spot on the globe every 35 days, but the maximum distance between two trajectories in the equator is only 80 km.

Increasingly efficient computational data systems now allow access to altimetric data in virtually real time. Its assimilation into models, in combination with other in-situ data and remote sensing imaging, has greatly contributed to the development of Operational Oceanography. Simultaneous operation of at least two altimetric satellites makes it possible to observe phenomena in mesoscale (from 100 to 300 km). For operational applications, three, or ideally four, altimetric satellites are required (Pascual et al. 2006).

It is important to note that each form of sea level measurement is related to some factor influencing this level. Floating-buoys, pressure recorders or radar tide gauges measure the relative sea level, i.e. surface variations relative to one ground reference; GPS (global positioning system) measurements allow detection of crust movements; measurements of gravimeters make it possible to map the geoid. Consequently, the use of the measurements of the mentioned tide gauges with GPS observations and gravimetric information makes it possible to measure the absolute sea level. On the other hand, measurements by satellite altimetry also correspond to the absolute level of the sea surface.

5.4 Sea Level Data Analysis and Periodicity Detection

The analysis of sea level time series generated by tide gauge considers the following steps given below. If there is access to the inclusion of data from GPS and gravimeter, the conclusions will be in relation to the absolute sea level; if no such measurements are available, the results of the analysis will be restricted to the relative sea level (Box et al. 1994; Godin 1972; Franco and Rock 1972; Franco 1997; Foreman 1977; Cartwright and Tayler 1971; Cartwright and Edden 1973).

1. The basic method of sea level data analysis corresponds to the time series statistical analysis, applied to a specific tide gauge record (possibly with the inclusion of GPS and gravity measurements). This analysis makes it possible to determine the basic statistical moments of this series – mean, median, standard deviation, minimum, maximum, kurtosis, asymmetry and distribution.
2. Next, the time series is submitted to spectral analysis, in order to determine the main periodicities present in the record; in this analysis, general initial information on the effects of tides is generally obtained, the behaviour of which is essentially periodic.
3. The third analysis applied to the time series of measurements is the tidal analysis, which constitutes a special formulation of the spectral analysis, in which the amplitudes and phases of the harmonic components are determined exactly in the tide frequencies. Besides astronomical corrections and applied, in view of determining the amplitudes and phases of the tide waves at the measurement site.
4. Given the amplitudes and phases of the tide components, tide predictions can be made for any period of interest, including for the period of the measurements that were analysed.
5. Subtracting the tide prediction series from the measurements series gives the residual sea level series (Fig. 5.8), i.e. variations in relative sea level associated with atmospheric effects (wind and air pressure), steric, halosteric and eustasy effects, as well as variations in sea level associated with earth's crust movements and variations in terrestrial gravity (if there were simultaneous measurements of GPS and gravity). The example of Fig. 5.8 shows a huge mean sea level increase that occurred in Santos from December 31, 1979 to January 02, 1980 – when the mean sea level exceeded 0.6 m and the spring tide had high tides above 0.5 m – so the sea level exceeded 1.25 m.
6. In general, the tide prediction series has mean zero (or very close to zero). The residual series has mean that corresponds to the mean sea level in the measurement period. On the other hand, in the residual series, averages can be calculated for time intervals of daily, weekly, monthly, seasonal, annual, decadal, etc. … periods – which makes it possible to obtain the corresponding mean sea level variations at these time intervals.
7. There is an alternative way of obtaining the residual sea level series – by applying filtering techniques to the time series of sea level measurements; these filters eliminate diurnal, semi-diurnal, ter-diurnal oscillations, etc., i.e. all frequencies above 1 day – which constitute the major part of the tide signal. The disadvantage of this alternative lies in the fact that it does not eliminate the components of long-term tides, with monthly, semi-annual, annual, etc. … periods,
8. Then, the residual series is subject to statistical and spectral analyses, in order to obtain various information of interest, especially in relation to extreme values, distribution function, more important periodicities, and especially, trends.

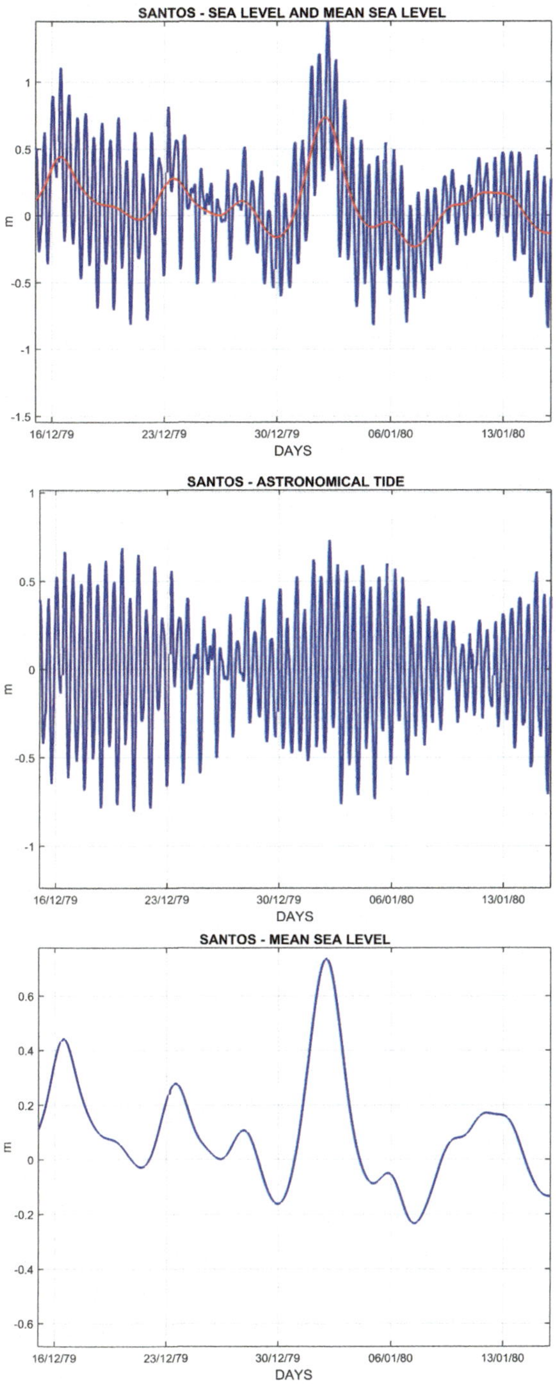

Fig. 5.8 Time series of sea level in Santos, from December 16, 1979 to January 15, 1980 in blue, with mean sea level values in red (upper); extraction of the tide signal (medium) and the mean sea level (lower)

In general, the residual series provides information on storm surges (exaggerated large residual level values), mean sea level elevation rates (through series linear adjustment), etc. The presented methodology therefore provides information on the time variation of sea level at the measurement site.

By applying this methodology to sea level measurements in a set of points appropriately distributed, it is possible to infer temporal and spatial variability of sea level – both for tides and for residual sea level. This is the case of satellite altimetry data, which are processed to obtain time series at points of interest, which can be chosen with distribution in the form of a grid (regular or not). It is also the case of sea level time series produced by hydrodynamic numerical models, again having the advantage of obtaining the analysis results for grid points. When applying the statistical, spectral and tidal analyses to grid points (regular or not), information on the temporal and spatial variability of sea level is obtained, for both its tidal and residual components.

5.5 Long Term Trends of the Sea Level and Their Consequences

One of the main risks associated with global warming due to climate change is the rise of the mean sea level, in particular the eustatic rise of the oceans related to the melting of glaciers and Antarctic ice sheets, besides the steric effect associated to the thermal expansion of sea water. Rising sea levels, currently at a rate above 2 mm/year, could accelerate substantially in the near future, endangering several coastal areas. An increase of only 1 m. at sea level can leave submerged islands in the Pacific and make uninhabitable vast areas of Bangladesh. In addition to the direct effect of rising sea level, there is also an increase in erosive power, especially in dune and beaches, and the impact on ports infrastructure and coastal defence against floods.

Sea-level rise is therefore a troubling issue to all human kind, directly responsible for the climate change that is now facing. Several scientific studies have been performed to monitor and explain the sea-level rise. In 10 years, between 1993 and 2003, the average global sea level, measured by the Topex/Poseidon and Jason-1 satellites, increased at a relatively constant rate of 3 mm/year. The IPCC (2007) report showed an increase of about 1.5 mm/year due to ocean water expansion (by warming), while 1.2 mm/year came from the loss of mass of ice caps and glaciers. Further, acceleration of the melting of terrestrial ice is responsible for the vast majority of sea-level rise in the period 2003–2008. To reach this conclusion, the researchers are based on data from the Jason-1 altimetry satellite, two satellites of the GRACE gravity space mission, and Argo buoys. Since 2003, there is still a fairly rapid increase in sea level (2.5 mm/year), but ocean warming has less effect, the steric contribution to sea level is only 0.4 mm/year.

Thermal expansion of the sea water can be computed by two independent methods: buoys and satellites. The Argo network of buoys transmit profiles of temperature and salinity in the global ocean; since 2003, the integration of all valid data in the first 900 m of the sea leads to a steric contribution of about 0.4 mm/year. This value was confirmed by space measurements, taking into account the difference between the sea-level rise observed by Poseidon/TOPEX and Jason-1 and the increase in the water supply in the oceans, as seen by GRACE. Satellites show a steric contribution of 0.3 mm/year, very close to the value obtained from Argo floats. Therefore, the increase in seawater mass, not its heat content, has been the main cause of the increase in its level since 2003. GRACE data were used to measure the mass variations of both the Arctic and Antarctica ice caps, contributing 1 mm/year to sea levels, or twice as much as in the previous decade. For glaciers, the most recent estimates indicate a glaciological contribution of 1.1 mm/year. Measurements indicate therefore that the water mass increases in the ocean, accounting for 80% of the increase in mean sea level, reaching about 4 mm/year. A study published by IPCC (2007) shows that sea level could rise by 1 m by 2100.

5.6 Computations for the Region of Santos

In the present study, sea-level data were used from:

1. Tide gauge in Torre Grande, at position 23.949°S 46.308°W, from 1945 to 1990.
2. Altimetric multi – satellite data of dynamic topography and sea level anomaly at position 23.875°S 46.375°W, from 1993 to 2014.

Sea-level data, both from tide-gauge and altimetry, have variability over several time scales, making it difficult to access a defined value of their long-term trend. Some considerations are important: first, the Fourier Transform (FFT) of the Torre Grande complete series (1945–1990) shows that the annual component has an amplitude of 5.4 cm (Fig. 5.9), but also important other components of long period, with periodicities of 7.7 years, 9.2 years, 11.5 years, 23.0 years and 4.6 years (similar to the periodicities of the phenomenon El-Niño – Southern Oscillation). As for the annual component, the normal monthly averages (means of all January, of all February, etc. …) have maximum in April–May and minimum in September–October, evidencing the steric effect at sea level. On the other hand, the FFT of the altimetric data (Fig. 5.10) also shows the annual component (with amplitude of 5.1 cm) and other important periodicities, such as 21.0 years and 2.6 years. Figures 5.9 and 5.10 therefore indicate that any long-term mean sea level assessment should consider a large number of complete years.

The analysis of the linear trend of the mean sea level data in Torre Grande, from 1945 to 1990, gave 0.13 ± 0.03 cm/year (Fig. 5.11), while the daily mean sea level data series from altimetry, for the period 1993–2013, had a linear trend of 0.27 ± 0.06 cm/year (Fig. 5.12). The computation of the linear trend of the mean sea

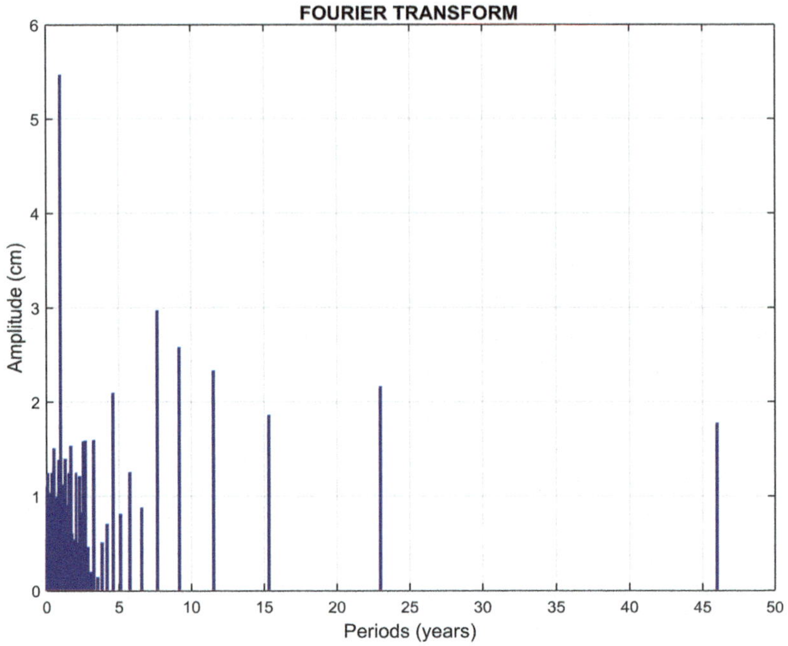

Fig. 5.9 FFT of sea level data of Torre Grande (1945–1990)

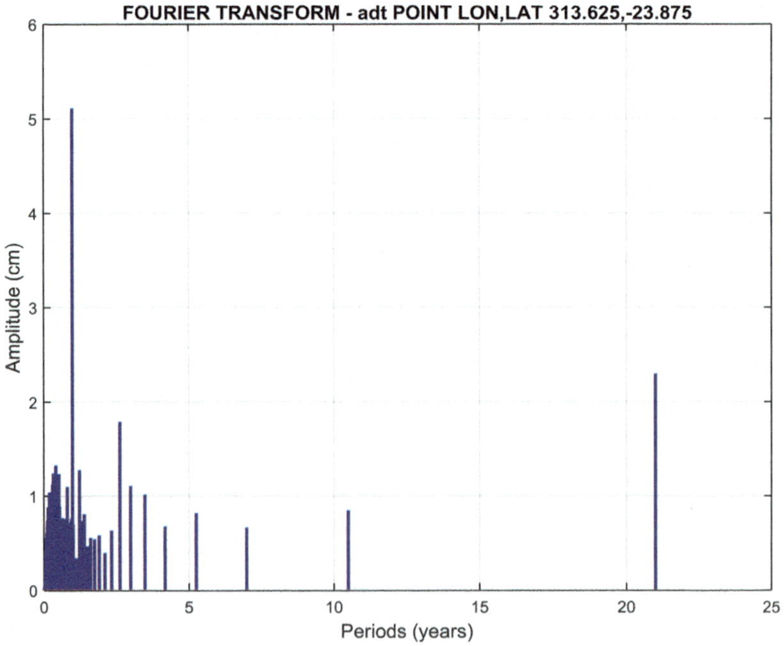

Fig. 5.10 FFT of altimetric data of dynamic topography (1993–2014)

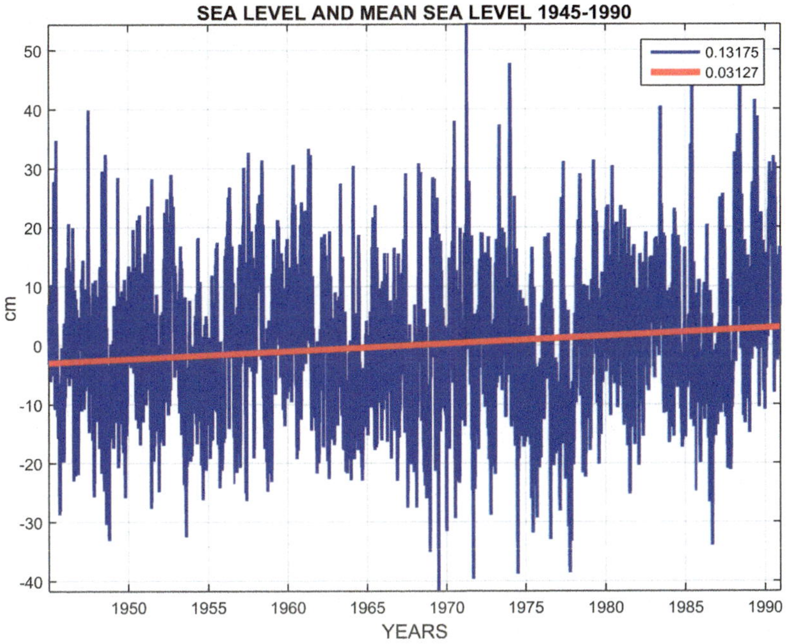

Fig. 5.11 Time series of the mean sea level of Torre Grande (1945–1990) and its linear trend

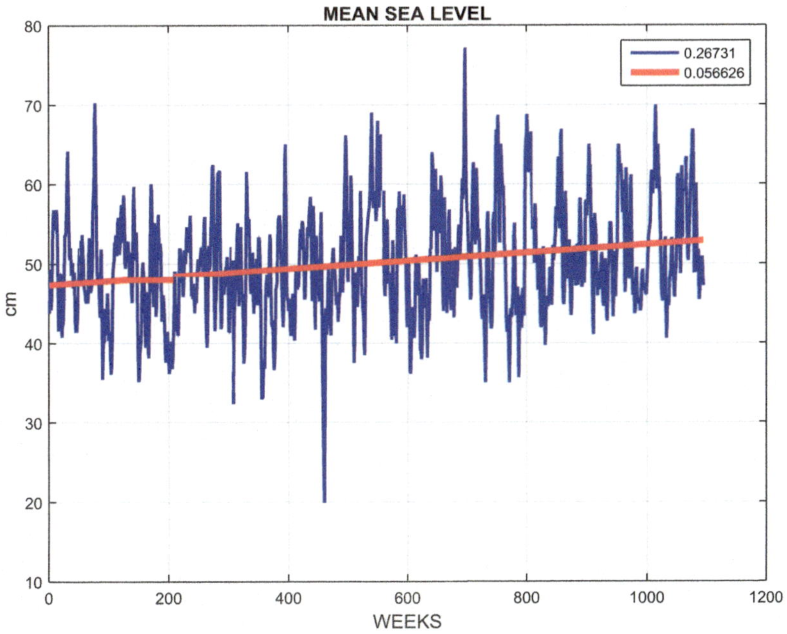

Fig. 5.12 Time series of mean sea level series from altimetry data (1993–2014) and its linear trend

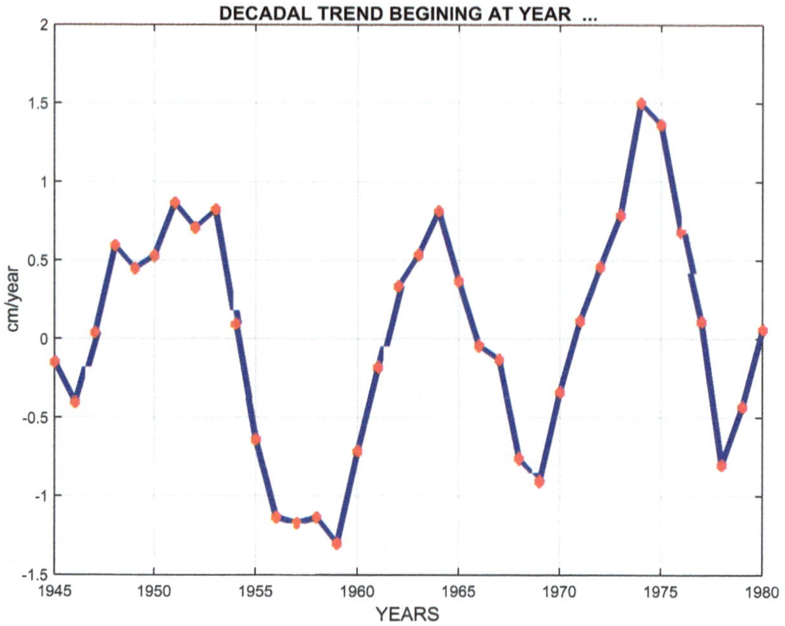

Fig. 5.13 Linear trend of the mean sea level for sub-series with 10-year extension in Torre Grande (beginning in 1945, 1946, …, 1980)

level for the two time series, considering subsections with a 10-year extension, leads to the results shown in Figs. 5.13 and 5.14.

Harari and Camargo (1995) analysed the data from Torre Grande tide-gauge, on monthly, seasonal and annual basis, leading to the following conclusions:

1. The tide components have very small annual trends, with −0.0162 cm/year for the main lunar semi-diurnal component (M2) and +0.013 cm/year for the main solar semi-diurnal component (S2);
2. Periodicities were detected in the annual series of amplitudes and phases of the tide components (of 7.67 years, 9.20 years and 11.50 years, mainly), which can be attributed to limitations in tidal analyses or may be due to long-term meteorological effects;
3. The rate of elevation of the mean sea level is much higher in the winter months (0.1823 cm/year) than in summer months (0.0301 cm/year).

Figure 5.15 shows the linear trends of the sea level time series of satellite altimetry calculated for the entire region 20°S–30°S 50°W–40°W, for the complete period 1993–2014.

The analyses lead to the following conclusions:

1. The mean sea level in Santos has great variability, so that a large number of complete years should be considered for assessing the real sea level trend, because several long period periodicities are present in the records (Figs. 5.9 and 5.10);

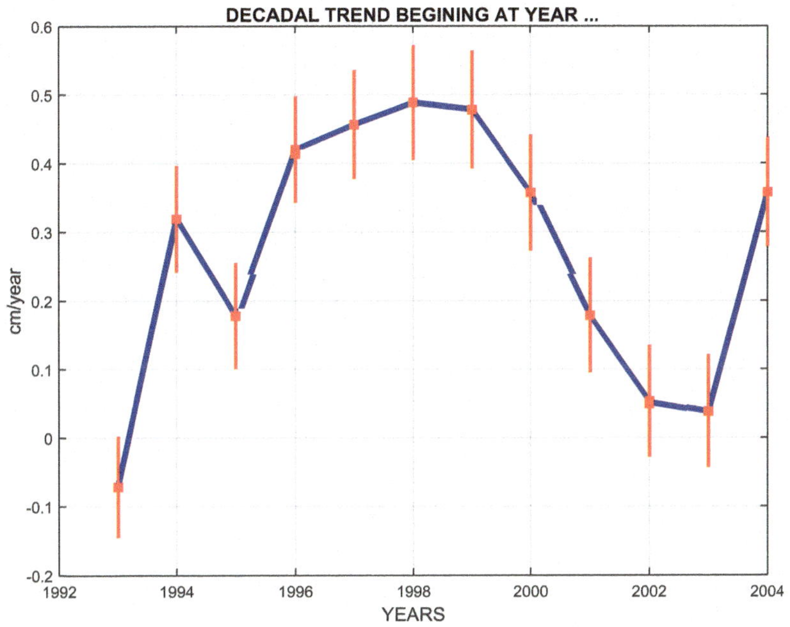

Fig. 5.14 Linear trend of the mean sea level for sub-series with 10-year extension of altimetry data (beginning in 1993, 1994, …, 2004)

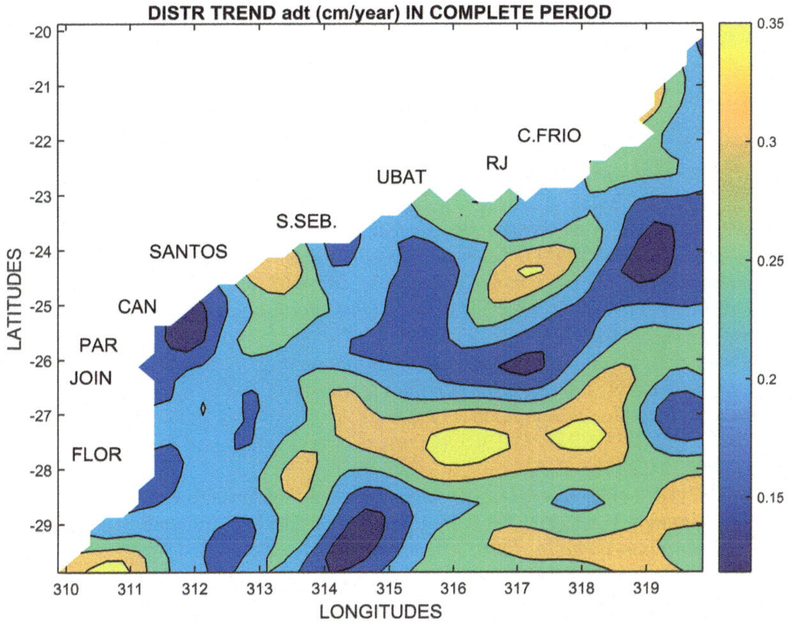

Fig. 5.15 Spatial distribution of mean sea level trends measured by satellite altimetry (in cm/year) for the region 20° S–30° S 50° W–40° W, in the period 1993–2014

2. The computed trends show great variations, and some decades have a definite decreasing trend of mean sea level (see Figs. 5.13 and 5.14);
3. Yet, earlier decades tend to have lower trends than recent ones, so the rise in mean sea level can be considered to accelerate over time;
4. In addition to the temporal variability, the mean sea level trend also presents great spatial variability, as shown in Fig. 5.15.

Of course, the analysis of the mean sea level trends in coastal regions presents many uncertainties; for example, lack of important complementary data (on crust movements, for example), discontinuities in the series, morphological changes of the coastal regions, among others. However, the major problem in these analyses is in the limited extension of records, which have intrinsic variabilities. This fact makes it difficult to obtain conclusive results.

References

Ablain, M., et al. (2009). A new assessment of the error budget of global mean sea level rate estimated by satellite altimetry over 1993–2008. *Ocean Science, 5*(2), 193–201.

Antonov, J. I., Levitus, S., & Boyer, T. P. (2002). Steric sea level variations during 1957–1994: Importance of salinity. *Journal of Geophysical Research, 107*(C12), 8013. https://doi.org/10.1029/2001JC000964.

AVISO. (2016). *SSALTO/DUACS user handbook: MSLA and (M)ADT near-real time and delayed time products*. CLS-DOS-NT-06-034 – Issue 5.0 – Date: 2016/08/20 – Nomenclature: SALP-MU-P-EA-21065-CLS.

Box, G., Jenkins, G., & Reinsel, G. (1994). *Time series analysis*, 3rd edn. Prentice Hall.

Carton, J. A., Giese, B. S., & Grodsky, S. A. (2005). Sea-level rise and the warming of the oceans in the Simple Ocean Data Assimilation (SODA) ocean reanalysis. *Journal of Geophysical Research, 110*, C09006. https://doi.org/10.1029/2004JC002817.

Cartwright, D. E., & Edden, A. C. (1973). Corrected tables of tidal harmonics. *Geophysical Journal of the Royal Astronomical Society, 33*, 253–264.

Cartwright, D. E., & Tayler, R. J. (1971). New computations of the tide-generating potential. *Geophysical Journal of the Royal Astronomical Society, 23*, 45–74.

Cazenave, A., & Remy, F. (2011). Sea level and climate: Measurements and causes of changes. *Interdisciplinary Reviews: Climate Change, 2*(5), 647–662. https://doi.org/10.1002/wcc.139.

Foreman, M. G. G. (1977). *Manual for tidal heights analysis and predictions*. Pacific Marine Science Report, 77–10, 97pp.

Franco, A. S. (1997). *Tides, fundamentals, analysis and prediction*. São Paulo: IPT Press, 286pp.

Franco, A. S., & Rock, N. J. (1972). *The fast fourier transform and its application to tidal oscillations*. Bol. do Instituto Oceanogr. de S.Paulo.

Fu, L. L., et al. (1994). TOPEX/POSEIDON mission overview. *Journal of Geophysical Research: Oceans, 99*(C12), 24369–24381.

Godin, G. (1972). *The analysis of tides*. Toronto: University of Toronto Press, 264pp.

Harari, J., & Camargo, R. (1995). Tides and mean sea level variabilities in Santos (SP), 1944 to 1989. *Relatório Interno do Instituto Oceanográfico da USP*, n° 36, 15 p.

IOC (2015). Manual on sea level measurement and interpretation. Manuals and Guides 14. Intergovernmental Oceanographic Commission. Volumes I–V. http://www.psmsl.org/train_and_info/training/manuals/.

IPCC. (2007). In B. Metz, O. R. Davidson, P. R. Bosch, R. Dave, & L. A. Meyer (Eds.), *Contribution of working group III to the fourth assessment report of the intergovernmental panel on climate change, 2007*. Cambridge/New York: Cambridge University Press.

Llovel, W., Guinehut, S., & Cazenave, A. (2010). Regional and interannual variability in sea level over 2002–2009 based on satellite altimetry, Argo float data and Grace ocean mass. *Ocean Dynamics, 60*, 1193–1204. https://doi.org/10.1007/s10236-010-0324-0.

Pascual, A., Faugère, Y., Larnicol. G., & Le Traon, P.Y. (2006). Improved description of the ocean mesoscale variability by combining four satellite altimeters. *Geophysical Research Letters, 33*(2).

Pavlis, N. K., Holmes, S. A., Kenyon, S. C., & Factor, J. K. (2012). The development and evaluation of the Earth Gravitational Model 2008 (EGM2008). *Journal of Geophysical Research: Solid Earth (1978–2012), 117*(B4) April 2012.

Schaeffer, P., Faugere, Y., Legeais, J.F., Ollivier, A., Guinle, T., & Picot, N. (2012). The CNES CLS11 global mean sea surface computed from 16 years of satellite altimeter data. *Marine Geodesy, 35*, 2012, Special Issue, Jason-2.

Souza, R. B. (2009). Oceanografia por satélites, Ed. Oficina de Textos, 2ª ed., 382 p., ISBN: 9788586238741.

Stewart, R. H. (2005). *Introduction to physical oceanography*. Department of Oceanography, Texas A and M University. 344 pp. Available at http://oceanworld.tamu.edu/resources/ocng_textbook/contents.html.

Tomczak, M., & Godfrey, J. S. (1994). *Regional oceanography: an introduction*. London: Pergamon Press. 391 pp. (available at http://www.es.flinders.edu.au/~mattom/regoc/pdfversion.html).

Tran, N., Labroue, S., Philipps, S., Bronner, E., & Picot, N. (2010). Overview and update of the sea state bias corrections for the Jason-2, Jason-1 and TOPEX missions. *Marine Geodesy, 33*(S1), 348–362.

Wolter, K., & Timlin, M. S. (2011). El Niño/Southern Oscillation behaviour since 1871 as diagnosed in an extended multivariate ENSO index (MEI.ext). *International Journal of Climatology, 31*, 1074–1087.

Chapter 6
Long Term Analysis of Meteorological-Oceanographic Extreme Events for the Baixada Santista Region

Celia Regina de Gouveia Souza, Agenor Pereira Souza, and Joseph Harari

Abstract This chapter presents a database of extreme events, including storm surges (SS) and coastal inundation/flooding (CI/F) that caused injuries and economic/environmental losses in cities from Baixada Santista Metropolitan Region between 1928 and 2016 (hemerographic method). A group of seven indicators describes the boundary conditions of each event: duration/evolution interval, lunar phase, meteorological tide height, precipitation, wind direction and intensity, significant wave height and direction, and ENSO phases. They were listed 115 SS (76.5% only in the current century) and 123 CI/F (47.2%). Around 76.5% of SS occurred between April and September, while 50.4% of CI/F between January and April. Spring tides influenced 52.2% of SS and 65.6% of CI/F. Accumulated rainfall volume during the duration interval was 227.1 mm in SS and 277.8 mm in CI/F. Maximum height of meteorological tides was 0.78 m for both types. Wind intensity reached 20.6 m/s in SS and 17 m/s in CI/F events, both with predominant SW-SSW directions. Significant waves reached 7 and 5.5 m respectively in SS and CI/F, being S-SSE directions predominant. ENSO phenomenon seems to control these extreme events, once 54.8% of SS and 46.3% of CI/F occurred during EN, and 40% of SS and 37.4% of CI/F during LN phases.

Keywords Storm surge · Coastal inundation · Flooding · Boundary conditions · Long-term series

---

C. R. G. Souza (✉)
Institute of Geology – Secretariat for the Environment
of the State of São Paulo (IG-SMA/SP), São Paulo, SP, Brazil

Post-Graduate Programme on Physical Geography – Faculty of Philosophy, Languages and Human Sciences, University of São Paulo (FFLCH-USP), São Paulo, SP, Brazil
e-mail: celia@sp.gov.br

A. P. Souza
Environmental consultant, São Paulo, São Paulo, Brasil

J. Harari
Institute of Oceanography, University of São Paulo, São Paulo, São Paulo, Brasil
e-mail: joharari@usp.br

L. H. Nunes et al. (eds.), *Climate Change in Santos Brazil: Projections, Impacts and Adaptation Options*, https://doi.org/10.1007/978-3-319-96535-2_6

6.1 Introduction

Coastal zones are influenced by oceanographic, atmospheric and continental processes and are therefore particularly sensitive to climate variations. In a climatic change scenario, sea-level rise and changes in the intensity, spatial distribution and temporal frequency of extreme meteorological-oceanographic events induced by wind, rain, waves and tides might produce significant impacts in these regions.

Much evidence suggest that climate changes have already modified the magnitude and frequency of extreme meteorological events around the world, though attributing isolated events to climate change is still difficult (IPCC 2012). Moreover, a general increase in extreme events is expected in future decades (IPCC 2014).

Brazil's coastal zone is very susceptible, and its inhabitants vulnerable, to climate change and its effects (Marengo et al. 2017a, b). Considering sea-level rise (SLR) alone, the consequences can be diverse, as they impact both natural and anthropogenic environments. In this regard, many Brazilian regions are already being affected by severe coastal erosion and coastal inundation linked to extreme storm and tidal surges (Souza et al. 2005; Muehe 2006; Nicholls 2006; Gasparro et al. 2008; Souza 2009a, b, 2010, 2011, 2012; Magrin et al. 2014; PBMC 2014; Marengo et al. 2017a, b).

A 61-year time span (1948–2008) of wave climate data reanalysis for the Latin American and Caribbean regions showed that many changes in wave height and the mean direction of the energy flux have occurred along the Brazilian coast, mainly in the South Region (Losada et al. 2013; Reguero et al. 2013; CEPAL 2016).

Reguero et al. (2013) and CEPAL (2016) pointed to the following results for the Brazilian coast: increase in mean annual significant height (Hs) by ca. 6 mm/year; increase in Hs12 (Hs exceeded for 12 h on average every year, thus related to the annual extreme values; Hs is closely linked to the closure depth of the beach profile and thereby to potential erosion), with mean values of around 3 cm/year; a moderate (1.5 mm/year) increase in surge extremes was noted on the southern Brazilian coast, with a similar reduction in the northern coast; and a clockwise rotation trend in the direction of the mean energy flux (DFE; related to sediment transport and pocket beach rotation). The clockwise rotation trend is probably related to the previously detected trend of rotation toward the poles in extratropical storm activity, meaning more storm activity at high latitudes and less at medium latitudes, something which is more pronounced in the South Hemisphere (SH). The southern Atlantic coast shows low positive correlation according to the northward displacement of the Atlantic intertropical convergence zone. The correlation pattern, however, has strong similarities to the pattern of the Tropical South Atlantic Index (measures sea-surface temperatures in the eastern tropical South Atlantic Ocean), because both indices are related to Atlantic warming. In view of these results, no apparent link exists between El Niño–Southern Oscillation (ENSO/SOI) and wave climate on the South Atlantic coast, for either wave height or direction.

Despite the limited mapping scale, the data from wave climate reanalysis (1948–2008) presented by Reguero et al. (2013) and CEPAL (2016) allows us to identify some trends specifically for the coast of São Paulo State, such as mean annual Hs of 1.5 m; maximum mean Hs of 2.5 during June-July-August, and 3.8 m during September-October-Novenber; Hs12 ranging between 3.3 and 4.3 m; DFE of SSE (135–180°); astronomical tide ranges of 0.5–1.0 m; meteorological tide (storm surge influence) of 0.3–0.4 m on average. Considering only extreme events (Hs + storm surge) and return periods of 50 years (period of reanalysis) and 500 years (reference year: 2010), $Hs_{(50)}$ = 5.5–6.0 m; $Hs_{(500)}$ = 7.0–8.0 m; meteorological tide (50) = 1.3–1.8 m. The long-term (61-year time span) trends indicate the following increases in average values: Hs = 0.1–0.3 cm/year; Hs12 = 0.7–1.3 cm/year; DFE = 0.13–0.22°/year clockwise (values also increase northward along the state coast).

According to Losada et al. (2013) and CEPAL (2016), the annual trends in storm surge (SS) extremes indicated that the zone with the greatest positive trend was Río de la Plata, with values of up to 5 mm/year between 1948 and 2008. This was also the area with the greatest surge extremes throughout all seasons. Trends decreased to 2 mm/year immediately north of the river inlet, extending northward to Brazil's southern coast. The increase is associated with the 100-year return period level, which is at its greatest in northern Argentina, Uruguay and southern Brazil, where the trend is dominated by SS extremes (note that these areas are poles of cyclogenesis). At the remaining points, the mean sea level (MSL) trend predominates. Moreover, long-term shifts in extreme probability density functions suggest that extreme values have become more frequent in recent decades.

These results imply that coastal flooding risk in low-lying areas may be increasing due to a combination of rising MSL and variations in extreme SS events. By itself, rising MSL may not cause flooding, but rising water levels have caused a decrease in the return periods of the extreme total water levels during the last five decades (Losada et al. 2013).

Concerning climate extreme projections for the Brazilian coastal zone, trends point toward an increase in magnitude and frequency of extreme rainfall events (5 days), as well as a decrease in their return periods for all regions, the Southeastern and Southern Regions in particular, and an increase in wind velocity for all regions, the Northeast in particular (e.g. Magrin et al. 2014; CEPAL 2016; Marengo et al. 2017a). These trends, in addition to an increase in the SLR rate for the entire Brazilian coast, will likely increase the number and magnitude of natural disasters related to meteorological-oceanographic events, such as coastal erosion, coastal inundation and flooding.

This chapter presents an analysis of an extensive time series of hazardous events and associated processes, such as storm surges and coastal inundation/flooding, that have caused injuries and widespread human, material, economic or environmental losses in many areas of the Baixada Santista Metropolitan Region (BSMR).

6.2 Main Concepts

Storm surge (SS) is the abnormal rise in seawater level during a storm on an open coast, measured as the height of the water above the normal predicted astronomical tide, resulting from the combined impact of wind setup due to wind stress on the water surface, the atmospheric pressure reduction (cyclonic storm), decreasing water depth, and the horizontal boundaries of the adjacent water (Pugh 1987; NOAA 2017; Mangor et al. 2017). The "surge" is the difference between the observed and the predicted (astronomical tide) levels, which can be either positive or negative and causes a rapid increase or decrease in sea level (SL), respectively (Pugh 1987). Therefore, it corresponds to the meteorological tide. Storm tide is the total observed seawater level during a storm, resulting from the combination of storm surge and astronomical tide (NOAA 2017).

In summary, high winds and low pressure created by the cyclonic storm cause water to accumulate at this centre, resulting in water piling up along the front of the storm during its migration across the ocean (Pugh 1987). Closer to the shore, water is pushed toward the beach, causing a rapid elevation in SL, with waves becoming higher and more powerful. Consequently, an SS occurs. However, this complex phenomenon is further influenced by many different factors. The amplitude of the SS at any given location depends on the orientation of the coastline relative to the storm track; the intensity, size, and speed of the storm; and the local bathymetry and underwater topography (NOAA 2017; Mangor et al. 2017). The effects associated with atmospheric pressure are less than 10% of the total, wind shear stress on the sea surface being the main component (Camargo and Harari 1994; Marone and Camargo 1994).

In the southern region of South America, atmospheric circulation is controlled by atmospheric perturbation, particularly the South Atlantic Tropical Anticyclone—a wet and stationary (westerly winds) high pressure centre—and the cold migratory anticyclones, such as the Atlantic Polar Anticyclone, which are responsible for the northward migration of extratropical cyclones (southeastern winds) and associated cold fronts (e.g. Satyamurti et al. 1998).

South America has four cyclogenesis centres (Gan and Rao 1991; Sinclair 1996; Parise et al. 2009; Reboita et al. 2010) responsible for the formation and migration, or meteorological patterns, of extratropical cyclones: (a) Pattern I – cyclogenesis in the southern Argentinian coast, with a displacement to the east and a trajectory between 47.5°S and 57.5°S; (b) Pattern II – cyclogenesis in the southern Uruguayan coast, with a displacement to the east and a trajectory between 28°S and 43°S; (c) Pattern III – cyclogenesis in the southern Uruguayan coast, with a displacement to the southeast and a trajectory between 32°S and 57.5°S; (d) Pattern IV – high-pressure centre generating easterly winds.

Machado et al. (2010) conducted a hindcast study of wave energy in deep water (100 m), using a wave model based on reanalysis of 29 years (1979–2008) of wind data for the coast of Rio Grande do Sul. They identified a total of 40 extreme events, 53.66% of which had a Pattern II trajectory, while 26.82% were associated with

Pattern III; collectively, these patterns represented 80% of all extreme events. Coastal erosion episodes were associated mostly with Pattern II, while Pattern III caused the highest surges.

Parise et al. (2009) studied extreme SS in southern Brazil and concluded that the highest sea level elevation events resulted from the action of SW winds, which blow parallel to the main NE-SW coastline orientation in the region, and may be explained by the Coriolis effect (i.e. Ekman transport) causing a build-up of water along the coast.

Another way to retrieve and analyse historical data on extreme meteorological-oceanographic events is through a survey of articles from local and regional news media. Paula et al. (2015) called this type of survey the "hemerographic method". Examples of this approach being used in Brazil can be found in the works of Bittencourt et al. (2008), Lins-de-Barros (2010), Pontes and Zee (2010), Paula et al. (2015), and Rudorff et al. (2014).

Chapter 8 also surveys news articles, as sources of information on the occurrence of mass movements.

6.3 Coastal Processes and Extreme Events in Santos

In general, on any ocean coastline, SS has the following results:

(a) Beach erosion – due to the transfer of sediments from the immersed/subaerial contour to the submerged contour of the beach (Bruun rule), and the temporary transverse/vertical migration of the beach contour in the direction of the mainland;
(b) Erosion of urban infrastructure and retention structures – due to the hydraulic and mechanical forces generated by the waves (overtopping) on these structures;
(c) Coastal inundation – due to the raised elevation of the SL and the overtopping of waves over urban structures.

Coastal erosion (of the entire coastline) can be classified as chronic or acute (e.g. Mangor et al. 2017). Normally, coastal erosion develops progressively and continuously over the years. This happens due to the deficit in the sedimentary balance of a given coastal cell. Acute erosion results from extreme events, such as storm surges (waves heights greater than normal) and positive meteorological tides (short-term sea level rise), whose effect is the migration of the beach profile toward the mainland and the displacement of large quantities of sand to the immersed contour of the beach (e.g. Souza 2009a, b, 2012).

The consequences of coastal erosion can be diverse – a reduction in the width of the beach and/or retrogradation of the coastline; disappearance of the backshore and even of the beach itself; loss of properties and assets along the coastline; destruction of man-made structures parallel and transverse to the coastline; problems with, or collapse of, the sanitation drainage system (underground systems and marine outfall); reduction of the recreational water quality of coastal waters; erosion in the region downstream of estuarine ecosystems, with possible alteration of estuarine circulation; loss of fishery resources; loss of scenic value of the beach and/or coastal region; loss of the property value of oceanfront homes; compromised tourism potential of the region; impairment of socioeconomic activities connected to beach-related tourism and leisure; artificial construction of the coastline (coast "protection" structures); increased expenses due to restoration of beaches and reconstruction of the waterfront (Souza et al. 2005; Souza 2009a, 2012).

In Santos, chronic erosion has affected Ponta da Praia since the beginning of the 1940s, triggered mainly by the construction of the avenue to the waterfront over the beach itself (Souza et al. 2012; Souza 2017). Further contributions to this erosion include destruction of the dunes, coastal strands and mangroves; alterations to the drainage network; increasing impermeability of the land close to the coastline; landfills above the estuarine channel; installation of structures transverse to the coastline; construction of retaining walls and stone bulkheads; dredging (mainland and Santos Bay); removal of sand from the beach; sea-level rise; and increased occurrence of extreme events (storm surges and meteorological tides).

On the short- and mid-term temporal scale, the morphodynamic processes of Praia de Santos are controlled by the occurrence of extreme meteorological-oceanographic events (Souza et al. 2016a, b; Souza 2017). In the east-southeast sector (Southeast Zone) of the beach (Fig. 6.1), these events cause acute erosion of the beach and destruction of urban infrastructure, and they intensify the chronic erosion of Ponta da Praia and its expansion toward Canal 4, as well as provoking coastal flooding in the Ponta da Praia region and, eventually, the waterfront of neighbourhoods adjacent to Canal 4 (Fig. 6.2).

Costal inundation has also affected the Northwest Zone of Santos (see Fig. 6.1). This area also suffers with flooding caused by intense rains and high concentration of surface runoff, influenced by the drainage network coming from the hills of Santos; the low declivity of the entire region; the increasing impermeability of the land (this region was largely built over old mangroves and wetlands); the inefficiency of the urban macro-drainage system; and the damming of water in the mouth of the basin (man-made channel) connected to the estuary, which is provoked by high tides, especially during spring tides. What occurs in the area is a sum of processes that trigger coastal inundation and flooding.

In addition to the costs of the restoration of Ponta da Praia and the continual maintenance of the sanitation canals, the greatest socioeconomic harm arising from these extreme events in Santos is related to loss and damage of the public and private natural heritage, due to coastal erosion; coastal inundation; flooding in lowlands; siltation in the navigation channel; damage to coastal protection works; structural or operational damage to Santos Port and terminals; damage to urban structures; flooding of the underground parking of buildings; structural damage or operational damage to sanitation and port infrastructure works; exposure of buried pipelines or structural damage to exposed pipelines; and serious disruption of the functioning of coastal communities, especially transportation and services.

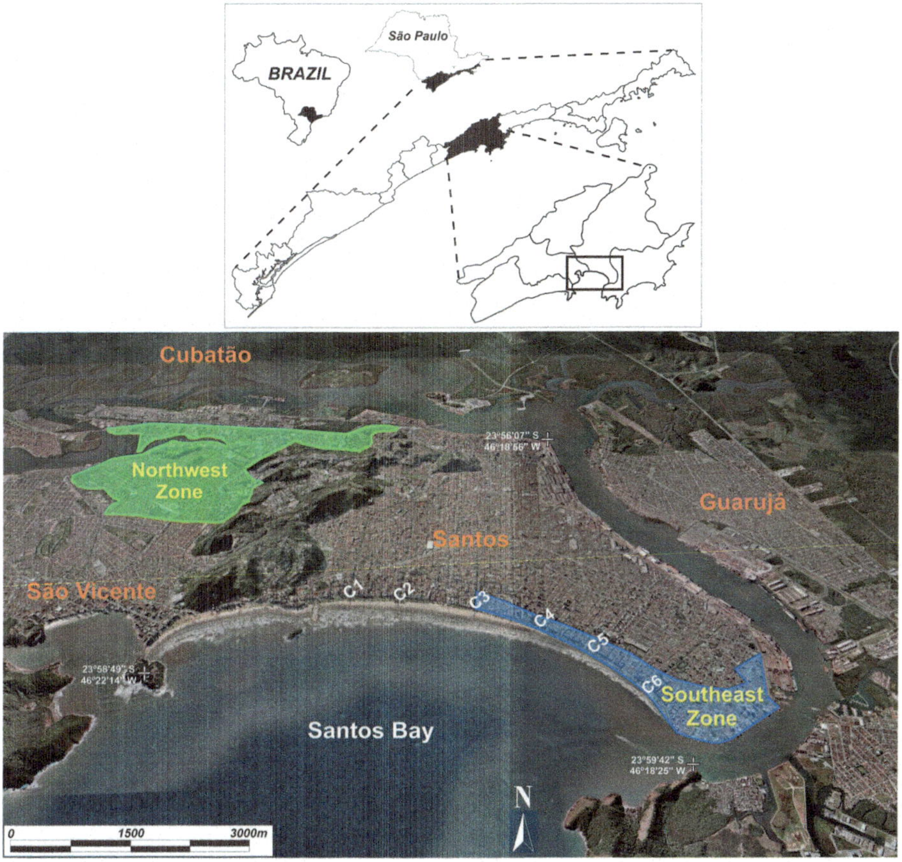

Fig. 6.1 Areas in Santos most affected by storm surges, causing coastal erosion and coastal inundation/flooding (C1–C6 = man-made drainage canals)

6.4 Data Bank Framework and Indicators of Boundary Conditions

The database presented here includes a historical analysis, using the hemerographic method, of the occurrence of extreme meteorological-oceanographic events that affected the region. These events are reported in the media as "*Ressacas*" (storm surges) and "*Marés Altas*" (high tides). In fact, they correspond to one type of hazard (storm surge) and its associated geodynamic processes, including coastal erosion, coastal inundation and flooding. In this case, flooding is caused by the combination of heavy rainfall and high tides, so that the waters from the watersheds discharges summed to the superficial continental runoff flowing down are blocked by the tidal waters rising into the existing drainage system (Marengo et al. 2017b).

Fig. 6.2 Some impacts caused by storm surges on 07/04/2015 (**a**), 06/08/2016 (**b**), 16-08/2017 (**c**) and 20/05/2018 (**d**) in Ponta da Praia region (Southeast Zone) (photos from the first author)

The research focused mainly on the largest regional newspaper, *A Tribuna de Santos*, through both its physical (since 1928) and online (www.atribuna.com.br/) archives. Other newspapers, such as *Diário do Litoral, Expresso Popular, Jornal da Tarde, D.O. Urgente*, served as complementary references. News and information on oldest historical events were also consulted on the website *Novo Milênio* (www.novomilenio.inf.br/).

Generally, these events are only reported on by the newspapers when they significantly alter the beaches and/or cause some disruption for the city. In other words, most of the reports refer to events of high intensity and magnitude; therefore, we refer to them here as "extreme events".

It must be noted that most of the news reports refer to Santos, home of the surveyed newspapers. As the urbanisation in Santos seaside dates back to the beginning of 1930s, we can assume that an absence of news is not due to a lack of urbanisation.

For comparison, we also consulted the extreme events registry – *Cadastro de Eventos Extremos* – developed by the Geological Institute for the period 1923–2010 (Gutjahr 2011), and the climate studies registry – *Cadastro do Laboratório de Estudos Climáticos* – of the Institute of Geosciences/Unicamp (LECLIG, for *Laboratório de Estudos Climáticos do Instituto de Geociências* in Portuguese), covering the period 1928–2014.

6.4.1 Methodological Approach

The meteorological-oceanographic events are influenced by a set of forcings linked to winds, waves, tides, lunar phase, precipitation, atmospheric circulation at the regional scale, and climate phenomena at the global scale, such as the El Niño Southern Oscillation (ENSO).

Regarding the atmospheric phenomena that influence these events in the state of São Paulo, the studies of Campos et al. (2010) were pioneering. The authors analysed a historical series (1951–1990) of data for wind fields, surface pressure, and sea level rise to identify SS in the Baixada Santista region and to compare them to some high-intensity occurrences that caused significant damages to Santos. The results showed that SS occurred due to the migration and persistence of a low-pressure system (extratropical cyclone) over the ocean, moving from southern South America, as well as the action of an anticyclone over the continent for a period of days, with winds above 8 m/s acting upon the ocean near the coast. The authors concluded that, in general, 2 days before an event of maximum sea level elevation is felt in Santos, an associated cold front is positioned over the coast of Paraná and São Paulo; one day before, the front passes over Santos, moving to Rio de Janeiro; and on the day of peak SS in Santos, the cold front reaches Espirito Santo. The period of greatest SS activity occurs between autumn and winter (April and August) and appears to depend on the intensification, size and position of the troughs over the ocean. In both the autumn and the winter, there are intense SW winds approximately parallel to the coast, a pattern determined by the troughs over the ocean in conjunction with high pressure over the mainland, increasing the number of SS events. In winter, severe intensification of the anticyclone over the mainland increases the pressure gradients responsible for intense SW winds; this period has the highest incidence of extreme cases. Conversely, in spring and summer, the troughs weaken, and the wind intensity diminishes, reducing the occurrence of events. For the analysed series, the distribution of events was as follows: 14.5% in spring; 13.4% in summer; 40.2% in autumn; and 30.8% in winter. Comparing the considered decades, the authors observed an increasing trend in the number of positive extremes between 1951 and 1980, of the order of 13%.

Thus, to determine the boundary conditions of each event registered by the data banking we selected seven indicator groups:

(a) Duration and evolution intervals of the events
(b) Lunar phase
(c) Maximum height of the meteorological tide
(d) Precipitation accumulated in the duration and evolution intervals of the event
(e) Direction and intensity of the winds in the region
(f) Significant wave height and direction in the region
(g) Correlation with phases of El Niño and La Niña

6.4.1.1 Duration and Evolution of Events

The "duration interval" of the event was defined as being the period between the actual start date of the event (not necessarily of the news about it) and last news report on the same event.

Considering the results of Campos et al. (2010), we defined the "evolution period" of the event, which encompasses 4 days prior to the beginning of the event in addition to the "duration interval". This period was used for synoptic analysis of the events and to obtain rainfall accumulations.

6.4.1.2 Lunar Phase

Record of the lunar phase for each event was obtained from websites (http://www.cosmobrain.com.br and http://magodosol.atspace.org/lua).

The phases of greatest sea level rise occurred during the full and new moons, i.e. during spring tides, while the least SLR occurred during neap tides. Some events can occur in transition phases or extend over more than one lunar phase. Thus, to represent the different phases, we adopted the scale shown in Table 6.1.

6.4.1.3 Meteorological Tide

The tidal regime on the coast of São Paulo is semi-diurnal with diurnal inequality, having two high tides and two low tides at approximately 6-h intervals, the nocturnal high tide generally being the highest.

To describe this indicator, the highest meteorological tide levels were used for the duration interval; these values were made available by the Institute of Oceanography of the University of São Paulo (historical series of the Torre Grande tide gauge and satellite altimetry data) and the Hydrodynamic Research Centre at the University of Santa Cecília (measuring instrument of the Praticagem as of 2015).

It must be stressed that, because the data is from different sources, the results should be interpreted with caution.

Table 6.1 Values attributed to lunar phases for showing in graphic representation

Lunar Phase	Value
Full/waning	0.5
Waning	1
Waning/new	1.5
New	2
New/waxing	2.5
Waxing	3
Waxing/full	3.5
Full	4

6.4.1.4 Rainfall Volume

Rainfall data were acquired from the pluviometric databases of Santos provided by the Caete station (DAEE E3-041, 23°53′00″S/46°13′00″W, 200-m high) and the Saboó station (DNAEE-02316279, originally located at 23°56′06″S/46°20′22″W, 60-m; currently at 23°55′12″S/46°20′40″W, 32-m), and from the rainfall index of the Santos Municipal Government (www.santos.sp.gov.br/indicepluviometrico). The Caete station covers the period 1937–2004, and the Saboó station, from 1939 onward.

The data from the Saboó station were also analysed in Chaps. 3 and 8.

Between the data from the two stations, we always selected the highest volume and accumulated totals for the duration interval and evolution period (duration interval + 4 days prior to the beginning of the event).

For the data analysis, the cumulative rainfall data were divided into range classes (Table 6.2), chosen according to previous studies on accumulated minimum rainfall for flooding/inundation development in other coastal regions of the state of São Paulo (Santana et al. 2004; Souza 2009c).

6.4.1.5 Waves, Winds and Synoptic Charts

The wind and wave data were compiled from databases provided by the following institutions: National Bank of Oceanographic Data of the Brazilian Navy (BNDO/DHN, for *Banco Nacional de Dados Oceanográficos – Diretoria de Hidrografia e Navegação*), National Institute for Space Research (INPE-CPTEC, for *Instituto Nacional de Pesquisas Espaciais – Centro de Previsão de Tempo e Estudos Climáticos*) and Hydrodynamic Research Centre of the University of Santa Cecília (Unisanta).

The data from BNDO/DHN (the *Meteoromarinha* bulletins) include the oldest historical series, starting in 1941, though it contains many gaps. Until November 1978, these bulletins presented descriptive (not numerical) information about the intensity of the winds, significant wave heights and general direction of wind and waves. We therefore transformed the information on wind intensity and wave height into numerical data based on the Beaufort and Sea State scales (available on the website of the DHN), respectively. The absence of wave period data until November

Table 6.2 Rainfall accumulation classes for duration interval and evolution period

Time	Duration interval	Evolution period
Rainfall (mm)	0–60	0–100
	61–120	101–200
	121–180	201–300
	181–240	>300
	>240	

1978 was the reason that this attribute was not used as a descriptor of the events' boundary conditions.

The data provided by the INPE-CPTEC derived from simulations made by the model WAVEWATCH, implemented in Brazil in March 2006. For the Baixada Santista, the model has a spatial resolution of 1 × 1°, over an area centroid of 24°S and 46°W.

The data provided by Unisanta is also on a regional scale and is based on SWAN wave models and NOAA meteorological models, with wave data starting in October 2014 and wind data since December 2015 (Ribeiro et al. 2016).

The values entered into the present database (for the event duration interval) were always the highest among all available data sources. For the wind and wave direction data, the azimuth notations were transformed into numerical values (0–360°), and then the average of the values was used.

It must be noted that, due to different data sources and types of collection/modelling, the presented results should be interpreted with caution.

The synoptic charts (BNDO and INPE-CPTEC) were used to analyse the synoptic evolution of the atmospheric systems and validate the news reports.

6.4.1.6 El Niño and La Niña

To characterise the active phases of ENSO – El Niño and La Niña – we consulted the INPE-CPTEC website for the historical series 1877–2010 (http://enos.cptec.inpe.br/) and the NOAA website for the period 1950–2017, based on the Oceanic Niño Index – ONI (http://www.cpc.ncep.noaa.gov/products/analysis_monitoring/ensostuff/ensoyears2011.shtml), as well as another compilation organised into annual intervals and based on the ONI (http://ggweather.com/enso/oni.htm).

The available data indicates the active year and the classification of the phenomenon, such as very strong to strong, medium and weak intensities (Table 6.3), which is based on the pattern and magnitude of anomalies of the Tropical Pacific Sea Surface Temperature.

Table 6.3 ENSO phases and attributed intensity values

Intensity	El Niño	La Niña
Very strong-strong	3	−3
Moderate	2	−2
Weak	1	−1
Neutrality	0	0

6.5 Storm Surges and Coastal Inundation/Flooding Events: Boundary Conditions

Table 6.4 presents a list of events registered between 25 April 1928 and 28 October 2016, as well as some earlier ones, in which *Storm Surge* and *Coastal Inundation/ Flooding* are referred to as SS and CI/F, respectively.

The first storm surge and high tide event registered in the region, and in Brazil, occurred in 1541, called the "tsunami" of de São Vicente in historical documents. The consequences were determinants of the region's history, with the near total destruction of then Vila de São Vicente, and the relocation of the port there to the opposite end of Ilha de São Vicente, where the Povoado do Porto de Santos and, later, the Vila de Santos were built. In 1580, another strong event occurred, destroying part of the then reconstructed Vila de São Vicente. There are records of strong storm surges in the years 1905, 1908, 1914, 1927 and 1946; however, the exact dates are unknown.

Most of the events compiled refer to the municipality of Santos, but may also involve other municipalities or all of the Baixada Santista, and may even include all of coastal São Paulo. News articles that do not cite Santos account for 37 events (13 SS and 24 CI/F), which correspond to 15.6% of the total. These events refer mainly to São Vicente, a contiguous neighbour located on the same island.

Some of the indicators that depend on measurements in the field, such as winds, waves, rain and real tides, present several gaps in their data series, especially prior to the mid-twentieth century. Hence, for these indicators, the statistical analyses do not represent the totality of registered events.

6.5.1 General Characteristics

Between 1928 and 2016, 238 extreme events were registered (see Table 6.4), among them 115 SS (48%) and 123 CI/F (52%). Fifty of these events, or 21%, were simultaneously SS and CI/F.

In the graphic showing the annual distribution of these events (Fig. 6.3), a large increase is evident in the number of extreme events per year, especially SS, starting in 1999. Between the 1920s and 1970s, there were 19 CI/F events and 9 SS events. In the 1980s, the number of events increased considerable, with 33 CI/F and 12 SS. In the 1990s, the number decreased, totalling 13 CI/F and 6 SS, 4 of which occurred in 1999 alone. In the 2000s, the number of SS jumped to 49, and CI/F events increased to 24. The year 2010 had the highest number of events, with 14 SS and 3 CI/F, followed by 2009 (11 SS and 5 CI/F) and 2016 (10 SS and 6 CI/F).

Also evident is an increase, starting in the 1980s, of the number of consecutive years in which both types of events occurred, eventually becoming yearly. The only years without any record of extreme events during the studied period were 1993,

Table 6.4 Historical series of Storm Surges (SS) and Coastal Inundation/Flooding (CI/F) records

ID	Type	Date	Affected area	ID	Type	Date	Affected area	ID	Type	Date	Affected area
	SS/CI/F	1541	São Vicente Village	**78**	CI/F	04/08/1992	São Vicente	**159**	CI/F	16/09/2008	Santos
	SS/CI/F	1580	Santos, São Vicente	**79**	CI/F	12/01/1994	São Vicente	**160**	SS	03/01/2009	Santos
	CI/F	1905, 1908	Santos	**80**	CI/F	28/01/1994	Baixada Santista	**161**	SS	25/02/2009	Baixada Santista
	SS	1914, 1927,194«	Santos/São Vicente	**81**	CI/F	06/02/1994	Baixada Santista	**162**	CI/F	25/02/2009	Baixada Santista
1	CI/F	25/04/1928	Santos	**82**	CI/F	09/03/1995	São Vicente	**163**	SS	11/04/2009	Santos/São Vicente
2	CI/F	10/01/1948	São Vicente	**83**	SS	12/02/1996	São Paulo Littoral	**164**	CI/F	11/04/2009	Santos/São Vicente
3	CI/F	01/03/1956	Santos	**84**	CI/F	12/02/1996	São Paulo Littoral	**165**	SS	16/04/2009	Santos
4	CI/F	27/03/1956	Santos	**85**	CI/F	04/03/1996	Santos/São Vicente	**166**	SS	03/06/2009	Santos
5	CI/F	30/09/1961	Santos	**86**	CI/F	20/03/1996	Baixada Santista	**167**	SS	20/07/2009	São Vicente
6	SS	30/09/1961	Santos	**87**	SS	07/04/1997	Santos/São Vicente	**168**	SS	24/07/2009	São Vicente
7	CI/F	08/03/1966	Santos	**88**	CI/F	07/04/1997	Santos/São Vicente	**169**	SS	30/07/2009	São Vicente
8	CI/F	20/11/1969	Santos	**89**	SS	18/04/1999	São Vicente	**170**	CI/F	30/07/2009	São Vicente
9	CI/F	16/12/1970	Santos	**90**	SS	21/05/1999	Santos	**171**	SS	21/08/2009	São Vicente
10	SS	01/01/1971	Baixada Santista	**91**	SS	31/05/1999	Baixada Santista	**172**	SS	08/09/2009	Santos
11	CI/F	01/01/1971	Baixada Santista	**92**	SS	15/08/1999	Guarujá	**173**	CI/F	08/09/2009	Santos
12	CI/F	16/02/1971	Baixada Santista	**93**	SS	16/07/2000	Santos	**174**	SS	24/09/2009	Santos
13	SS	26/02/1971	Santos	**94**	CI/F	16/07/2000	Santos	**175**	CI/F	02/12/2009	São Vicente
14	CI/F	26/02/1971	Santos	**95**	CI/F	12/01/2001	Santos	**176**	SS	26/02/2010	Santos/São Vicente
15	CI/F	16/01/1973	Baixada Santista	**96**	SS	05/05/2001	São Vicente	**177**	CI/F	26/02/2010	Santos/São Vicente
16	CI/F	23/02/1975	Baixada Santista	**97**	CI/F	05/05/2001	São Vicente	**178**	SS	17/03/2010	São Vicente
17	CI/F	28/02/1975	Santos	**98**	SS	21/06/2001	Santos/Guarujá	**179**	CI/F	17/03/2010	São Vicente
18	CI/F	28/01/1976	Santos	**99**	SS	27/07/2001	Santos	**180**	SS	08/04/2010	Santos/São Vicente
19	CI/F	25/03/1976	Santos	**100**	SS	17/09/2001	São Paulo Littoral	**181**	SS	09/05/2010	Santos
20	CI/F	24/12/1976	Baixada Santista	**101**	CI/F	17/09/2001	São Paulo Littoral	**182**	SS	13/07/2010	Santos
21	CI/F	02/04/1977	São Vicente	**102**	SS	16/01/2002	Santos	**183**	SS	27/07/2010	Santos

ID	Type	Date	Affected area	ID	Type	Date	Affected area	ID	Type	Date	Affected area
22	SS	17/05/1977	Baixada Santista	**103**	CI/F	16/01/2002	Santos	**184**	SS	02/08/2010	Baixada Santista
23	SS	09/03/1978	Santos	**104**	SS	21/03/2002	Santos/São Vicente	**185**	SS	11/08/2010	Santos
24	CI/F	09/03/1978	Santos	**105**	CI/F	29/03/2002	Santos	**186**	SS	05/09/2010	Santos
25	SS	31/05/1978	Santos	**106**	SS	22/05/2002	Santos/São Vicente	**187**	SS	18/09/2010	Santos
26	SS	18/07/1978	Santos	**107**	SS	17/06/2002	Santos	**188**	SS	08/10/2010	Baixada Santista
27	SS	22/07/1979	Santos	**108**	SS	10/07/2002	São Vicente	**189**	CI/F	08/10/2010	Baixada Santista
28	SS	02/01/1980	Baixada Santista	**109**	SS	02/09/2002	Santos	**190**	SS	18/10/2010	Santos
29	CI/F	02/01/1980	Baixada Santista	**110**	SS	07/11/2002	Santos/São Vicente	**191**	SS	11/11/2010	Santos
30	CI/F	15/03/1980	São Vicente	**111**	CI/F	21/01/2003	Santos	**192**	SS	13/12/2010	Santos, Guarujá
31	CI/F	01/08/1980	São Vicente	**112**	CI/F	18/02/2003	Santos	**193**	SS	14/03/2011	Santos
32	SS	17/08/1980	Santos	**113**	SS	21/03/2003	Santos	**194**	CI/F	14/03/2011	Santos
33	CI/F	22/12/1980	Santos	**114**	CI/F	21/03/2003	Santos	**195**	SS	03/05/2011	Baixada Santista
34	CI/F	22/01/1981	Santos	**115**	SS	12/04/2003	Santos	**196**	CI/F	03/05/2011	Baixada Santista
35	CI/F	17/03/1981	São Vicente/Cubatão	**116**	CI/F	12/04/2003	Santos	**197**	SS	28/05/2011	Santos/São Vicente
36	SS	08/06/1981	São Vicente	**117**	SS	25/05/2003	Santos	**198**	SS	08/06/2011	Santos
37	CI/F	08/06/1981	São Vicente	**118**	CI/F	25/05/2003	Santos	**199**	SS	20/08/2011	Santos
38	CI/F	16/10/1981	Santos/São Vicente	**119**	SS	12/07/2003	Santos/São Vicente	**200**	CI/F	19/07/2013	Santos
39	CI/F	14/07/1982	São Vicente	**120**	CI/F	09/01/2004	Santos/Guarujá	**201**	CI/F	17/01/2014	Santos
40	SS	19/09/1982	Santos/São Vicente	**121**	SS	29/02/2004	Baixada Santista	**202**	CI/F	03/03/2014	Santos
41	CI/F	19/09/1982	Santos/São Vicente	**122**	CI/F	06/03/2004	Santos	**203**	SS	14/04/2014	Santos
42	SS	28/09/1982	São Vicente	**123**	SS	07/05/2004	Baixada Santista	**204**	CI/F	14/04/2014	Santos
43	CI/F	28/09/1982	São Vicente	**124**	SS	12/05/2004	Santos	**205**	CI/F	26/04/2014	Santos
44	SS	14/10/1982	São Vicente	**125**	SS	21/05/2004	Santos	**206**	CI/F	06/05/2014	Santos
45	CI/F	14/10/1982	São Vicente	**126**	SS	28/05/2004	Santos	**207**	CI/F	25/07/2014	Santos
46	CI/F	02/02/1983	Santos	**127**	CI/F	28/05/2004	Santos	**208**	SS	28/08/2014	Santos

(continued)

Table 6.4 (continued)

ID	Type	Date	Affected area	ID	Type	Date	Affected area	ID	Type	Date	Affected area
47	CI/F	03/06/1983	Santos/São Vicente	**128**	SS	12/06/2004	Santos/São Vicente	**209**	CI/F	28/08/2014	Santos
48	CI/F	11/06/1983	Baixada Santista	**129**	SS	22/03/2005	Santos	**210**	CI/F	22/12/2014	Santos/São Vicente
49	CI/F	12/07/1983	Santos	**130**	SS	26/04/2005	Santos	**211**	CI/F	24/01/2015	Santos
50	CI/F	25/10/1983	São Vicente	**131**	SS	21/05/2005	Santos/São Vicente	**212**	CI/F	30/01/2015	Santos
51	SS	28/06/1984	Santos	**132**	SS	06/07/2005	Santos	**213**	SS	06/04/2015	Santos
52	CI/F	28/06/1984	Santos	**133**	SS	11/02/2006	Santos	**214**	CI/F	06/04/2015	Santos
53	CI/F	06/04/1985	Santos	**134**	CI/F	27/03/2006	Santos	**215**	SS	20/04/2015	Santos
54	SS	07/06/1985	Santos/São Vicente	**135**	CI/F	17/05/2006	Santos	**216**	SS	19/06/2015	Santos
55	CI/F	07/06/1985	Santos/São Vicente	**136**	SS	28/06/2006	Santos	**217**	SS	25/06/2015	Santos
56	CI/F	26/03/1986	Santos	**137**	CI/F	28/06/2006	Santos	**218**	SS	23/07/2015	Santos
57	SS	15/04/1986	Santos	**138**	SS	30/07/2006	Guarujá	**219**	SS	02/09/2015	Santos
58	CI/F	15/04/1986	Santos	**139**	CI/F	30/07/2006	Guarujá	**220**	CI/F	19/09/2015	Santos
59	SS	21/07/1986	Santos	**140**	SS	21/08/2006	Guarujá	**221**	SS	17/10/2015	Santos
60	CI/F	21/07/1986	Santos	**141**	SS	04/09/2006	Santos/Guarujá	**222**	SS	07/11/2015	Santos
61	CI/F	02/12/1986	Cubatão	**142**	CI/F	04/09/2006	Santos/Guarujá	**223**	SS	21/04/2016	Santos
62	CI/F	17/03/1987	Santos	**143**	CI/F	05/10/2006	Baixada Santista	**224**	SS	27/04/2016	São Paulo Littoral
63	CI/F	23/01/1988	Baixada Santista	**144**	CI/F	16/10/2006	Santos	**225**	CI/F	27/04/2016	São Paulo Littoral
64	CI/F	06/02/1988	Baixada Santista	**145**	SS	09/11/2006	Santos	**226**	SS	16/05/2016	Santos
65	CI/F	21/02/1988	São Paulo Littoral	**146**	CI/F	09/11/2006	Santos	**227**	SS	11/06/2016	Santos
66	CI/F	03/04/1988	Santos	**147**	SS	27/04/2007	Guarujá	**228**	SS	18/07/2016	Santos
67	CI/F	02/06/1988	São Vicente	**148**	SS	25/05/2007	Santos	**229**	CI/F	18/07/2016	Santos
68	CI/F	30/06/1988	São Vicente	**149**	SS	28/05/2007	Santos	**230**	SS	28/07/2016	Santos
69	SS	10/08/1988	Guarujá/São Vicente	**150**	CI/F	28/05/2007	Santos	**231**	SS	11/08/2016	Santos
70	CI/F	10/08/1988	Guarujá/São Vicente	**151**	SS	28/07/2007	Santos	**232**	CI/F	11/08/2016	Santos
71	CI/F	20/04/1989	Santos	**152**	CI/F	28/07/2007	Santos	**233**	SS	21/08/2016	São Paulo Littoral

72	SS	17/09/1989	Santos/São Vicente	**153**	CI/F	16/10/2007	Santos	**234**	CI/F	21/08/2016	São Paulo Littoral
73	SS	23/05/1990	Santos	**154**	SS	17/11/2007	Santos	**235**	SS	15/09/2016	São Paulo Littoral
74	CI/F	23/05/1990	Santos	**155**	CI/F	26/02/2008	Santos	**236**	CI/F	15/09/2016	São Paulo Littoral
75	CI/F	22/06/1990	São Vicente	**156**	SS	16/06/2008	Santos/São Vicente	**237**	SS	28/10/2016	São Paulo Littoral
76	CI/F	20/07/1990	São Vicente	**157**	CI/F	16/06/2008	Santos/São Vicente	**238**	CI/F	28/10/2016	São Paulo Littoral
77	CI/F	07/10/1991	Baixada Santista	**158**	CI/F	26/07/2008	Santos/São Vicente				

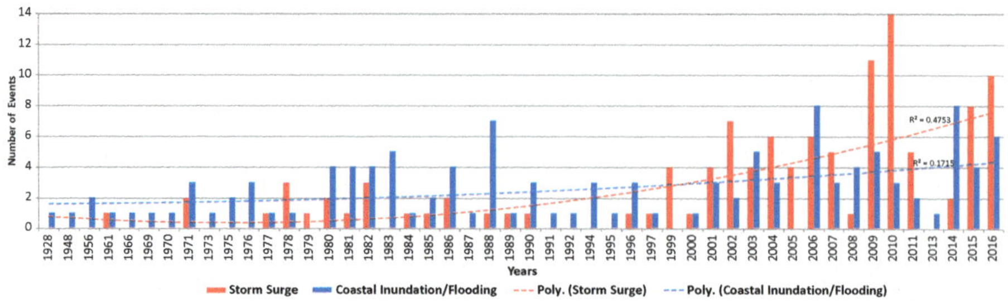

Fig. 6.3 Annual distribution of storm surges and coastal inundation/flooding events

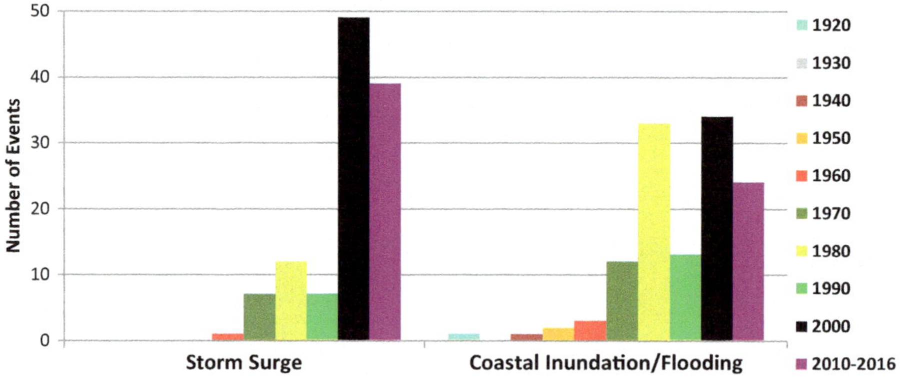

Fig. 6.4 Decadal distribution of storm surges and coastal inundation/flooding events

1998 and 2012. The 1970s seem to have been a transition, with SS becoming ever more frequent and in greater numbers per year.

The distribution of events per decade (Fig. 6.4) shows that the largest number occurred in the 2000s, which had 83 occurrences (35% of the 238 events), comprising 49 SS (42.6% of the 115 registered events) and 34 CI/F (27.6% of the 123 registered). The 2010s also look promising, since 65 events had already occurred in only half of the decade (39 SS and 24 CI/F). Therefore, in the last 17 years alone (2000–2016), there have been 146 events (61.3% of the total), of which, 88 (76.5%) are SS and 58 (47.2%) are CI/F. The 1930s was the only decade without registered events. SS has appeared only from the 1960s onward.

Another interesting fact is the occurrence of combined events, whose number increased to 10 in the 1980s, compared to 1 in the 1960s and 3 in the 1970s (and none before 1960). In the 1990s, the number of occurrences decreased again, occurring only three times. However, the number jumped to 19 in the 2000s, and there were 14 between 2010 and 2016. As these combined events are of greater magnitude, affecting all of Baixada Santista and, generally, other sectors of coastal São Paulo, we may conclude that the number of very strong events increased starting in the 2000s.

Table 6.5 Distribution of events along the centuries and rates of increase

Century (years)	SS (no)	CI/F (no)	ALL (no)	% SS (115)	% CI/F (123)	Rate SS (no/year)	Rate CI/F (no/year)	Rate ALL (no/year)
XX (1928–1999)	27	65	92	23.5	52.8	0.4	0.9	1.3
XXI (2000–2016)	88	58	146	76.5	47.2	5.2	3.4	8.6
Rate of Increase	3.3	0.9	1.6	226.0	10.8	13.6	3.8	6.7

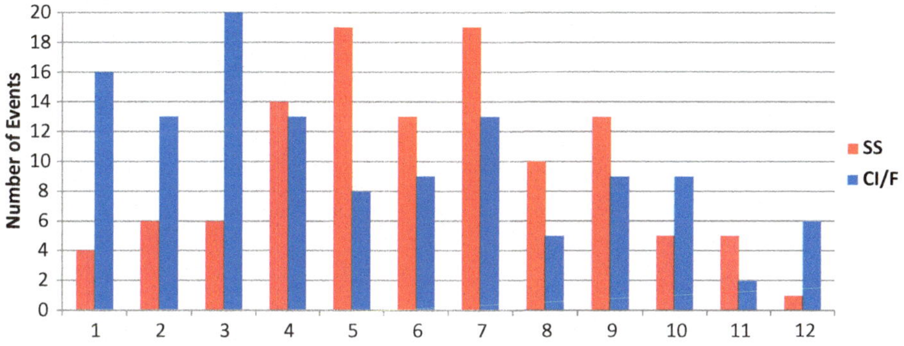

Fig. 6.5 Monthly distribution of occurrence of storm surges (SS) and coastal inundation/flooding (CI/F) events

Analysis of SS event distribution per century (Table 6.5) shows that the twentieth century, specifically 1928–1999 (72 years), had 27 events (23.5% of the total), which represents an average of 0.4 event/year. In the twenty-first century, in only the first 17 years (2000–2016), there have been 88 events (76.5% of the total), which corresponds to an average rate of 5.2 events/year. This indicated an increase in the number of events by 3.3 times (226%) in the twenty-first century, or an increase in the rate of events/year by 13.6 times, relative to the twentieth century. For CI/F, 65 events happened in the twentieth century (52.8%), indicating a rate of 0.9 event/year; while in the twenty-first century, 58 events have occurred (47.2%) at a rate of 3.4 events/year. Therefore, the increase in CI/F this century is 0.9 times the number of events and 3.8 times the rate of events/year.

In terms of the monthly distribution of events (Fig. 6.5), although both types occurred in all months, some key trends are highlighted above.

July was the month with the most events, totalling 32 (19 SS and 13 CI/F), followed by April (14 SS and 13 CI/F) and May (19 SS and 8 CI/F), both with 27 events each, while March had 26 (6 SS and 20 CI/F). The months with the fewest occurrences were November (5 SS and 2 CI/F) and December (1 SS and 6 CI/F), with seven events each.

The distribution of SS is most concentrated between April and September (cyclone/SS season, autumn/winter), totalling 76.5% (88 occurrences); May and

July had the highest number (19 each). These results are similar to those obtained by Campos et al. (2010), which showed that 71% of SS events occurred in autumn/winter (in a time series for 1951–1990).

CI/F occurs mainly between January and April, making up 50.4% of these events, with most happening in March (20), followed by January (16). The summer months alone (January–March) had 39.8% of events (49), while the autumn/winter had 46.3% (57). These results show that the CI/F are controlled by the more intense rains during the summer, often triggering flooding, and then by the effects of the cyclone/SS season and the steric effect (warming of the ocean during the summer) in autumn/winter, when the main process is coastal inundation. This assertion is supported by reports of combined events (those of greater magnitude), since among those 50 events, 11 occurred in the summer months, and 39 in the autumn/winter months.

December and January had the fewest SS events, and November and August had the fewest CI/F.

6.5.2 Duration of Events

The time span of each event's duration interval varied from 1 to 8 days (Fig. 6.6). For SS, the average duration interval was 1.9 days (variation of 1–8 days, with standard deviation of 1.4) and for CI/F, 1.8 days (variation of 1–5 days, with standard deviation of 1.1).

Around 49.2% of all events had a 1-day duration (4.6% of SS and 52% of CI/F). Events of 2- or 3-day duration totalled 40.3%; 9.2% lasted 4 or 5 days, and only 1.3% of SS events lasted for 7–8 days.

In the 2000s and 2010s, the SS events were of notably longer duration, particularly in the years 2007, 2010, 2015 and 2016.

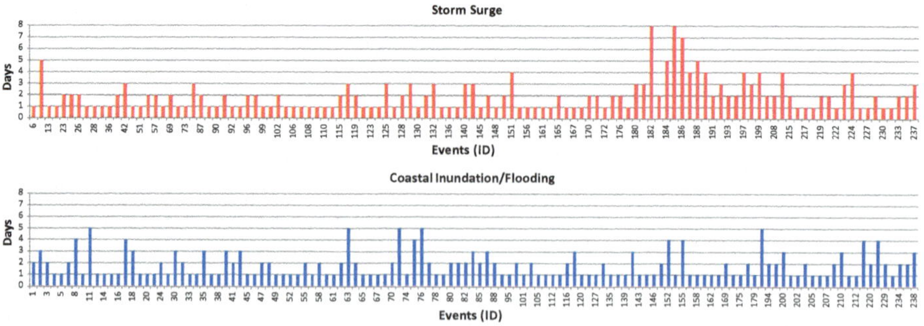

Fig. 6.6 Duration interval for storm surge and coastal inundation/flooding events

6.5.3 Lunar Phase

Figure 6.7 and Table 6.6 show that both types of events occurred during all lunar phases, including the intermediate phases, when the effects of one or the other are offset. However, most of the events occurred during the full moon (30.7%) and new moon (28.6%) phases, i.e. during spring tides, constituting 59.2% of the total (141 events).

For SS, the dominant phases are new moon (29.6%), waxing moon (23.5%) and full moon (22.6%), the spring tides predominating (52.2%). During the CI/F, as expected, full moon (38.2%) and new moon (27.6%) phases predominate, accounting for almost 66%.

However, the results show that factors other than the influence of the moon also contribute, since around 48% of SS events and 34% of CI/F occurred during neap tides and intermediate phases.

6.5.4 Meteorological Tide Heights

The meteorological tide heights varied from 0.01 to 0.78 m for both SS and CI/F (Fig. 6.8). The maximum height was reached in only one combined event, which occurred in 07 June 1985. Another seven events had meteorological tide heights ≥0.70 m (17 May 1977; 02 January 1980; 02 June 1988; 07 April 1997; 16 July 2000; 05 May 2001; 25 March 2003), four of them being combined events.

The overall average height for all the events (178 events with information) was 0.36 m; for SS, 0.35 m (84 events), and for CI/F, 0.38 m (94 events).

The predominant heights were 0.41–0.60 m, making up 36% of all events, 35.7% of the SS, and 36.2% of the CI/F (Table 6.7). For SS, that class was followed by the interval 0.01–0.20 m, in 34.5% of events. For CI/F, the second most frequent class was 0.21–0.40 m, in 31.9% of events. These results agree with those presented by Reguero et al. (2013) and CEPAL (2016), which found that the mean meteorological tide reaches 0.3–0.4 m at the São Paulo coast.

The monthly pattern of maximum meteorological tide heights (Fig. 6.9) show clearly the influence of the cyclone/storm surge season on both types of events, as expected. Only in January was the behaviour anomalous in relation to the summer

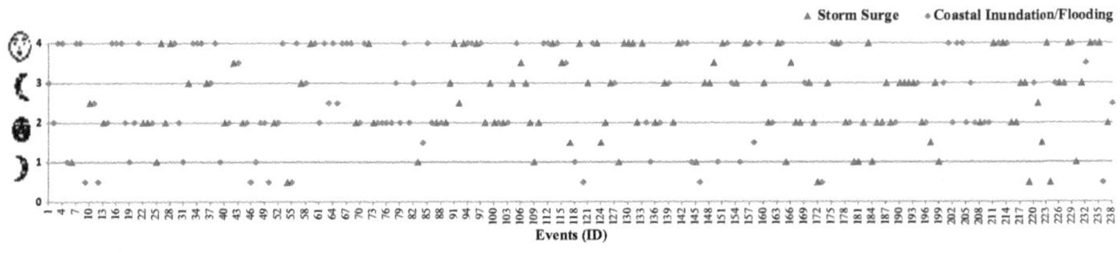

Fig. 6.7 Lunar phase

Table 6.6 Distribution of lunar phases

ALL			SS			CI/F		
Moon	**N° events**	**%**	**Moon**	**N° events**	**%**	**Moon**	**N° events**	**%**
0.5	16	6.7	**0.5**	5	4.3	**0.5**	11	8.9
1	23	9.7	**1**	13	11.3	**1**	10	8.1
1.5	3	1.3	**1.5**	2	1.7	**1.5**	1	0.8
2	68	28.6	**2**	34	29.6	**2**	34	27.6
2.5	7	2.9	**2.5**	3	2.6	**2.5**	4	3.3
3	40	16.8	**3**	27	23.5	**3**	13	10.6
3.5	8	3.4	**3.5**	5	4.3	**3.5**	3	2.4
4	73	30.7	**4**	26	22.6	**4**	47	38.2

See Table 6.1 and Fig. 6.9 above for lunar phase references
ALL all events (238), *SS* Storm Surge (115), *CI/F* Coastal Inundation/Flooding (123)

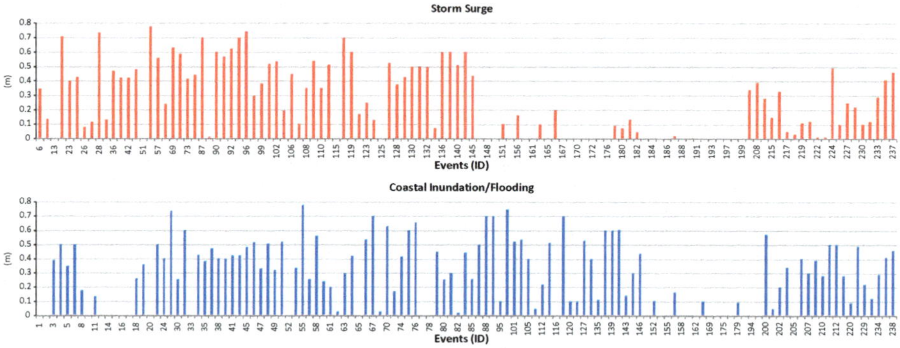

Fig. 6.8 Meteorological tide height for storm surge and coastal inundation/flooding events (the absence of bars signifies a gap in the data)

Table 6.7 Distribution of meteorological tides heights

Height	ALL		SS		CI/F	
Classes (m)	n°	%	n°	%	n°	%
0.01–0.20	50	28.1	29	34.5	21	22.3
0.21–0.40	46	25.8	16	19.0	30	31.9
0.41–0.60	64	36.0	30	35.7	34	36.2
0.61–0,78	18	10.1	9	10.7	9	9.6

months, due to the event of 02 January 1980 (0.73 m). The greatest meteorological tide height occurred in the month of June and corresponds to the combined event of 07 June 1985 (0.78 m). The month of May follows (0.74 m) with the combined event of 05 May 2001. The lowest heights occurred in December (0.44 m for CI/F and 0.28 m for SS).

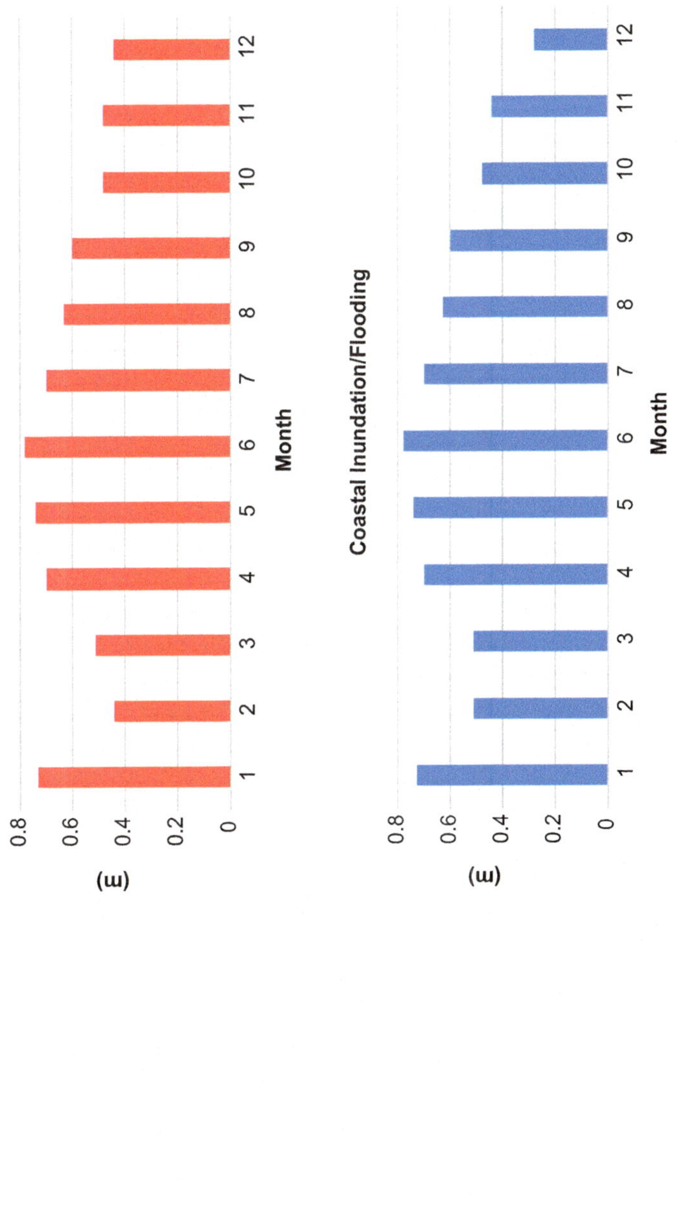

Fig. 6.9 Maximum meteorological tide height by month during storm surge and coastal inundation/flooding events

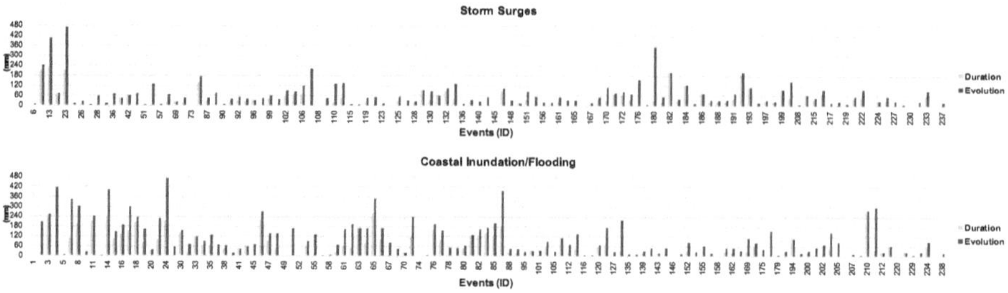

Fig. 6.10 Rainfall index distribution during duration interval and evolution period for storm surge and coastal inundation/flooding events

Table 6.8 Rainfall index classes for duration interval and evolution period of the events

	Rainfall	ALL (237)		SS (115)		CI/F (122)	
	Classes	n°	%	n°	%	n°	%
Duration interval	0–60	185	78.1	98	85.2	87	71.3
	61–120	29	12.3	13	11.3	16	13.1
	121–180	9	3.8	0	0,0	9	7.4
	181–240	11	4.6	5	4.3	7	5.7
	>240	3	1.3	0	0,0	2	1.6
Evolution period	0–100	172	72.6	96	83.5	76	62.3
	101–200	40	16.9	14	12.2	26	21.3
	201–300	15	6.3	3	2.6	12	9.8
	>300	10	4.2	3	2.6	7	5.7

6.5.5 Rainfall Volume

The pattern of rainfall accumulation in the event duration intervals and evolution periods are shown in Fig. 6.10.

For the event duration intervals, the greatest accumulations were 227.1 mm for SS and 277.8 mm for CI/F. For the evolution period, the greatest accumulation was 468.4 mm for both SS and CI/F.

Among all events, the average rainfall accumulation in the duration interval was 40.9 mm, and for the evolution period, 84.0 mm. In SS events, the averages were 33.1 mm for the duration interval and 63.9 mm for the evolution period. In CI/F events, the averages were 51.3 mm for the duration interval and 103.0 mm for the evolution period.

In the duration intervals of all events, the predominant rainfall accumulations were between 0 and 60 mm (78.1%), representing 85.2% of SS and 71.3% of CI/F events (Table 6.8). During the evolution period, rainfall accumulations between 0 and 100 mm occurred in 72.6% of events—83.5% of SS and 62.3% of CI/F.

These conclusions are also shown in the graphs of the monthly rainfall accumulation averages, which are controlled largely by seasonality (Fig. 6.11).

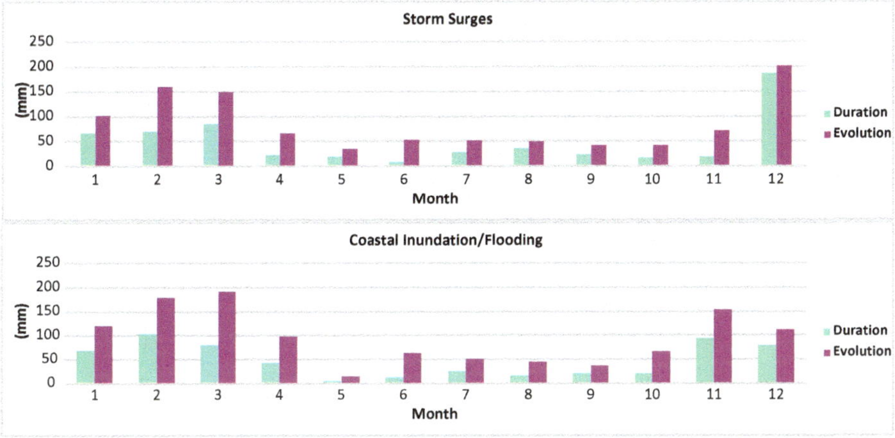

Fig. 6.11 Monthly distribution of mean rainfall index for storm surges and coastal inundation/flooding events

As expected, both types of events have the highest rainfall accumulations in the summer months (December to March). In the autumn/winter months (the SS season), lower accumulations prevail. For SS, the anomalous pattern in December is due to an isolated event (13 December 2010).

In the case of CI/F, seasonality clearly controls the type of process involved: in the spring/summer months (the rainiest), floods predominate, and during the cyclone season, the autumn/winter (least rainy), coastal inundations predominate.

The results also indicate a greater influence of rainfall on CI/F events than on SS events. This suggests low concomitance of a passing cold front and the peak of an SS event, which corroborates the studies of Campos et al. (2010).

6.5.6 Winds

The wind velocity varies between 1 (Light air) and 20.6 (Gale) m/s (Fig. 6.12). The average velocity for all 214 events was 6.63 m/s—for SS, 6.23 m/s, and for CI/F, 7.03 m/s; i.e., all were of the moderate Breeze class. The maximum velocities were 20.6 m/s (Gale) for SS and 17.0 m/s (near Gale) for CI/F.

In most events, the conditions were gentle to moderate Breeze (3.4–7.9 m/s), with very close distribution between the two classes, making up 73.4% of all events, 75.2% of the SS and 62.9% of the CI/F (Table 6.9). Winds between 8 and 20.6 m/s (fresh Breeze, strong Breeze, near Gale and Gale) occurred in 18.3% of SS and 32.4% of CI/F (though Gale winds did not occur during CI/F).

The results show predominant and average velocities slightly lower than those found by Campos et al. (2010), who concluded that during the events, winds near the coast are stronger than 8 m/s.

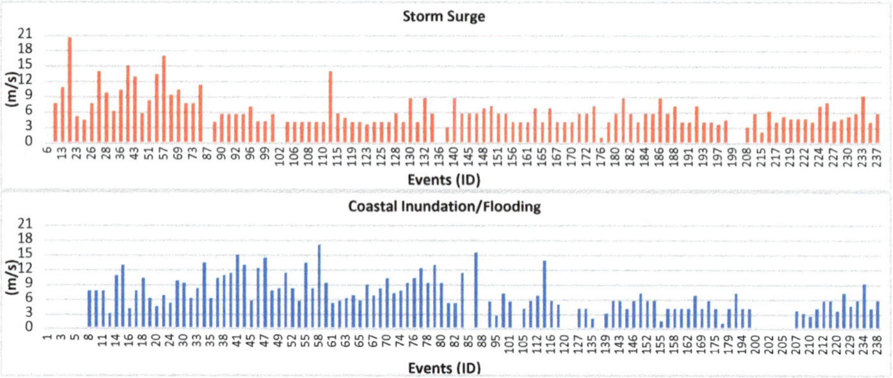

Fig. 6.12 Wind intensity for storm surge and coastal inundation/flooding events

Table 6.9 Wind intensity frequency according to the Beaufort scale

Beaufort scale (m/s)	Light air		Light breeze		Gentle breeze		Moderate breeze		Fresh breeze		Strong breeze		Near Gale		Gale	
	1–1.5		1.6–3.3		3.4–5.4		5.5–7.9		8–10.7		10.8–13-8		13.9–17.1		17.2–20.7	
Distribution	n°	%	n°	%	n°	%	n°	%	n°	%	n°	%	n°	%	n°	%
ALL	3	1.4	9	4.2	79	36.9	78	36.4	28	13.1	16	7.5	9	4.2	1	0.5
SS	1	0.9	6	5.5	42	38.5	40	36.7	11	10.1	4	3.7	4	3.7	1	0.9
CI/F	2	1.9	3	2.9	28	26.7	38	36.2	17	16.2	12	11.4	5	4.8	0	0.0

Regarding the wind direction (Fig. 6.13), values varied between 2° and 358° (N), the average of all events being 198° (SSW). During the SS events, winds varied between 2° and 358° (N), with an average of 204° (SSW). In CI/F, winds varied between 2° (N) and 340° (NNW), and the average was 191° (S-SSW).

The distribution of directions according to the classes and quadrants (Table 6.10) reveals a broad spectrum of winds blowing from all quadrants, notwithstanding a predominance of the S-W quadrant with similar percentages in all cases: 42.2% among all events, 42.3% among SS, and 42.1% among CI/F. These results corroborate the results of Campos et al. (2010).

SW (14.7%) and SSW (13.3%) were the predominant directions for all events as well as for CI/F (15% and 15.9%, respectively). For SS, SW (14.4%), S (12.6%) and WSW (12.6%) directions predominated.

6.5.7 *Significant Waves*

The significant wave heights (Hs) varied between 1 and 7 m (Fig. 6.14), the total average being 2.5 m (213 events). For SS, the average was 2.8 m (1.5–7 m; standard deviation 0.91), and for CI/F, 2.2 m (0.5–5.5 m; standard deviation 0.85).

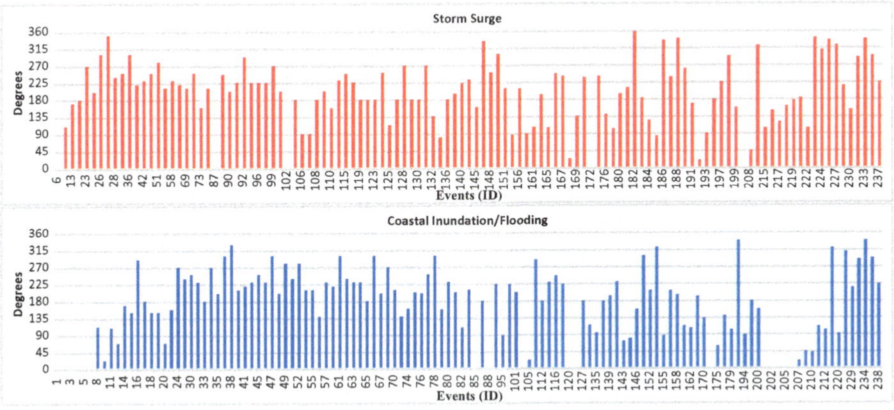

Fig. 6.13 Wind direction for storm surge and coastal inundation/flooding events

These results show heights greater than the mean annual Hs of 1.5 m obtained by Reguero et al. (2013) for the coast of São Paulo State.

The trend lines in both graphs suggest a slight increase in the Hs height over time. In the case of SS, if the purged value of event 22 (17 May 1977) were reduced, the slope of the line would increase considerably.

These trends agree with results from Reguero et al. (2013) and CEPAL (2016), which found increases in mean annual significant height (Hs) of ca. 6 mm/year; in Hs12, with mean values of around 3 cm/year; and in surge extremes, of 1.5 mm/year.

The distribution of significant wave height frequencies (Table 6.11) show that the intervals 1.1–2.0 and 2.1–3.0 m predominated in all situations, totalling 75.6% of all events, 70.3% of SS, and 81.4% of CI/F.

However, considering SS separately, the 2.1–3.0 m class (46.8%) predominated over the 1.1–2.0 m class (23.4%), contrary to CI/F, for which the 1.1–2.0 class (47.1%) predominated (as expected) over the 2.1–3.0 class (34.3%).

The next most frequent heights are 3.1–4.0 m, with 17.4% of the total; for SS, this class corresponds to 23.4%, same as the 1.1–2.0 m class. For the CI/F, as expected, the frequency is much lower, less than 11%.

No SS event had waves under 1.5 m, unlike the CI/F (nine events). Only seven events presented waves higher than 4 m, five of which were in SS (17 May 1977, 08 June 1981, 08 April 2010, 03 May 2011, 11 June 2016) and two in CI/F (08 June 1981, 03 May 2011). Note that two CI/F events were combined.

In relation to the monthly distributions of Hs (Fig. 6.15), the highest averages were concentrated in the autumn/winter months for both types of events, as expected, and varied between 2.6 m (September) and 3 m (May) during SS, and between 2.2 m (April) and 2.7 m (May/June) during CI/F. The month of November, however, fell outside this pattern, with averages of 3.1 m for SS and 3.3 m for CI/F. The lowest averages occurred in the summer months and December (2.3–2.7 m for SS and 1.7–2.1 m for CI/F).

Table 6.10 Frequency of wind directions

Quadrant	*N – E*				*E – S*				*S – W*				*W – N*		
(Degrees)	<36	36–55	56–80	81–100	101–120	121–145	146–170	171–190	191–210	211–235	236–260	261–280	281–310	311–335	>335
Direction	**N-NE**	**NE**	**NE-E**	**E**	**E SE**	**SE**	**SSE**	**S**	**SSW**	**SW**	**WSW**	**W**	**WNW**	**NW**	**NNW**
ALL (n°)	7	4	6	11	17	8	18	22	29	32	21	10	18	6	9
SS (n°)	3	1	1	6	8	4	9	14	12	16	14	5	8	3	7
CI/F (n°)	4	3	5	5	9	4	9	8	17	16	7	5	10	3	2
ALL (%)	12.8				29.8				42.2				15.1		
SS (%)	9.9				31.5				42.3				16.2		
CI/F (%)	15.9				28.0				42.1				14.0		

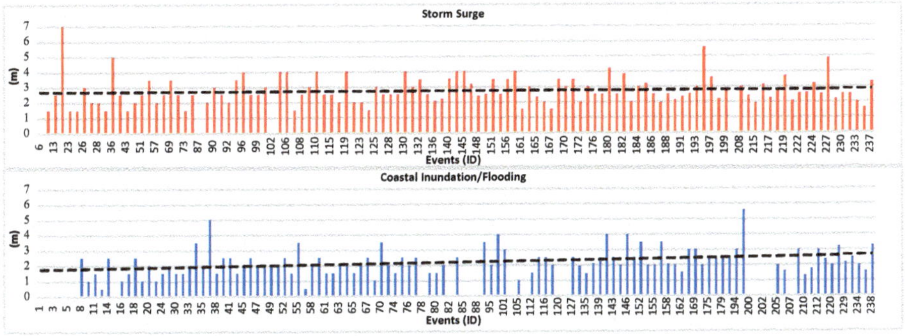

Fig. 6.14 Hs height for storm surge and coastal inundation/flooding events

Table 6.11 Frequency of significant waves (Hs) classes

Hs (m)	0.5–1.0		1.1–2.0		2.1–3.0		3.1–4.0		4.1–5.0		5.1–7.0	
Distribution	N°	%	N°	%	N°	%	N°	%	N°	%	N°	%
ALL	8	3.8	74	34.7	87	40.8	37	17.4	4	1.9	3	1.4
SS	0	0.0	26	23.4	52	46.8	26	23.4	3	2.7	2	1.8
CI/F	8	7.8	48	47.1	35	34.3	11	10.8	1	1.0	1	1.0

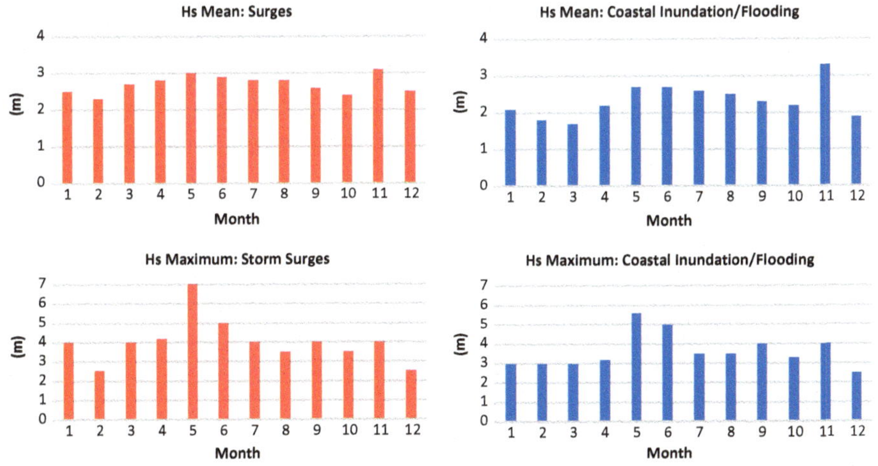

Fig. 6.15 Monthly mean and maximum Hs for storm surge and coastal inundation/flooding events

The pattern of maximum heights is less regular, though the month of May presented the greatest heights for both types of events (7 m for SS and 5.6 for CI/F), followed by the month of June (5 m for both). However, according to the studies of Reguero et al. (2013), these values obtained for May would be quite anomalous and comparable to projections for the return periods of 50 years (5.5–6.0 m) and 500 years (7.0–8.0 m).

Regarding seasonality, the result obtained here, although from another sample universe, are similar to those in Reguero et al. (2013) (maximum mean Hs of 2.5 m for J-J-A, and 3.8 m for S-O-N)—for J-J-A, we found an overall average of 2.7 m and an average maximum of 4 m; for S-O-N, the overall average was 2.7 m, and the average maximum, 3.8 m; for combined events, the overall average was 2.7 m.

In relation to wave direction, note the more consistent behaviour in the SS events than in CI/F events (Fig. 6.16). The directions varied between 23° (NNE) and 300° (NW), the average among all events being 170° (SSE-S). During SS, waves varied from 90° (E) to 300° (NW), with an average of 177° (S). During CI/F, the direction varied from 23° (NNE) to 280° (W), with an average of 162° (SSE).

The distribution of directions according to classes and quadrants reveal that waves originate mainly from the E-S quadrant, with similar percentages in all cases: 70.7% among all events, 75.5% during SS, and 65.7% during CI/F (Table 6.12).

The predominant direction was S, followed by SSE, with the following distribution, respectively: 34.4% and 21.4% in relation to all events; 41.8% and 25.5% for SS; and 26.7% and 17.1% for CI/F.

6.5.8 *El Niño and La Niña*

Among the 238 registered events, 212 (89.1%) occurred in phases of ENSO activity, 50.4% of them under El Niño (EN) and 38.7% during La Niña (LN) (Table 6.13).

Most of the events happened during phases of moderate to weak EN (37.8%) and moderate to weak LN (33.6%); the phases of strong/very strong EN/LN totalled 17.6% of all events. Thus, only 10.9% of the events occurred in phases of neutrality.

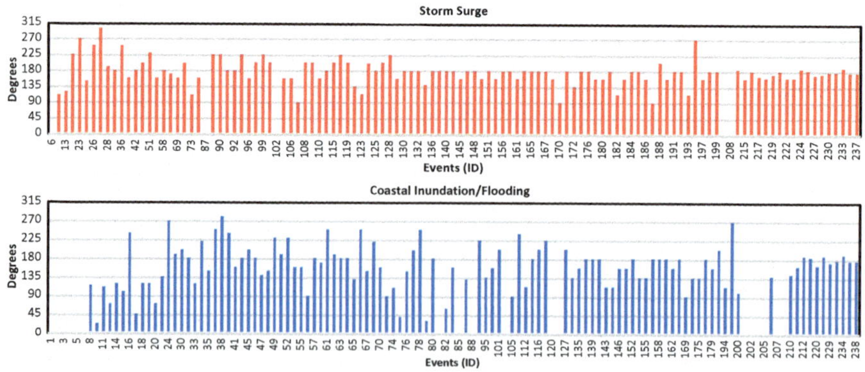

Fig. 6.16 Hs direction for storm surge and coastal inundation/flooding events

Table 6.12 Frequency of Hs directions

Quadrant	*N – E*				*E – S*				*S – W*				*W – N*
(Degrees)	<36	36–55	56–80	81–100	101–120	121–145	146–170	171–190	191–210	211–235	236–260	261–280	281–305
Direction	N-NE	NE	NE-E	E	ESE	SE	SSE	S	SSW	SW	WSW	W	WNW
ALL (n°)	1	3	3	9	17	15	46	74	18	14	9	5	1
SS (n°)	0	0	0	3	6	3	28	46	11	8	2	2	1
CI/F (n°)	1	3	3	6	11	12	18	28	7	6	7	3	0
ALL (%)	7.4				70.7				21.4				0.5
SS (%)	2.7				75.5				20.9				0.9
CI/F (%)	12.4				65.7				21.9				0.0

Table 6.13 Frequency of ENSO phases and intensities

Events	Intensity / Distribution	Strong/very strong n°	Strong/very strong %	Moderate n°	Moderate %	Weak n°	Weak %	Total n°	Total %
ALL	El Niño	30	12.6	46	19.3	44	18.5	120	50.4
	La Niña	12	5.0	40	16.8	40	16.8	92	38.7
	Neutrality							26	10.9
SS	El Niño	16	13.9	24	20.9	23	20,0	63	54.8
	La Niña	2	1.7	25	21.7	19	16.5	46	40.0
	Neutrality							6	5.2
CI/F	El Niño	14	11.4	22	17.9	21	17.1	57	46.3
	La Niña	10	8.1	15	12.2	21	17.1	46	37.4
	Neutrality							20	16.3

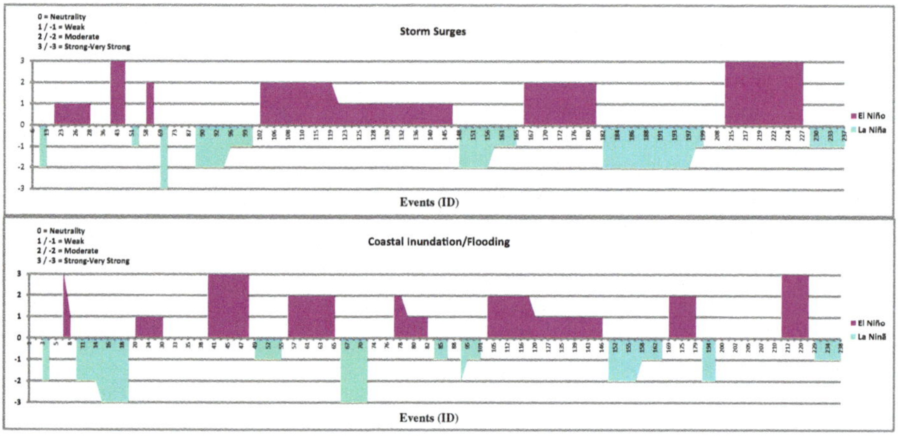

Fig. 6.17 ENSO phases and intensities for storm surge and coastal inundation/flooding events

SS events had the following distribution: 54.8% under EN activity, moderate and weak intensities predominating, with similar percentages of about 20% each; 40% of events happened during LN, with predominantly moderate intensity (21.7%); only 5.2% of the events occurred during neutrality phases.

For CI/F events, the trends were somewhat similar: 46.3% occurred under EN activity, with predominantly moderate to weak intensities and similar percentages of around 17% each; 37.4% of the events happened in LN, with predominantly weak intensity (17.1%); and during the neutrality phases, 16.3% of the events occurred.

The temporal distribution of the events, illustrated in the graphs of Figs. 6.17 and 6.18, show the increasing trend in the number of events with EN and LN activity, starting in the 2000s.

When compared to the years of global ENSO occurrence (see, for example: http://ggweather.com/enso/oni.htm), these results underline the conditioning of the

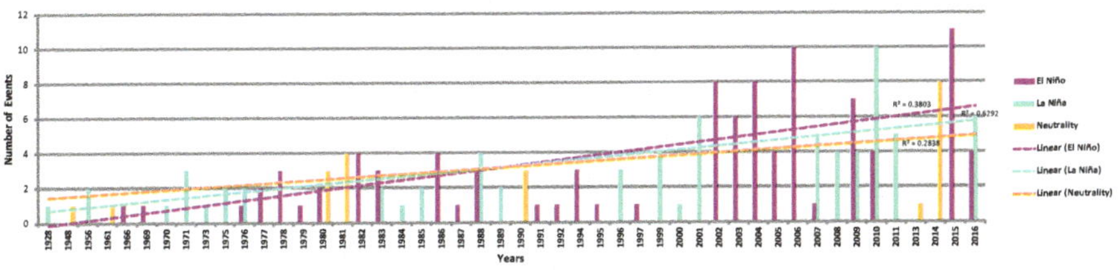

Fig. 6.18 ENSO and neutrality phases in an annual distribution of SS and CI/F events (conjugated events of SS and CI/F were considered as only 1 event)

extreme climatic events at the global scale, but they also suggest that other factors must be controlling these events, among them the regional climatic variability and the behaviour of the atmospheric systems preceding ENSO itself, as well as other possible forcings not analysed here.

However, as seen in Table 6.13, the intensity of the ENSO does not seem to influence the number of events per year, because in the years with strong El Niño and La Niña activity, the number of events did not increase, with exception of 2015–2016; however, these years still follow the decadal trend.

The control exerted by other forcings is clear when the following are observed: (a) the occurrence of 26 events (10.9%) during neutrality phases, 20 CI/F and 6 SS, and distributed throughout all seasons of the year; (b) the reduced number of events in the 1950s and 1960s, because, since the 1950s at least, the number of years with global ENSO seems not to have varied much (8 in the 1950s, 6 in the 1960s, 1980s, 1990s e 2000s, 9 in the 1970s, and 4 between 2011 and 2016); (c) there is no apparent link between El Niño–Southern Oscillation (ENSO/SOI) and wave climate in the South Atlantic coast, either for wave heights or for directions, as postulated by Reguero et al. (2013).

6.6 Final Remarks

The volume of manipulated data in the database exceeded 7000.

A synthesis of the main conditions surrounding the 238 events is exhibited in Table 6.14.

The two events considered most intense, given the proportion of their impacts on the Santos coastline, were combined; their characteristics are presented in Table 6.15. It is interesting to note that neither of them presented very high Hs, meteorological tide or wind intensity values.

This and all the results presented here suggest that, as mentioned previously, other factors are also very important triggers of impacts on Santos. One of the main factors must be the direction of the waves entering the Bay of Santos and the changes they undergo until they reach the coastline. The local winds, in turn, can also affect the behaviour of these waves and of the longshore drift currents during extreme

Table 6.14 Summary of the boundary conditions for SS and CI/F events

Indicator	Storm Surge (115 events)	Coastal Inundation/Flooding (123 events)
Distribution	Highest n° of events: 2000s (49 = 42.6%); year 2010 (14); 33% in May and July; 76.5% between April and September	Highest n° of events: 2000s (34 = 27.6%); years 2006 and 2014 (8 each); 16.3% in March; 50.4% between January and April
	21st century (2000–2016): 76.5% of the total; increased by 3.3 times relative to the 20th century	21st century: 47.2% of the total; reduction by 0.9 or constant relative to the 20th century
Duration	Average: 1.9 days (1–8 days)	Average: 1.8 days (1–5 days)
Lunar phase	52.2%: during spring tides	65.6% during spring tides
	34.8%: during neap tides	18.7%: during neap tides
Meteorological tide height	Average 0.35 m (maximum: 0.78 m); predominant 0.41–0.60 m (35.7%); greatest heights (>0.7 m) in January, May and June	Average 0.38 m (maximum: 0.78 m); predominant 0.41–0.60 m (36.2%); greatest heights (>0.7 m) in January, May and June
Rainfall volume	Duration interval: average 33.1 mm; maximum 227.1 mm; 0–60 mm accumulated = 84.3%	Duration interval: average 51.3 mm; maximum 277.8 mm; 0–60 mm accumulated = 72.1%
	Evolution period: average 63.9 mm; maximum 468.4 mm; 0–100 mm accumulated = 83.5%	Evolution period: average 103.0 mm; maximum 468.4 mm; 0–100 mm accumulated = 62.3%
Winds	Intensity: average 6.2 m/s (moderate breeze); maximum 20.6 m/s (gale); 75.2% between 3.4 and 7.9 m/s (gentle-moderate breeze)	Intensity: average 7.0 m/s (moderate breeze); maximum 17.0 m/s (near gale); 62.9% between 3.4 and 7.9 m/s (gentle-moderate breeze)
	Direction: average 204° (SSW) (2–358°); SW/S/WSW predominate (37.8%)	Direction: average 191° (S-SSW) (2–340°); SW/SSW predominate (30.9%)
Significant Waves	Height: average 2.8 m; maximum 7 m; 75.7% between 1.5 and 3.0 m; 45% between 2.5 and 3.0 m; highest average in May (3.0 m)	Height: average 2.2 m; maximum 5.5 m; 80.4% between 1.5 and 3.0 m; 50% between 1.5 and 2.0 m; highest average in May–June (2.7 m)
	Direction: average 177° (S) (90–300°); S/SSE predominate (41.8 and 25.5%)	Direction: average 162° (SSE) (23–280°); S/SSE predominate (26.7 and 17.1%)
ENSO	El Niño: 54.8%; moderate to weak intensities predominate (40.9%)	El Niño: 46.3%; moderate to weak intensities predominate (35%)
	La Niña: 40%; moderate intensity predominates (21.7%)	La Niña: 37.4%; weak intensity predominantes (17.1%)
	Neutrality: 5.2%	Neutrality: 16.3%

events. Furthermore, the morphodynamic state of the beach preceding an extreme event is also a determinant of how the beach receives and readjusts to the impacts of storm waves. Beaches in a state of chronic erosion and beaches affected by consecutive SS (acute erosion) will present morphodynamic disequilibrium and negative

Table 6.15 Characteristics of the most impactful events for Santos coastline

Date	Duration (day)	Lunar phase	Meteorological tide (m)	Rainfall volume: event duration; evolution period (mm)	Winds: intensity; direction	Significant Waves: height direction	ENSO
26/04/2005	1	Full	0.50	26.6; 85.4	8.7 m/s; 180°	4.0 m; 180°	Weak EN
27/04/2016	4	Full/ waning	0.49	26.4; 26.4	7.2 m/s; 310°	3.2 m; 187°	Strong EN

sedimentary balance, which will result in greater erosion during SS, facilitating the incursion of water over the mainland and increasing coastal inundation.

Therefore, there is clearly a need to gather local monitoring data to better understand the behaviour of these very complex extreme events.

In any respect, the data set obtained for this 89-year historical series of extreme events of SS and CI/F in the Baixada Santista region, today constitutes an important tool for the municipal contingency[1] plans for storm surges and anomalous high tides (currently under development in most cases) as well as city planning within the context of climate change and its effects.

Although the socioeconomic impacts caused by these events are difficult to quantify, the damages and restoration expenses in many cases were on the order of millions of Brazilian reals, especially after large-magnitude events.

Acknowledgements The authors thank Oceanographer Eduardo G. Rosa, Environmental Analyst Johann C. Lima and Geographer Graziella S. R. Rodrigues for conducting part of the work of compiling the data and making some graphic treating. We also thank Dr. Mirian R. Gutjahr (Instutute of Geology-SMA/SP) and Professor Dr. Luci H. Nunes (LECLIG-University of Campinas) for making available their disaster registries for São Paulo coastal zone; the BNDO – National Bank of Oceanographic Data (Captain Vladimir C. Maluf in particular), the Centre for Hydrodynamic Research at the University Santa Cecília (Profs. Renan B. Ribeiro MSc. and Dr. Alexandra F. P. Sampaio), as well as the Santos Civil Defence (Eng. Ernesto K. Tabuchi and Geologist Marcos P. Bandini in particular) for kindly making their databases available. Finally, our thanks to the State of São Paulo Docas Company (2010–2011), the IG-SMA/SP, and the São Paulo Research Foundation (Fapesp n° 2012/51876-0 and 2015/08192-1) for the various types of financial support for this research.

[1] In 2016, São Paulo State created the "*Plano Preventivo de Defesa Civil para Erosão Costeira, Inundações Costeiras e Enchentes/Alagamentos causadas por Eventos Meteorológicos-Oceanográficos Extremos como Ressacas do Mar e Marés Altas*" (a preventative plan for coastal erosion, coastal inundation and flooding caused by meteorological-oceanographic extremes such as storm surges and high tides), in which one of the main instruments is the municipal contingency plan, *Plano Municipal de Contingência* (*Resolução CMIL 17-610 – Cedec, de 28-11-2016*).

References

Bittencourt, A. C. S. P., Medeiros, K. O. P., Dominguez, J. M. L., Guimarães, J. K., & Dutra, F. R. L. S. (2008). Severe coastal erosion hotspots in the city of Salvador, Bahia, Brazil. *Shore and Beach, 76*, 8–14.

Camargo, R., & Harari, J. (1994). Modelagem numérica de ressacas na plataforma sudeste do Brasil a partir de cartas sinóticas de pressão atmosférica na superfície. *Boletim do Instituto Oceanográfico USP, 42*(1), 19–34.

Campos, R. M., Camargo, R., & Harari, J. (2010). Caracterização de eventos extremos do nível do mar em Santos e sua correspondência com as re-análises do modelo do NCEP no Sudoeste do Atlântico Sul. *Revista Brasileira de Meteorologia, 25*, 175–184.

CEPAL-Comisión Económica para America Latina y el Caribe. (2016). *Efectos del cambio climático en la costa de América Latina y el Caribe: Dinámicas, tendencias y variabilidad climática.* Santiago de Chile: Naciones Unidas. Available https://www.cepal.org/es/publicaciones/3955-efectos-cambio-climatico-la-costa-america-latina-caribe-dinamicas-tendencias. Accessed May 2017.

Gan, M. A., & Rao, B. V. (1991). Surface ciclogenesis over South America. *Monthly Weather Review, 119*, 293–302.

Gasparro, M. R., Sousa, E. C. P. M., Giordano, F., Argentino-Santos, R. C. (2008). *Occupation history of the Santos estuary, perspectives on integrated coastal zone management in South America.* In: R. Neves, et al. (Eds.), IST Press. Available https://www.unisanta.br/arquivos/CoastalZoneManagementFinal.pdf. Accessed May 2017.

Gutjahr, M. R. (coord) (2011). *Banco de Dados: Estudos históricos relacionados a eventos climáticos na Baixada Santista – SP.* Instituto Geológico–SMA/SP. Available http://www.igeologico.sp.gov.br. Accessed May 2010.

IPCC-Intergovernmental Panel on Climate Change. (2012). Managing the risks of extreme events and disasters to advance climate change adaptation. In C. B. Field et al. (Eds.), *Special report of working groups I and II of the Intergovernmental Panel on Climate Change.* Cambridge: Cambridge University Press. Available https://www.ipcc.ch/pdf/special-reports/srex/SREX_Full_Report.pdf. Accessed May 2017.

IPCC-Intergovernmental Panel on Climate Change. (2014). Summary for policymakers. In C. B. Field et al. (Eds.), *Climate change 2014: impacts, adaptation and vulnerability. Part A: Global and sectoral aspects. Contribution of working group II to the fifth Assessment report of the Intergovernmental Panel on Climate Change.* Cambridge: Cambridge University Press. Available https://www.ipcc.ch/pdf/assessment-report/ar5/wg2/ar5_wgII_spm_en.pdf. Accessed May 2017.

Lins-de-Barros, F. M. (2010). *Contribuição metodológica para análise local da vulnerabilidade costeira e riscos associados: Estudo de caso da Região dos Lagos.* Tese de Doutorado: Universidade Federal do Rio de Janeiro.

Losada Rodríguez, I. J., Reguero, B. G., Méndez, F. J., Castanedo, S., Abascal, A. J., & Mínguez, R. (2013). Long-term changes in sea-level components in Latin America and the Caribbean. *Global and Planetary Change, 104*, 34–50.

Machado, A. A., Calliari, L. J., Melo, E., & Klein, A. H. F. (2010). Historical assessment of extreme coastal sea state conditions in southern Brazil and their relation to erosion episodes. *Pan-American Journal of Aquatic Sciences, 5*(2), 277–286.

Magrin, G. O., Marengo, J. A., Boulanger, J. P., Buckeridge, M. S., Castellanos, E., Poveda, G., Scarano, F. R., & Vicuña, S. (2014). Central and South America. In V. R. Barros et al. (Eds.), *Climate change: Impacts, adaptation, and vulnerability. Part B: Regional aspects, Contribution of working group ii to the fifth Assessment report of the Intergovernmental Panel on Climate Change* (pp. 1499–1566). Cambridge: Cambridge University Press. Available https://www.ipcc.ch/pdf/assessment-report/ar5/wg2/WGIIAR5-Chap27_FINAL.pdf. Acessed May 2017.

Mangor, K., Drønen, N. K., Kærgaard, K. H., & Kristensen, S. E. (2017). *Shoreline management guidelines* (4ª ed.). Horsholm: DHI Water & Environement. Available https://www.dhigroup.

com/upload/campaigns/shoreline/assets/ShorelineManagementGuidelines_Feb2017-TOC.pdf. Accessed May 2017.

Marengo, J. A., Scarano, F. R., Klein, A. F., Souza, C. R. G., & Chou, S. C. (2017a). Impacto, vulnerabilidade e adaptação das cidades costeiras brasileiras às mudanças climáticas. In J. A. Marengo & F. R. Scarano (Eds.), *Relatório especial do Painel Brasileiro de Mudanças Climáticas (PBMC)*. Rio de Janeiro: COPPE-UFRJ. https://doi.org/10.13140/RG.2.2.36042.16329.

Marengo, J. A., Nunes, L. H., CRG, S., Harari, J., Hozokawa, E. K., & Tabuchi, E. K. (2017b). Vulnerability in Brasilian coastal comunities: An integrated framework to analyse local decision making and adaptation to sea-level rise in Santos, São Paulo – Brazil. In V. Marchezini et al. (Eds.), *Redution of vulnerability to disasters: From knowlegde to action* (Vol. 1, 1st ed., pp. 397–408). São Carlos: Rima Editora.

Marone, E., & Camargo, R. (1994). Marés meteorológicas no litoral de estado do Paraná: o evento de 18 de agosto de 1993. *Revista Nerítica, 8*(1–2), 73–85.

Muehe, D. (org) (2006). *Erosão e progradação do litoral brasileiro*. MMA/PGGM, Brasília, DF, p 476.

Nicholls, R. J. (2006). Storm surges in coastal areas. In M. Arnold et al. (Eds.), *Natural disaster hotspots – case studies, Disaster Risk Management Ser nº 6* (pp. 79–108). Washington, DC: The World Bank Hazard Management Unit.

NOAA-National Oceanic and Atmospheric Administration. (2017). *What is storm surge*? Available https://oceanservice.noaa.gov/facts/stormsurge-stormtide.html. Accessed May 2017.

Parise, C. K., Calliari, L. J., & Krusche, N. (2009). Extreme storm surges in the south of Brazil: atmospheric conditions and shore erosion. *Brazilian Journal of Oceanography, 57*(3), 175–188.

Paula, D. P., Morais, J. O., Ferreira, O., & Dias, J. A. (2015). *Análise histórica das ressacas do mar em Fortaleza (Ceará, Brasil): origem, características e impactos*. In: D. V. Paula, & J. A. Dias (orgs) Ressacas do mar/temporais e gestão costeira, pp 173–201.

PBMC-Painel Brasileiro de Mudanças Climáticas. (2014). Base científica das mudanças climáticas. In T. Ambrizzi & M. Araújo (Eds.), *Contribuição do grupo de trabalho 1 do Painel Brasileiro de Mudanças Climáticas ao primeiro Relatório da avaliação nacional sobre mudanças climáticas* (p. 464). Rio de Janeiro: COPPE, Universidade Federal do Rio de Janeiro.

Pontes, N. Z., & Zee, D. M. W. (2010). *Mudanças climáticas globais e seus reflexos nas praias oceânicas do município do Rio de Janeiro*. In: Expanded Abstracts of the 4° Congresso Brasileiro de Oceanografia, Rio Grande, RS, October 2010.

Pugh, D. T. (1987). *Tides, surges and mean sea level. A handbook for engineers and scientists* (p. 472). New York: Wiley.

Reboita, M. S., Rocha, R. P., & Ambrizzi, T. (2010). South Atlantic ocean cyclogenesis climatology simulated by regional climate model (RegCM3). *Climate Dynamics, 35*, 1331–1347.

Reguero, B. G., Méndez, F. J., & Losada, I. J. (2013). Variability of multivariate wave climate in Latin America and the Caribbean. *Global and Planetary Change, 100*, 70–84.

Ribeiro, R. B., Leitão, J. M. C. F. L., Leitão, P. M. C. F. L., Puia, H. L., & Sampaio, A. F. P. (2016). *Integration of high-resolution metocean forecast and observing systems at Port of Santos*. In: Proceedings of the IX PIANC-COPEDEC, Conference on Coastal and Port Engineering in Developing Countries, Rio de Janeiro, October 2016.

Rudorff, F. M., Bonetti Filho, J., Moreno, D. A., Oliveira, C. A. F., & Murara, P. G. (2014). Maré de tempestade. In M. L. P. Herrmann (Ed.), *Atlas de Desastres Naturais do Estado de Santa Catarina: Período de 1980 a 2010* (pp. 151–154). Florianópolis, Santa Catarina, Brasil: Cadernos Geográficos.

Santana, C. L., Souza, C. R. G., & Harari, J. (2004). *Correlação de dados pluviométricos, fluviométricos e maregráficos em eventos de enchentes/inundações no baixo Rio Ribeira de Iguape (SP)*. In: Proceedings of the I Simpósio Brasileiro de Desastres Naturais – SIBRADEN, Florianópolis, September 2004.

Satyamurti, P., Nobre, C., & Dias, P. L. S. (1998). South America. In D. J. Karoly & D. J. Vincent (Eds.), *Meteorology of the Southern Hemisphere* (pp. 119–139). Boston: American Meteorological Society.

Sinclair, M. R. (1996). A climatology of anticyclones and blocking for the Southern Hemisphere. *Monthly Weather Review, 124*, 245–264.

Souza, C. R. G. (2009a). A erosão costeira e os desafios da gestão costeira no Brasil. *J Integr Coast Manag, 9*, 17–37.

Souza, C. R. G. (2009b). *A erosão nas praias do estado São Paulo: causas, consequências, indicadores de monitoramento e risco*. In: Bononi VLR, Santos Júnior NA (org) Memórias do conselho científico da Secretaria do Meio Ambiente: A síntese de um ano de conhecimento acumulado. Instituto de Botânica-SMA/SP, São Paulo, pp 48–69.

Souza, C. R. G. (2009c). Flood risk assessment in coastal drainage basins through a multivariate analysis within a GIS-based model. *Journal of Coastal Research, SI, 56*(1), 900–904.

Souza, C. R. G. (2010). *Impactos das mudanças climáticas no litoral do Estado de São Paulo (Sudeste do Brasil)*. In: Proceedings of the VI Seminário Latino Americano de Geografia Física e II Seminário Ibero Americano de Geografia Física, Coimbra. Available http://www.uc.pt/fluc/cegot/VISLAGF/actas/tema4/celia_regina. Accessed May 2107.

Souza, C. R. G. (2011). *Os ecossistemas costeiros frente às mudanças climáticas no Brasil: efeitos da elevação do nível do mar*. In: Expanded abstracts of the XIV Congresso Latino-Americano de Ciências do Mar – COLACMAR, Balneário Camboriú, October–November 2011.

Souza, C. R. G. (2012). Oceanic sandy beaches of São Paulo (Brazil): Synthesis of knowledge on morphodynamics, sedimentology, sediment transport and coastal erosion. *J Dep Geogr USP*, n. 2012, Spec Vol 30 yers, pp 308–371. https://doi.org/10.7154/RDG.2012.0112.0015.

Souza, C. R. G. (2017). Balanço sedimentar de longo e curto termos da Praia de Santos e causas da erosão na Ponta da Praia. In: Proceedings of XVI Congresso Associação Brasileira de Estudos do Qaternário (ABEQUA), Bertioga (São Paulo), 21–27/10/2017. http://www.abequa.org.br/trabalhos/254_resumo.PDF. (ISSN: 2318–0986). Accessed May 2017.

Souza, C. R. G., Souza Filho, P. W. M., Esteves, S. L., Vital, H., Dillemburg, S. R., Patchineelam, S. M., & Addad, J. E. (2005). Sandy beaches and coastal erosion. In Souza et al. (Eds.), *Quaternário do Brasil* (pp. 130–152). Ribeirão Preto: Holos Editora.

Souza, C. R. G., Luna, G. C., & Souza, A. P. (2012). *Causas da erosão na Ponta da Praia de Santos (São Paulo, Brasil)*. In: Abstracts of the II Workshop Antropicosta Iberoamerica, Montevideo, 2012.

Souza, C. R. G., Gouveia, M. L., & Souza, A. P. (2016a). *Balanço sedimentar da Praia de Santos antes, durante e após as obras de dragagem de aprofundamento do canal do Porto de Santos (São Paulo, Brasil)*. In: Abstracts of the VII Congresso Latinoamericano de Sedimentologia and XV Reunión Argentina de Sedimentologia, Santa Rosa, October 2012. Available http://www.isbn.org.ar/cal/laimg/21314/488040.pdf. Accessed May 2017.

Souza, C. R. G., Souza, A. P., & Gouveia, M. L. (2016b). *Identificação de processos sedimentares em praias por meio da variabilidade temporal de células de deriva litorânea*. In: Abstracts of the VII Congresso Latinoamericano de Sedimentologia and XV Reunión Argentina de Sedimentologia, Santa Rosa, October 2012. http://www.isbn.org.ar/cal/laimg/21314/488040.pdf. Accessed May 2017.

Chapter 7
First Approach of a Storm Surge Early Warning System for Santos Region

Renan Braga Ribeiro, Alexandra Franciscatto Penteado Sampaio, Matheus Souza Ruiz, José Chambel Leitão, and Paulo Chambel Leitão

Abstract Historically, extratropical cyclones associated with frontal systems cause storm surges in Santos city. Although there are no fatality records, these events cause several socio-economic loss, especially in vulnerable regions including the Port of Santos. Accepting the impossibility of eliminating the risks, adapting to natural phenomenon can be accomplished through adequate emergency procedures. Early warnings are a key tool to minimize events impact on human activities and prevent economic loss. The present chapter describes an early warning system developed for Santos region, focusing on an intense storm surge event that occurred in October 2016. Executing and sending detailed bulletins once intense events are predicted, reporting the probability of significant impacts on vulnerable areas, is an important feature of the system. Since May 2016, more than 50 comprehensive bulletins were sent, in which at least 20% have warned the possibility of flooding and impacts on urban infrastructure. Currently, the Civil Defense Action Plan is based on this system, notifying the population each time an extreme event is forecast, through a variety of media and by sending cell phone messages, thus enabling residents to be prepared.

Keywords Storm surge · Early warning · Forecast · Santos city · Coastal flood

R. B. Ribeiro (✉) · A. F. P. Sampaio · M. S. Ruiz
Núcleo de Pesquisas Hidrodinâmicas – Universidade Santa Cecília (NPH-UNISANTA), Santos, SP, Brazil
e-mail: renanribeiro@unisanta.br; canastra@unisanta.br; matheusruiz@unisanta.br

J. C. Leitão · P. C. Leitão
HIDROMOD, Porto Salvo, Portugal
e-mail: jcleitao@hidromod.com; paulo.chambel@hidromod.com

L. H. Nunes et al. (eds.), *Climate Change in Santos Brazil: Projections, Impacts and Adaptation Options*, https://doi.org/10.1007/978-3-319-96535-2_7

7.1 Introduction

Storm surge is a sea-level rise usually associated with strong winds and an atmospheric pressure gradient, particularly from extratropical or tropical cyclones such as hurricanes and typhoons (Resio and Westerink 2008). Generally, tropical cyclones cause the most extreme events, but extratropical cyclones can also produce high sea levels, mainly when they coincide with high tide (Pugh 1996). Extreme sea-level rise causes coastal flood events and can potentially cost human lives, destroy homes and civil infrastructure, and disrupt navigation in coastal and port regions.

The city of Santos is located in the Baixada Santista Metropolitan Region (BSMR), a coastal region characterised by intense socioeconomic activities such as beach tourism and the largest port in Latin America. BSMR is a hub for petrochemical industries and home to more than 1.7 million inhabitants, 420 thousand of whom reside in Santos. Historically, Santos has been susceptible to the occurrence of extreme meteorological and oceanographic events causing floods, landslides, lightning and storm surges (Chaps. 4, 6, 8, and 12). In this region, the main cause of a storm surge event is an extratropical cyclone associated with a frontal system, characterised by high-intensity southwest winds blowing parallel to the coast.

In this context, climate change puts additional pressure on coastal zones: changes in the frequency and intensity of climatic phenomena have consequential impacts on human activities established in coastal zones such as the Santos region. Accepting that, in some cases, it is not possible or desirable to eliminate risks of flooding or shoreline recession, adaptation can be accomplished through adequate provision for emergency procedures (Tompkins 2005).

In this regard, operational models that simulate the circulation in oceans and in coastal and estuarine regions are important tools that support decisions related to preventing the effects of extreme events and to contingency planning. With advancements in technology and knowledge, the operational systems for predicting mid and large-scale circulation are widespread. Some examples of systems applied to coastal and estuarine regions are the various Operational Forecast Systems (OFS) implemented by the National Oceanic and Atmospheric Administration (NOAA 2017).

To simulate circulation and wave propagation at the local scale, numerical modelling is commonly used, most often applied in case studies, toward specific objectives and simulating past conditions (e.g. Leitão et al. 2005; Dube et al. 2009; Bunya et al. 2010). Numerical modelling may also be applied, through an operational system for the prediction of oceanographic conditions (e.g., Verlaan et al. 2005; Alfieri et al. 2012; Murty et al. 2017; Valchev et al. 2018). However, despite several applications of numerical modelling in the Santos region (e.g., Harari and Camargo 2003; Sampaio et al. 2008; Cassiano et al. 2012; Roversi et al. 2016), there are few operational models of appropriate scale to simulate the hydrodynamic circulation linked to wave propagation in the region of Santos Bay and Estuary, and to produce results through a user-friendly interface that can assist local administrators in decision-making.

In Santos, an integrated system of observation, prediction and early warning alerts of sea conditions in the bay and estuary region is being used as one of the preventative measures against the impacts of natural disasters. This chapter describes the characteristics of this operational system, implemented initially with the broad objectives of supporting the prediction of recreational water quality at regional beaches (Leitão et al. 2015; Ribeiro et al. 2017) and supporting navigation (Ribeiro et al. 2016). However, the reliability of wave and sea level forecasts, as well as empirical knowledge of regional characteristics, enabled early warning alerts for storm surge events that occurred in 2016.

7.2 Warning System Framework

In 2014, the hydrodynamics research group *Núcleo de Pesquisas Hidrodinâmicas da Universidade Santa Cecília* (NPH-UNISANTA), in partnership with Mercoshipping, implemented AQUASAFE, a platform developed by Hidromod and honoured by the International Water Association (IWA) in 2012. This platform enables management of different data sources and high-resolution numerical models to provide forecasts; control of the interface to disseminate the generated data to users through various client devices (desktop, web, mobile); and automated transmission of reports and alerts.

The following sections describe this platform (named AQUASAFE Santos), the real-time monitoring network, the implementation of high-resolution operational numerical modelling, and the platform's integrated user interface.

7.3 AQUASAFE Platform

The AQUASAFE platform is widely applied in the fields of water resources and information management. It enables the automated management and integration of different data sources and numerical models for the provision of data—obtained in real-time from a variety of sensors (e.g., tide gauges, directional wave buoys, high-frequency radars, meteorological stations, precipitation radars, rain gauges)—high-resolution predictions, and interfaces for different categories of users, including desktop, web and mobile app. AQUASAFE also automatically generates and sends reports and alerts, transforming data organisation, processing and cross-checking into relevant information for a variety of users.

This platform is based on a server-client philosophy, as demonstrated in Fig. 7.1. The server is responsible for all information management, from automatic downloads of external data and information (e.g., from real-time monitoring and from global and regional numerical models), to pre-processing and execution of high-resolution numerical models, data synchronisation, and automated alerts and reports. The server also makes data and information available through a web service,

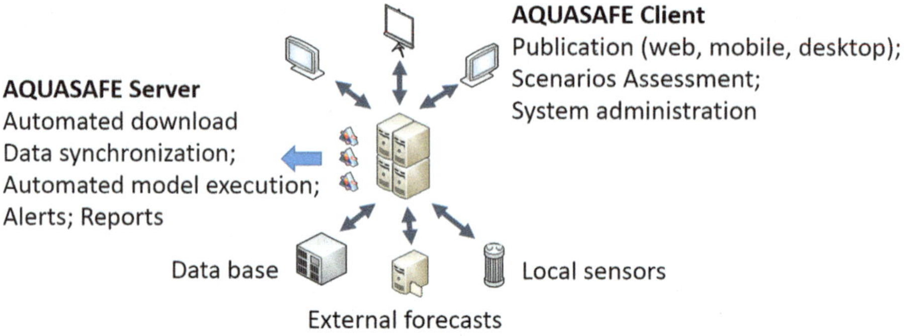

Fig. 7.1 Schematic representation of the AQUASAFE platform, showing the server-client philosophy and the connection to external sources

enabling connection and integration with external systems such as other software, platforms and websites. The AQUASAFE *client* allows access to all the information and data generated and synchronised by the server. The primary focus of its user-friendly desktop interface is to disseminate real-time data and forecasts to centres of operations and the personal computers of administrators and decision makers. There are three levels of access to the interface from the standpoint of functionalities and constraints:

- User – allows information from the server, in the form of graphs, tables, dashboards and GIS, among others, to be visualised in a workspace that is predefined by the following categories.
- Power-User – in addition to the visuals in the predefined workspaces, this user category allows workspaces to be edited and newly created based on information provided by the server.
- Admin – Users with administration privileges are permitted to configure the server. Among the functionalities are the creation and management of users, workspaces and automated tasks; configuration and management of numerical models and new data sources; and information dissemination, exporting data and forecasts in various formats (e.g., netcdf, hdf, images, pdf and excel reports) via e-mail, ftp, web service, etc.

Below, we present the workspaces and automatic reports developed for the early warning system.

7.4 Real-Time Monitoring Network

Currently, public and private real-time data from various stations are connected to the AQUASAFE Santos system.

Meteorological data are obtained from three different sources:(1) Meteorological network of the Brazilian Aeronautical Command (REDEMET—*Rede meteo-*

Table 7.1 Real-time monitoring network connected to the AQUASAFE Santos platform

Type	Name/location	Parameter	Data interval	Source
Real-time Tide gauge	Ilha das Palmas	Water level	10 min	SP Pilots
	Praticagem			
	Capitania			
	Ilha Barnabé			
	COSIPA			
Real-time Moored ADCP	Ilha das Palmas	Current speed and direction, pressure and water temperature	20 min	SP Pilots
	Praticagem			
	Capitania			
	Alemoa			
Real-time Moored ADCP	Ilha das Palmas	Significant wave height and peak period	20 min	SP Pilots
Real-time Meteorological station	Ilha das Palmas*	Wind speed and direction, visibility *includes: air temperature, air pressure and precipitation	<5 min	SP Pilots
	Praticagem*			
	Capitania			
	Alemoa			
	COSIPA			
Real-time Meteorological station	Santos Air Base	Wind speed and direction, visibility, air temperature, air pressure and precipitation	1 h	REDEMET
Real time Rain gauge	34 rain gauges	Precipitation	1 h	CEMADEN

rológica do Comando da Aeronáutica), with data from the Santos Air Base; (2) the station network of the Centre for Traffic Coordination, Communications and Operations (C3OT—*Centro de Coordenação, Comunicações e Operações de Tráfego*) of the São Paulo Pilots (*Praticagem de São Paulo*), with data from five stations; and (3) a network of automatic rain gauges belonging to the Centre for Natural Disaster Monitoring and Alert (CEMADEN—*Centro Nacional de Monitoramento e Alertas de Desastres Naturais*), with data from 34 stations. With respect to oceanographic data, all data originate from the C3OT from the São Paulo (SP) Pilots and include sea level elevation, speed and direction of currents, significant wave height, and peak wave period. A summary and the locations of these stations are presented below (Table 7.1 and Fig. 7.2).

7.5 Global and Regional Numerical Modelling Network

On a daily basis, the AQUASAFE Santos platform downloads the forecasts of five global/regional models and makes a sea level prediction using a global tidal solution based on harmonic constants. A summary of this modelling network is presented in Table 7.2.

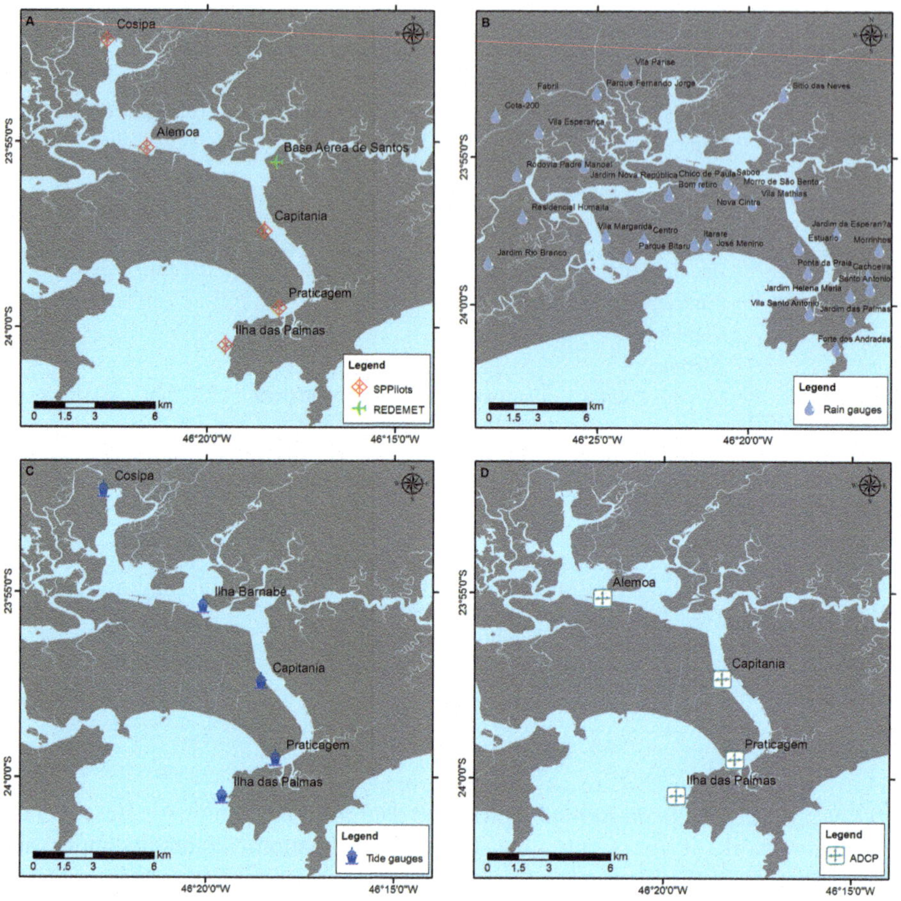

Fig. 7.2 Real time stations connected to AQUASAFE Platform: (**a**) meteorological stations; (**b**) rain gauges; (**c**) tide gauges; and (**d**) moored acoustic doppler current profilers (ADCPs)

The meteorological models serve both to predict meteorological conditions (air temperature, wind direction and intensity, precipitation, etc.) and to be used directly in numerical models (hydrodynamic, wave) as boundary conditions (forcing). Currently in use are three external-source models: (1) the Global Forecast System (GFS) model; (2) the 15 km Eta model; and (3) the 5 km Brazilian Regional Atmospheric Modeling System (BRAMS).

The GFS, available from the National Oceanic and Atmospheric Administration (NOAA), is a meteorological prediction model produced by the National Centers for Environmental Prediction (NCEP). Dozens of atmospheric and soil variables are available from this dataset, including temperature, wind, precipitation, soil humidity and atmospheric ozone concentrations. The GFS is a coupled model composed of four distinct models—atmospheric, oceanic, land/soil, and sea ice—that that work together to provide a precise picture of meteorological conditions.

Table 7.2 Global and regional numerical modelling network connected to AQUASAFE Santos platform

Type	Model/horizontal resolution	Parameter	Data interval	Source
Meteorological model	GFS (~50 km)	Wind speed and direction, relative humidity, air temperature, air pressure and precipitation	3 h	NOAA
Meteorological model	ETA (15 km)	Wind speed and direction, relative humidity, air temperature, air pressure and precipitation	3 h	CPTEC/INPE
Meteorological model	BRAMS (5 km)	Wind speed and direction, relative humidity, air temperature, air pressure and precipitation	6 h	CPTEC/INPE
Hydrodynamic model	CMEMS (~8.3 km)	Current speed and direction, water level, water temperature and salinity	24 h	CMEMS
Wave model	WW3 (~50 km)	Significant wave height, mean direction and peak period	3 h	NOAA
Tide model	FES2012 (~6.25 km)	Water level (astronomic tide)	<1 h	AVISO

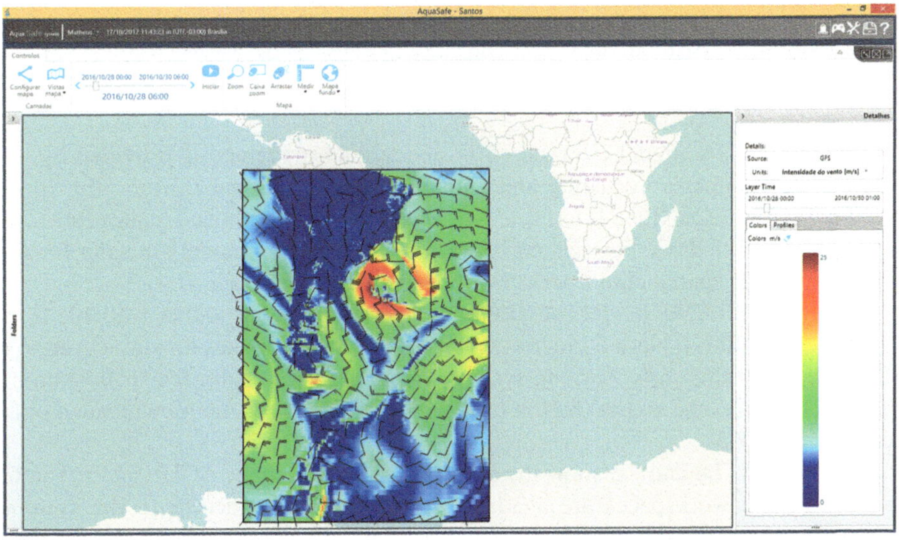

Fig. 7.3 Results of the GFS model for 28 October 2016 at 06:00 (UTC-3) for the South America region, visualised in the AQUASAFE Santos platform. The colour scale represents the wind intensity (varying from 0 to 25 m/s), and the direction is indicated by the wind barbs

For this early warning system, a spatial cutout for the South America region (Fig. 7.3) is downloaded once daily, with the results of the GFS model for the global numerical grid at a horizontal resolution of approximately 50 km. Using those results, the AQUASAFE Santos platform provides forecasts every 3 h for up to 7 days.

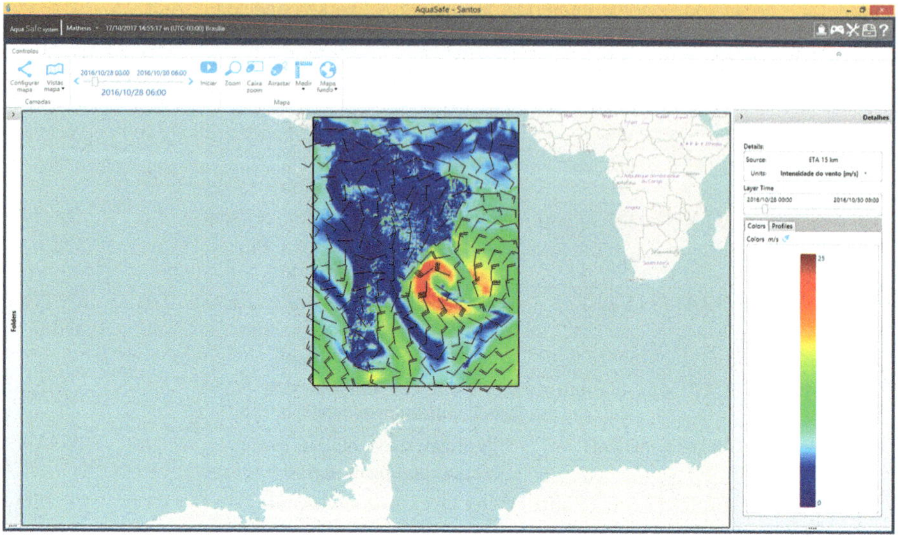

Fig. 7.4 Results of the 15 km Eta model for 28 October 2016 at 06:00 (UTC-3) for the South America region, visualised in the AQUASAFE Santos platform. The colour scale represents the wind intensity (varying from 0 to 25 m/s), and the wind barbs indicate the direction

The 15 km Eta model is executed in operational mode and is provided by the Centre for Weather Forecasting and Climate Change/National Institute for Space Research (CPTEC/INPE). It has a 15 km horizontal resolution and makes forecasts every 3 h for 11 days. For this early warning system, the model's results for the entire available spatial cutout are downloaded once per day (Fig. 7.4).

The 5 km BRAMS model is executed in operational mode and is available from the CPTEC/INPE. It has a 5 km horizontal resolution and makes forecasts every 6 h for 3 days. For this early warning system, the model's results for the spatial cutout, which comprises the coastal region of the state of São Paulo, are downloaded once per day (Fig. 7.5).

Figures 7.3, 7.4 and 7.5 show the workspaces created in the AQUASAFE Santos platform. The workspaces are available to the users with access to the system, allowing them to check global and regional meteorological forecasts, presented as maps and animations, as well as past results from the data sources stored in the platform. The illustrated cases show the results of the three meteorological models for 28 October 2016, at 06:00 (UTC-3), depicting the action of an extratropical cyclone associated with a frontal system. The 50 km GFS and 15 km Eta models predicted winds of 25 m/s (90 km/h) in the ocean region, and the three meteorological models predicted winds of 15.8 m/s (56.9 km/h) in the coastal region adjacent to Santos, parallel to the coastline and from the southwest quadrant. These conditions are typical of storm surge and, in this specific case, caused a strong event, which is presented in the following items.

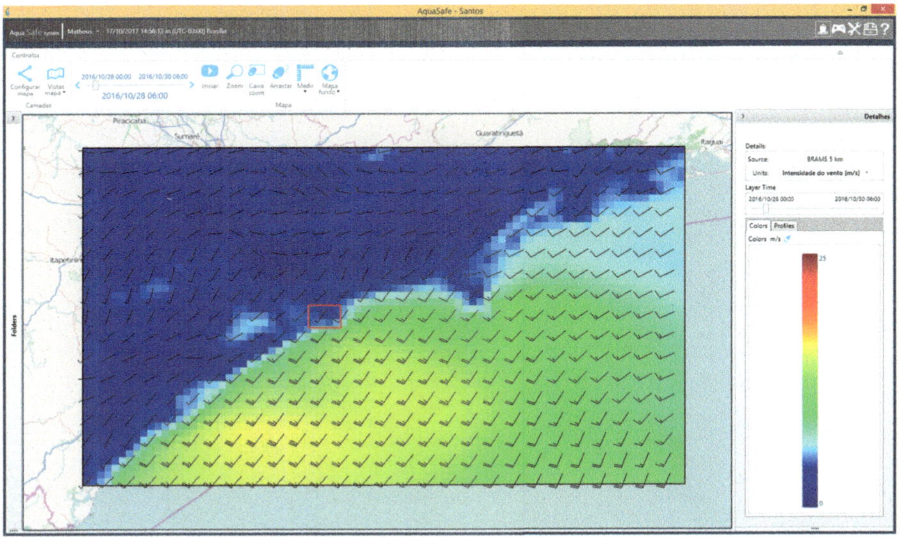

Fig. 7.5 Results of the 5 km BRAMS model for 28 October 2016 at 06:00 (UTC-3) for the coastal region of the state of São Paulo, visualised in the AQUASAFE platform with the region of Santos highlighted in red. The colour scale represents the wind intensity (varying from 0 to 25 m/s), and the wind barbs indicate the direction

As an external source of information on wave propagation at the global scale, results are downloaded from an operational system executed and provided by the National Weather Service/NOAA/NCEP. This system uses the wave model WAVEWATCH III® (WW3) and the operational products of the NCEP as forcing. The model is run four times per day, providing forecasts for up to 180 h (7.5 days).

For this early warning system, data from the NOAA operational model—significant wave height, peak wave period and wave direction—are downloaded once per day for a spatial cutout of the ocean region of Brazil, Uruguay and Argentina, corresponding to the global numerical grid with a horizontal resolution of approximately 50 km. Forecasts for up to seven days are also available (Fig. 7.6).

During the October 2016 event, the NOAA's WW3 model predicted a significant wave height of 6.5 m in the ocean region, reflecting the action of an extratropical cyclone associated with propagation of a frontal system.

In addition to the external data source of this global model, the global-scale WW3 model is run internally for use as a forcing in small-scale grids, simulating wave propagation at the scale of Santos Bay, as described in the following item.

As an external source of information about the daily average tide on a global scale and information on low-frequency sea level variations (on the order of days), also called meteorological tide, the results of an operational product from Copernicus Marine Environment Monitoring Service (CMEMS) are downloaded.

A variety of products are available from CMEMS. The product used in this early warning system is the GLOBAL_ANALYSIS_FORECAST_PHY_001_024, which

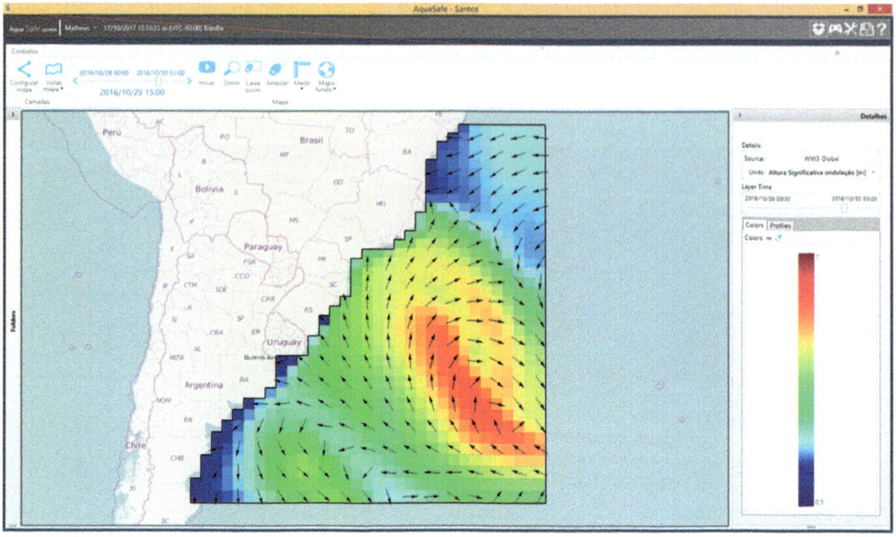

Fig. 7.6 Results for the WW3 model for 29 October 2016 at 15:00 (UTC-3) for the ocean region of Brazil, Uruguay and Argentina, visualised in the AQUASAFE Santos platform. The colour scale represents the significant wave height (varying from 0.5 to 7.0 m), and the vectors indicate the direction

provides analyses and forecasts of a global model with a horizontal resolution varying by approximately 9 km in the equatorial region, 7 km at mid-latitudes, and up to 2 km in the direction of the Ross and Weddell Seas. For vertical discretisation, this model considers 50 layers, with the resolution varying from 1 m in the top layers to 450 m in the bottom layers. A detailed description of this product can be found in Lellouche and Regnier (2015).

Once a day, AQUASAFE Santos downloads a spatial cutout similar to the wave model. The following example (Fig. 7.7) presents the sea-level rise for 29 October 2016, at 11:00 (UTC-3), showing a 0.90 m rise in sea level in the coastal region of Santa Catarina, Paraná and São Paulo states, reflecting the activity of a cyclone with intense winds in the southwest quadrant, generated by the advancement of a frontal system.

7.5.1 High-Resolution Operational Numerical Modelling

Two downscaling approaches, with hydrodynamic and wave models, were developed to provide high-resolution (~50 m horizontal) forecasts for the Santos region (Ribeiro et al. 2016).

Hydrodynamic downscaling uses the MOHID Water Modelling System—a set of four nested numerical grids (Fig. 7.8) using similar methodology as Leitão et al.

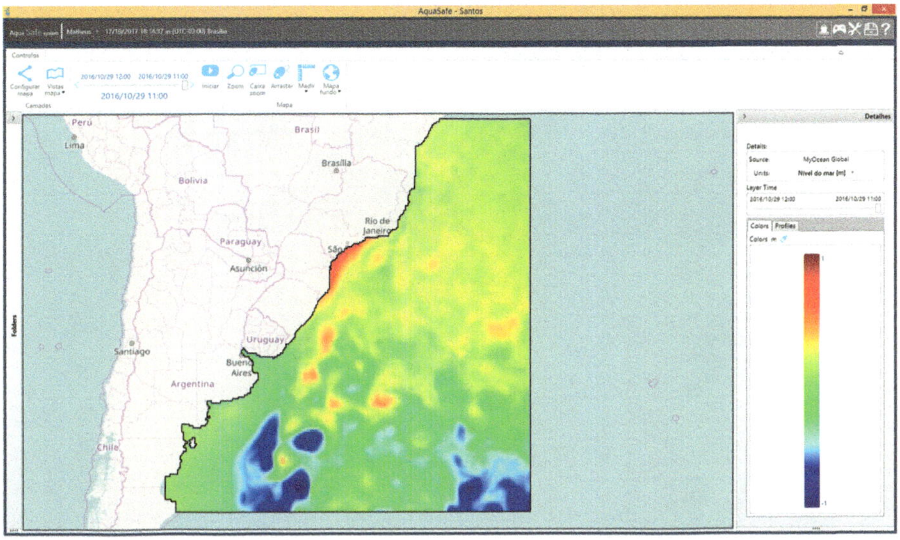

Fig. 7.7 Results of the CMEMS hydrodynamic model for 29 October 2016 at 11:00 (UTC-3), for the ocean region of Brazil, Uruguay and Argentina, visualised in the AQUASAFE Santos platform. The colour scale represents the sea level (varying from −1.0 to 1.0 m)

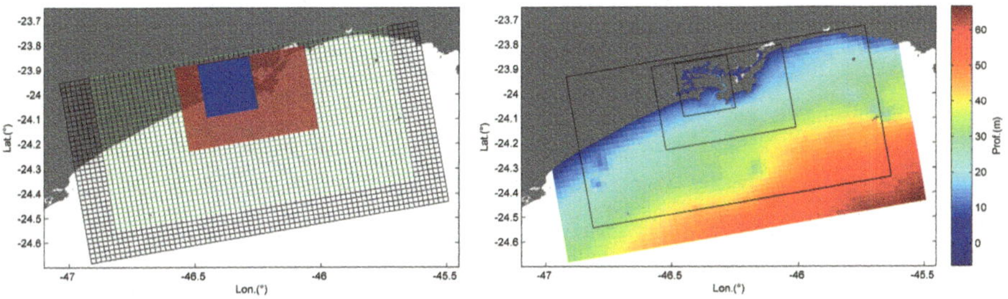

Fig. 7.8 Hydrodynamic numerical grids (left) and their bathymetry interpolated in numerical grids (right); the colour scale represents the depth in meters

(2005). The first numerical grid, called Level 1, is a 2D barotropic model, as are the three subsequent grids. Level 1 ocean boundary was forced by the spatial variable "astronomical tide", calculated from the harmonics of the global tide model FES2012 (Carrère et al. 2013), provided by the Archiving, Validation and Interpretation of Satellite Oceanographic data (AVISO). The second grid is forced at the ocean boundary by the conditions from the Level 1 grid (high frequency) and low frequency tides coming from the CMEMS model. The conditions generated by the Level 2 grid are imposed at the boundary of the Level 3 grid, which provides boundary conditions for the Level 4 grid. At the surface boundary, all four grids use the results from the metrological model GFS (50 km) provided by NOAA.

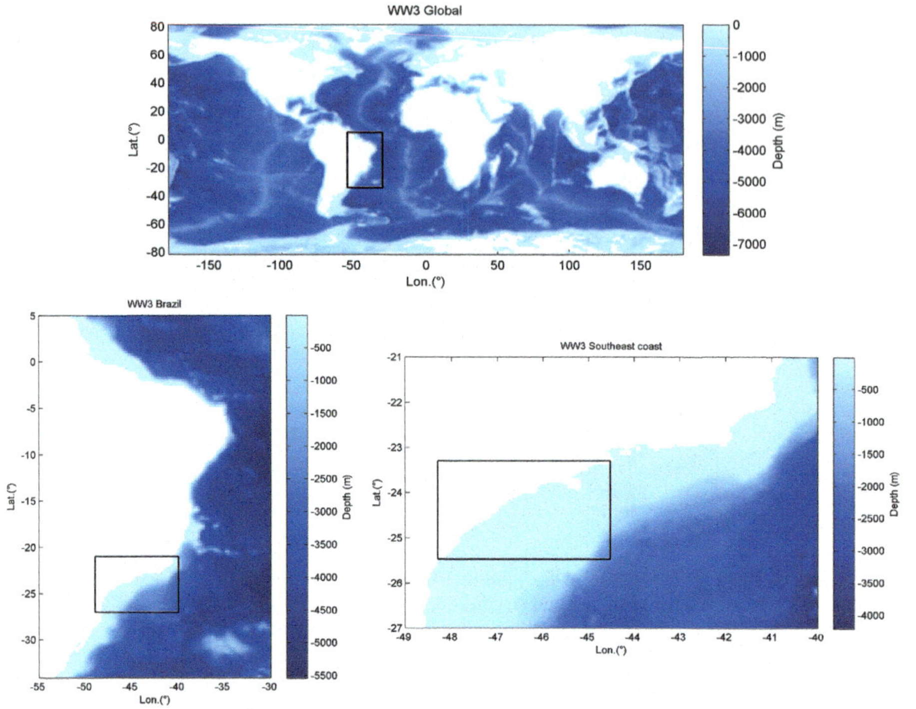

Fig. 7.9 WW3 grids: Global (top), Brazil (bottom left), and Southeast coast (bottom right). The colour scale represents the depth in meters, and the black polygons indicate the next level grids

Regarding the grids' characteristics, Level 1 has a horizontal resolution of 0.02° and 37 × 72 points; Level 2 has 31 × 61 points and a resolution of 0.02°; Level 3 has a resolution of 0.004° and 85 × 130 points; and the last grid (Level 4) has a resolution of 0.0005° (~50 m), with 432 × 416 points. The bathymetry was obtained by scanning the nautical charts of Directorate of Hydrography and Navigation (DHN) of the Brazilian Navy—Nos 1701; 1711; 23,100; and 23,200—and by recent surveys conducted by the SP Pilots and NPH-UNISANTA database.

The wave downscaling system uses two different numerical models. First, the model WAVEWATCH III® (WW3) is used to represent the generation and propagation of waves in deep water. This model comprises a set of three numerical grids, including a global grid with a horizontal resolution of 1.0° that generates boundary conditions for a second grid with a resolution of 0.25° at the Brazilian coast. This second grid generates boundary conditions for the third WW3 grid, with 0.05° of horizontal resolution and a domain covering the southeast coast of Brazil (Fig. 7.9). Next, the SWAN model is used to represent the propagation of waves in shallow waters. This model has two numerical grids: The first covers the coast of São Paulo State, with a resolution of 0.01°, and is forced at the ocean boundary by the last WW3 results. The second grid covers Santos Bay, with a horizontal resolution of

0.0005° (~50 m), and is used to provide forecasts with the appropriate level of detail for the region (Fig. 7.10).

Both the hydrodynamic and wave systems have been calibrated and are continuously validated through comparison of the wave and sea level data measured at various SP Pilots stations. Automated validation reports are sent weekly via the platform (Fig. 7.17). Additionally, a semiannual validation, which considers all the data collected by the platform, is performed using the following statistical indicators: (a) linear correlation coefficient (R); (b) normalised root mean square error (NRMSE); (c) root mean square error (RMSE); (d) Mean error (Bias); and (e) Skill score.

Based on the most recent validation, for the period from 01 July 2016 to 01 January 2017 (Table 7.3), the linear correlation coefficient for the stations with sea level data was 0.90–0.95. The prediction of the tidal tables, which uses only the harmonic components, presented coefficients between 0.70 and 0.84, representing a gain of about 20% with the use of this system of models. The significant wave height showed good linear correlation (0.88).

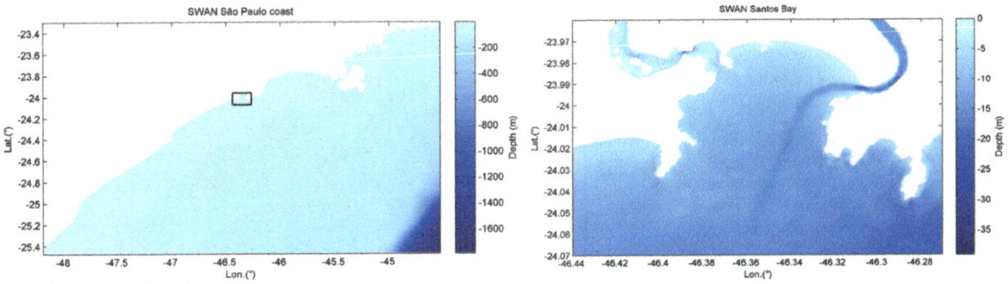

Fig. 7.10 SWAN grids (São Paulo coast at left and Santos Bay at right). The colour scale represents the depth in meters, and the black polygon indicates the next level grid

Table 7.3 Statistical comparison between the measured wave and sea level data, the tidal table prediction, and the results of the models for the period from 01 July 2016 to 01 January 2017

Parameter/station	Model/source	R	NRMSE	RMSE	BIAS	Skill
Water level Capitania	MOHID Level 4 (50 m)	0.94	9.6%	16.0 cm	−6.1 cm	0.96
	Tidal tables	0.86	14.3%	23.8 cm	−10.0 cm	0.90
Water level Cosipa	MOHID Level 4 (50 m)	0.90	13.9%	20.1 cm	−4.5 cm	0.94
	Tidal tables	0.70	20.1%	29.0 cm	−6.7 cm	0.83
Water level Ilha de Barnabé	MOHID Level 4 (50 m)	0.92	9.8%	17.8 cm	−4.0 cm	0.96
	Tidal tables	0.86	14.1%	25.5 cm	−9.1 cm	0.89
Water level Ilha das Palmas	MOHID Level 4 (50 m)	0.95	8.2%	13.0 cm	2.7 cm	0.97
	Tidal tables	0.83	13.8%	22.3 cm	−3.8 cm	0.90
Water level Praticagem	MOHID Level 4 (50 m)	0.95	12.3%	18.9 cm	−14.6 cm	0.94
	Tidal tables	0.84	17.3%	26.5 cm	−16.8 cm	0.86
Significant wave height Ilha das Palmas	SWAN (~50 m)Santos Bay	0.88	15.5%	31.7 cm	−17.5 cm	0.86
Peak period Ilha das Palmas	SWAN (~50 m) Santos Bay	0.47	27.4%	2.6 s	−0.4 s	0.69

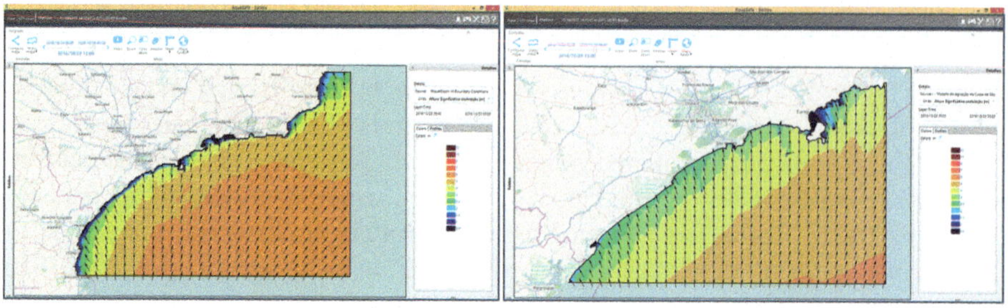

Fig. 7.11 Results of the models WW3 Southeast coast (left) and SWAN São Paulo coast (right) for 29 October 2016 at 15:00 (UTC-3), visualised in the AQUASAFE Santos platform. The colour scale represents the significant wave height (varying from 0.5 to 12.0 m), and the vectors indicate the direction

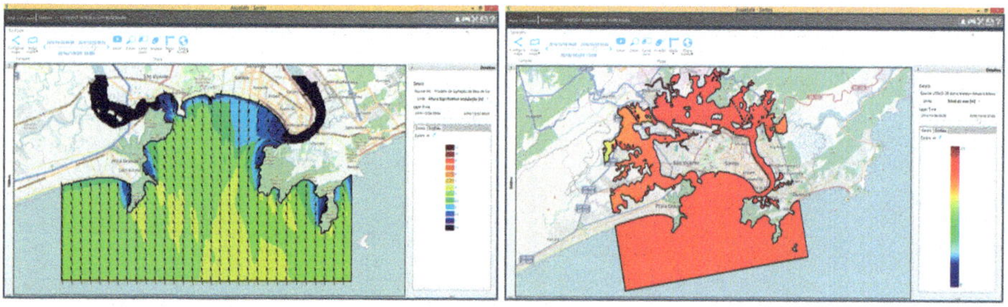

Fig. 7.12 Model results for 29 October 2016 at 15:00 (UTC-3), visualised in the AQUASAFE Santos platform: SWAN Santos Bay (left), with colour scale representing the significant wave height (varying from 0.5 to 12.0 m), and vectors indicating the propagation direction. MOHID Level 4 (right), with colour scale representing the sea level (varying from 0 to 2.5 m)

For the storm surge event of October 2016, this set of models adequately represented the wave and tide propagation at the scale of Santos Bay. The expected significant wave height was 4–6 m in the coastal region (shown as orange tones in Fig. 7.11) and 2–4 m at the entrance of Santos Bay (shown as green and yellow tones in Fig. 7.12). The expected sea level was over 2.1 m in the bay region and over 2.3 m within the estuary (Fig. 7.12).

7.6 AQUASAFE Santos Client Interface

As described previously, the graphic interface of the AQUASAFE *client* allows access to all the information and data generated and synchronised by the server. This tool was made available to users in early 2015, and ten different institutions currently have access to this tool. For use of the platform as an early warning system for storm surge, the five Civil Defences of the municipalities located in the Santos

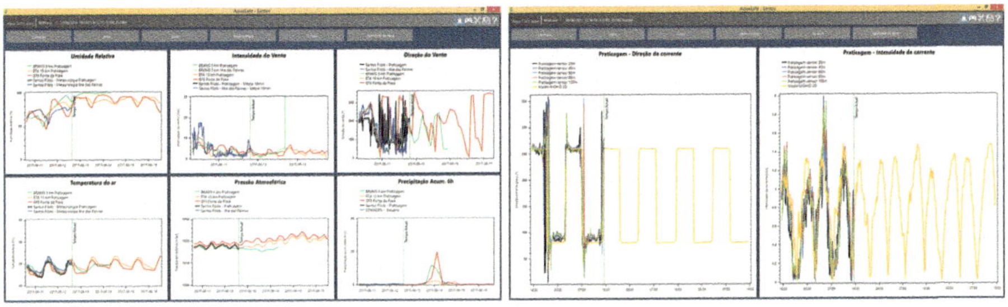

Fig. 7.13 Workspace showing the results of the three meteorological models, with data measured by the local meteorological stations, displayed in a time series format (left). Workspace showing the sea current data (direction and intensity), measured by the ADCP located at the SP Pilots station, and the results of the hydrodynamic model (right)

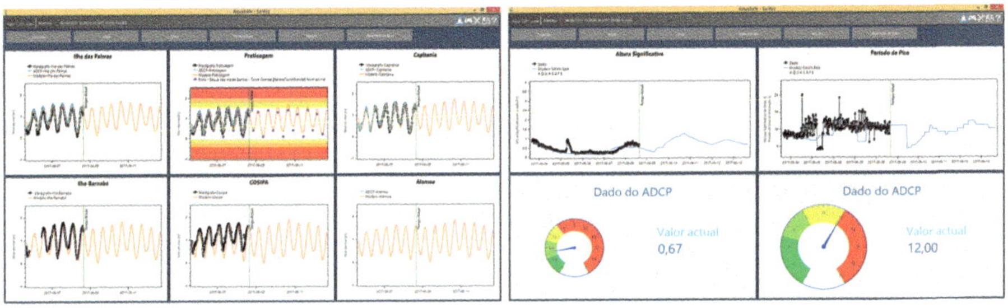

Fig. 7.14 Workspace showing the data measured by the six tide gauges and the model's results for sea level (left). Workspace showing the results of the wave model and the measured data on significant height and peak period (right)

region, as well as the SP Pilots, have access to this interface. The other five institutions with access to the interface focus on other uses related to water quality for swimming and sanitation.

To understand the needs of the system's users, various workspaces were created and organised through a customised menu. This menu separates the workspaces according to overall themes: Currents; Tides; Waves; Meteorology; Storm surge; and Water quality. Each menu also has sub-menus containing information pertaining to the data measured by the various stations (tide gauges, meteorological stations, rain gauges, SDCPs, etc.), as well as to the results of the numerical models, in the form of animated spatial maps (as previously shown) and time series.

Below (Figs. 7.13, 7.14, and 7.15) are some examples showing these real-time data and their comparison to the model predictions. The models agree with the data, and the real-time visualisation of this comparison adds to their reliability, which becomes relevant during the decision-making processes of final users.

Fig. 7.15 Workspace showing the data measured by the SP Pilots and the oceanographic prediction of determinant parameters of a storm surge event (significant height and peak direction of waves; sea level), displayed in a time-series format

7.7 Comunication and Warnings

According to United Nations Office for Disaster Risk Reduction (UNISDR 2009), an early warning system is the set of capacities needed to generate and disseminate timely and meaningful warning information that enables individuals, communities and organisations threatened by a hazard to prepare and to act appropriately and in sufficient time to reduce the possibility of harm or loss.

An effective response to the warnings depends on a range of factors, and a people-centred early warning system necessarily comprises four key elements:

1. Knowledge of the risks;
2. Monitoring, analysis and forecasting of the hazards;
3. communication or dissemination of alerts and warnings; and
4. Local capabilities to respond to the warnings received.
5. In this case, a common expression is "end-to-end warning system", which emphasises that the system needs to encompass all steps, from hazard detection through to community response (UNISDR 2009).

Initially, the effort to implement this early warning system was focused on validation of the numerical models and subsequent analysis and forecasting of the storm surge events.

Currently, the platform permits integrated and easy access to a large quantity of data. Moreover, as shown, an important function of the system should be

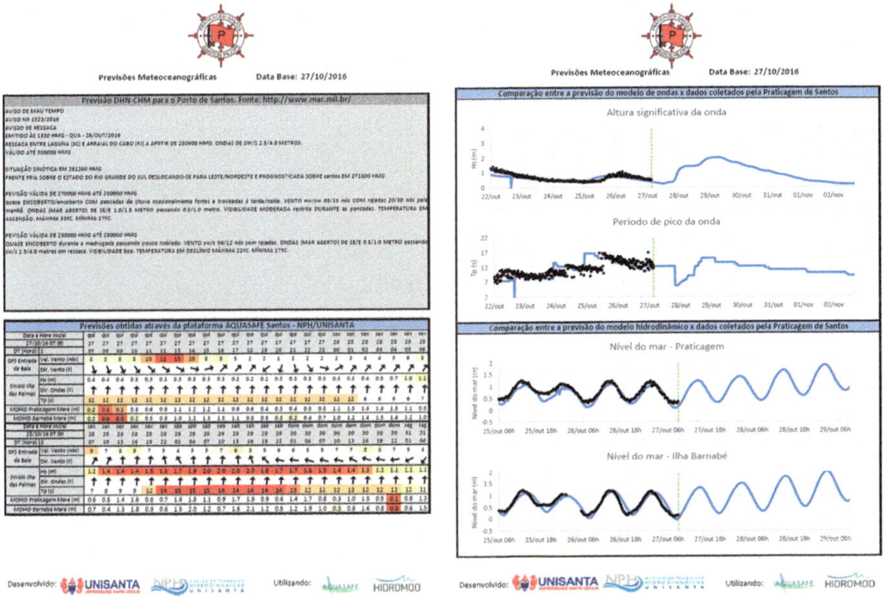

Fig. 7.16 This example shows the two-page daily report issued by the platform on 27 October 2016, with meteoceanographic forecasts for the following days

communication with users, so that this large quantity of data can be transformed into useful information for administrators, broadening dissemination and increasing their capacity to respond to the alerts.

In this regard, two separate reports, prepared in portable document format (pdf), are sent daily or weekly to a large number of people/administrators who do not necessarily have access to the graphic interface of the platform, such as coordinators and technicians for the municipal civil defences, technicians for the municipal secretariats of environment and of urban development, and technicians for the sanitation company.

The first report (Fig. 7.16) is focused mainly on the meteorological and oceanographic predictions generated by the platform. It is sent daily to a list of e-mail addresses. The second report (Fig. 7.17) is sent monthly by e-mail and details the validation of the hydrodynamic model and wave model, comparing the predictions with the data measured by equipment over the 15 days preceding the report's release.

It should be noted that the first report, presented as an example in Fig. 7.17, shows not only the weather forecasts issued by the Brazilian Navy, but also the results of the system's predictions, in table and graph formats, with the periods of increased magnitude and arrival of intense waves and intense winds, as well as elevated tide, indicated by a red on the colour scale. This report was issued on 27 October 2016, 36 h prior to the occurrence of the strongest storm surge event of 2016 and one of the strongest events in recent years. The bulletin from the previous days had also indicated the occurrence of this event, as discussed below.

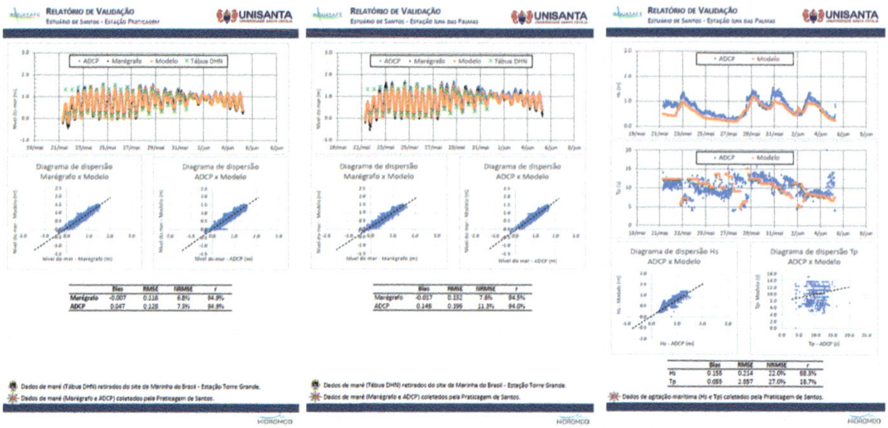

Fig. 7.17 This example shows the three-page monthly report referring to 05 June 2017, with the results of the validation of the hydrodynamic and wave models for the 15 previous days

Nevertheless, when these reports were first being disseminated, beginning in March 2015, there was a lack of local capability to respond to the warnings received, probably due to the diverse activities of the actors involved and to the limited or lack of knowledge of the meteoceanographic phenomena responsible for storm surge events and intense waves. Consequently, the numerical information and alerts, classified on a colour scale, were difficult to interpret.

As a result, and due also to the occurrence of significant storm surge events in recent years, the NPH-Unisanta began to issue warnings, called the *Boletim Informativo* (Information Bulletin), based on the predictions of large-magnitude events, describing these predictions and indicating the possibility of risk for flooding of roads and houses and impacts on urban infrastructure, among others, in the most susceptible areas within the city of Santos.

The systematic transmission of this information bulletin began in May 2016; thus, for the October 2016 event emphasised in this chapter, the authorities and the population could receive detailed warnings. Since then, as of August 2017, at least 36 information bulletins have been issued to the municipal civil defences and directly to some communications outlets. They contain descriptive and detailed information about storm surge forecasts, elevated tide and/or intense waves in the region, as well as warnings about the possibility of impacts on the urban perimeters. Twenty percent of the bulletins have warned of the possibility of flooding and strong impacts on urban infrastructure (Fig. 7.18).

Figure 7.18 shows the information bulletin issued on 26 October 2016, at least 60 h before the occurrence of the expected tidal and wave peak; the following information is provided:

> ***Intense storm surge should reach the Baixada Santista region starting Friday (28/10), peaking on Saturday (29/10), says the Núcleo de Pesquisas Hidrodinâmicas da Unisanta***

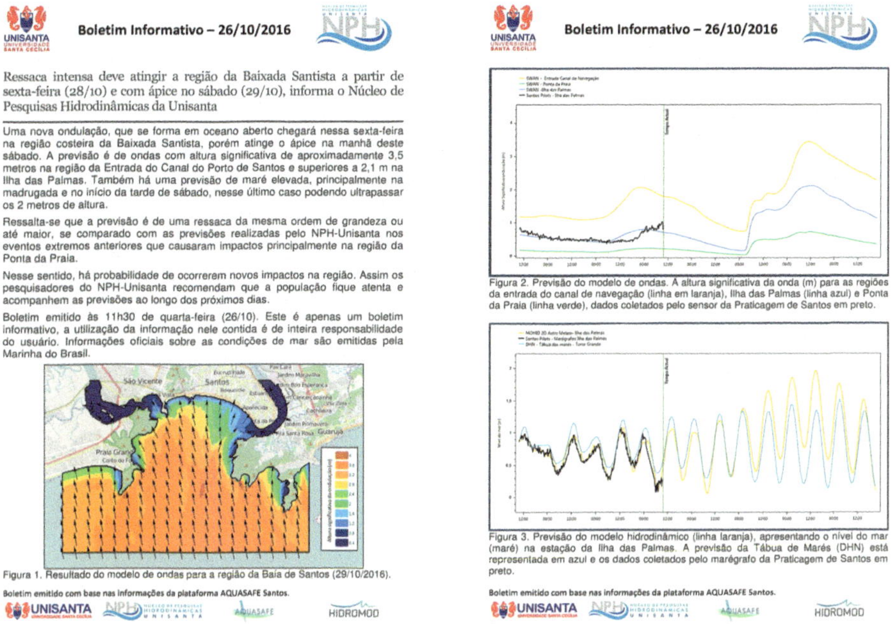
Boletim Informativo – 26/10/2016

Ressaca intensa deve atingir a região da Baixada Santista a partir de sexta-feira (28/10) e com ápice no sábado (29/10), informa o Núcleo de Pesquisas Hidrodinâmicas da Unisanta

Uma nova ondulação, que se forma em oceano aberto chegará nessa sexta-feira na região costeira da Baixada Santista, porém atinge o ápice na manhã deste sábado. A previsão é de ondas com altura significativa de aproximadamente 3,5 metros na região da Entrada do Canal do Porto de Santos e superiores a 2,1 m na Ilha das Palmas. Também há uma previsão de maré elevada, principalmente na madrugada e no início da tarde de sábado, nesse último caso podendo ultrapassar os 2 metros de altura.

Ressalta-se que a previsão é de uma ressaca da mesma ordem de grandeza ou até maior, se comparado com as previsões realizadas pelo NPH-Unisanta nos eventos extremos anteriores que causaram impactos principalmente na região da Ponta da Praia.

Nesse sentido, há probabilidade de ocorrerem novos impactos na região. Assim os pesquisadores do NPH-Unisanta recomendam que a população fique atenta e acompanhem as previsões ao longo dos próximos dias.

Boletim emitido às 11h30 de quarta-feira (26/10). Este é apenas um boletim informativo, a utilização da informação nele contida é de inteira responsabilidade do usuário. Informações oficiais sobre as condições de mar são emitidas pela Marinha do Brasil.

Figura 1. Resultado do modelo de ondas para a região da Baía de Santos (29/10/2016).

Boletim emitido com base nas informações da plataforma AQUASAFE Santos.

Boletim Informativo – 26/10/2016

Figura 2. Previsão do modelo de ondas. A altura significativa da onda (m) para as regiões da entrada do canal de navegação (linha em laranja), Ilha das Palmas (linha azul) e Ponta da Praia (linha verde), dados coletados pelo sensor da Praticagem de Santos em preto.

Figura 3. Previsão do modelo hidrodinâmico (linha laranja), apresentando o nível do mar (maré) na estação da Ilha das Palmas. A previsão da Tábua de Marés (DHN) está representada em azul e os dados coletados pelo marégrafo da Praticagem de Santos em preto.

Boletim emitido com base nas informações da plataforma AQUASAFE Santos.

Fig. 7.18 This example shows the two-page special information bulletin issued by the NPH-Unisanta on 26 October 2016, announcing the forecast of strong storm surge over the following days

> *A new ocean swell, which forms in open ocean, will arrive this Friday in the coastal region of Baixada Santista, though it will reach its apex this Saturday morning. The forecast is for significant wave heights of approximately 3.5 meters in the region of the Entrada do Canal do Porto de Santos, and over 2.1 m in the Ilha das Palmas. Elevated tide has also been predicted, mainly during the dawn and in the early afternoon on Saturday, which, in the latter case, may exceed 2 meters in height.*
>
> *Note that the forecast is of a storm surge of the same or larger magnitude as those predicted by NPH-Unisanta for previous extreme events that caused impacts in the Ponta da Praia region in particular.*
>
> *In this regard, new impacts are likely to occur in the region. Therefore, the researchers of the NPH-Unisanta recommend that the population stays alert and follows the forecasts over the coming days.*

Currently, the Civil Defence Action Plan is based on this system, and with each prediction of an extreme event, this institution makes an announcement through a variety of media (printed newspapers, television news programmes, internet and social medias) and sends cell phone messages to the registered population that resides in the regions most susceptible to impacts, thus enabling the population to prepare. In this regard, the system increases the local capacity to respond to the warnings.

The transmission of these special bulletins enabled the civil defences of the region, and of the municipality of Santos in particular, to monitor the storm surge

A-6 Cidades | A TRIBUNA www.atribuna.com.br | Sexta-feira 16 setembro de 2016

Efeitos de ressaca 'prevista' são minimizados

Trabalho preventivo é exaltado em Santos

SANDRO THADEU
DA REDAÇÃO

O trabalho integrado de previsão das ressacas e as ações de prevenção de transtornos à população, promovidos pela Defesa Civil de Santos e o Núcleo de Pesquisas Hidrodinâmicas (NPH) da Universidade Santa Cecília (Unisanta), foram considerados positivos.

O fenômeno registrado por volta das 14 horas de ontem ocorreu um pouco acima do esperado (2,03 metros contra 1,85 metro). A água tomou pra- chance para fazer uma selfie e estava convencido de que as fortes ondas do último dia 21 não ocorreriam novamente.

Moradora de Guarujá, a professora Eunice Pinheiro tinha uma consulta médica marcada, mas chegou com antecedência, na esperança de ser atendida mais cedo, com medo das ondas na hora de ir embora do consultório. "Estou com medo que as ondas estejam fortes, porque elas podem atrapalhar meu retorno para casa".

Prefeitura e Núcleo de Pesquisas Hidrodinâmicas, da Unisanta, ressaltam o trabalho integrado; ondas chegaram a 2,03 metros

Fig. 7.19 Page in the local newspaper showing an article praising the preventative work in Santos. (Source: Jornal A Tribuna from 16 September 2016)

predictions for the Santos region, and consequently, it enabled the publication of warnings and information about risks to the population susceptible to these events. Such publications include headlines in the local media (Fig. 7.19) that praise the preventative work undertaken in response to the predictions of these extreme events.

7.8 Final Considerations

Since March 2015, the AQUASAFE Santos platform has provided a large quantity of information from several monitoring networks as well as from global and regional numerical models. The high-resolution models were calibrated and, having demonstrated good correlation with the measured data, are run operationally and validated using the network of data measured in real time. The detailed results are disseminated through this same platform to the civil defences of five municipalities and to SP Pilots, the sanitation company (SABESP) and the environmental agency (CETESB). Complementally, the results of the early warning system are available on the NPH-Unisanta website (http://nph.unisanta.br) so that a greater number of people have access to the information. In the last 28 months, more than 50 storm surge and strong wave warnings have been published, reporting on the possibility of flooding and impacts on urban infrastructure.

Furthermore, the scenarios of sea-level rise (Chap. 1) and its consequences for the urban area of Santos (Chaps. 8 and 12; also reported by Berzin and

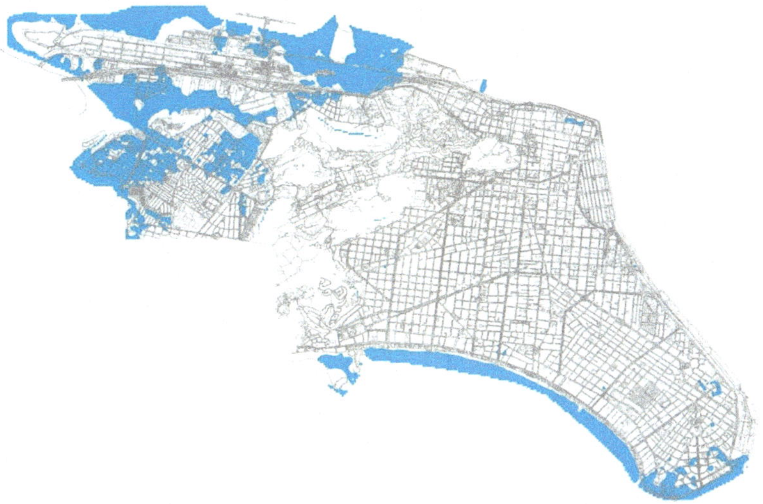

Fig. 7.20 Areas of the city of Santos (in blue) that are susceptible to flooding, given a tide of 2.48 m. (Source: Berzin and Ribeiro 2010)

Ribeiro 2010) show that some areas of the city are extremely susceptible to the rising sea level.

In the storm surge event of October 2016, considering both the surge and the astronomical tide, the total sea level reached 2.40 m, according to the tide gauge of the SP Pilots, and caused disruptions and economic losses for the city and its residents. This condition was very close to the sea-level rise scenario simulated by Berzin and Ribeiro (2010): in this work, the authors showed the flooded areas of the city of Santos, given a tide of 2.48 m (Fig. 7.20).

This and other extreme events demonstrate that the region currently faces problems caused by storm surges (Chap. 6) and intense waves, and with the predicted climate change scenarios, whether of sea-level rise or changes in the frequency and/or intensity of climatic phenomena (Chap. 1), the tendency is for the impacts on urban areas to grow.

Therefore, a plan to deal with these impacts is essential, and although the early warning systems do not prevent the occurrence of this type of extreme event, they are an adaptation measure to extreme events, and to climate changes, and are intended to minimise their impacts, particularly those felt by the population.

The system implemented and presented in this chapter is still a preliminary approach. However, with a view to becoming an end-to-end warning system, this vision can only be achieved through the integration of the various actors involved, particularly the municipal Civil Defences.

Acknowledgement This study was made possible in part by the Project n°2013-BS_COB-5, called "Implantação de sistema de monitoramento e previsão da qualidade da água por meio de modelagem numérica ambiental e desenvolvimento de base de dados na Bacia Hidrográfica do Estuário de Santos – SV", and supported by FEHIDRO (Fundo Estadual de Recursos Hídricos).

References

Alfieri, L., Salamon, P., Pappenberger, F., Wetterhall, F., & Thielen, J. (2012). Operational early warning systems for water-related hazards in Europe. *Environmental Science and Policy*. https://doi.org/10.1016/j.envsci.2012.01.008.

Berzin, G., & Ribeiro, R. B. (2010, August 10–12). O que os engenheiros precisam saber sobre a elevação do nível do mar e seus efeitos na *Baixada Santista*. Paper presented at XXI Encontro Técnico AESABESP, São Paulo.

Bunya, S., Dietrich, J. C., Westerink, J. J., Ebersole, B. A., Smith, J. M., Atkinson, J. H., … Roberts, H. J. (2010). A high-resolution coupled riverine flow, tide, wind, wind wave, and storm surge model for Southern Louisiana and Mississippi. Part I: Model development and validation. *Monthly Weather Review, 138*(2):345–377. https://doi.org/10.1175/2009 MWR2906.1.

Carrère, L., Lyard, F., Cancet, M., Guillot, A., & Roblou, L. (2013, September 24–29). FES 2012: a new global tidal model taking advantage of nearly 20 years of altimetry. In Ouwehand L (Ed.), *20 Years of Progress in Radar Altimetry*, Venice, Italy.

Cassiano, G. F., Ribeiro, R. B., & Yassuda, E. A. (2012, November 4–7). Acquisition of wave data and modeling in Santos Bay, São Paulo, Brazil. In *Proceedings of the 10th international conference on hydroscience & engineering*, Orlando, Florida, USA. Available via http://mdi-de.baw.de/icheArchive/documents/2012/41238372-Final.pdf. Accessed 14 May 2018.

Dube, S. K., Jain, I., Rao, A. D., & Murty, T. S. (2009). Storm surge modelling for the Bay of Bengal and Arabian Sea. *Natural Hazards, 51*, 3–27. https://doi.org/10.1007/s11069-009-9397-9.

Harari, J., & de Camargo, R. (2003). Numerical simulation of the tidal propagation in the coastal region of Santos (Brazil, 24°S 46°W). *Continental Shelf Research, 23*(16), 1597–1613. https://doi.org/10.1016/S0278-4343(03)00143-2.

Leitão, P., Coelho, H., Santos, A., & Neves, R. (2005). Modelling the main features of the Algarve coastal circulation during July 2004: A downscaling approach. *Journal of Atmospheric and Ocean Science, 10*(4), 421–462. https://doi.org/10.1080/17417530601127704.

Leitão, P., Leitão, J., Ribeiro, R., Sampaio, A., Galvão, P., Ribeiro, J., & Silva, A. (2015, October 14–16). Serviços de previsão de alta resolução de condições meteo-oceanográficas e de eventos de poluição costeira. In Coelho C et al (Eds.), VIII Congresso sobre Planeamento e Gestão das Zonas Costeiras dos Países de Expressão Portuguesa, Aveiro, Portugal. Available via http://www.aprh.pt/ZonasCosteiras2015/pdf/5A2_Artigo_080.pdf. Accessed 14 May 2018.

Lellouche, J. M., & Regnier, C. (2015). *GLOBAL ocean sea physical analysis and forecasting products*. Product user manual. Available via http://marine.copernicus.eu. Accessed 14 May 2018.

Murty, P. L. N., Padmanabham, J., Srinivasa Kumar, T., Kiran Kumar, N., Ravi Chandra, V., Shenoi, S. S. C., & Mohapatra, M. (2017). Real-time storm surge and inundation forecast for very severe cyclonic storm 'Hudhud'. *Ocean Engineering, 131*, 25–35. https://doi.org/10.1016/j.oceaneng.2016.12.026.

National Ocean and Atmospheric Administration (NOAA). (2017). *Operational forecast system*. Available via http://tidesandcurrents.noaa.gov/models. Accessed 14 Dec 2017.

Pugh, D. T. (1996). *Tides, surges and mean sea-level* (reprinted with corrections) (p. 472). Chichester: .Wiley. Available via https://eprints.soton.ac.uk/19157/1/sea-level.pdf. Accessed 14 May 2018.

Resio, D. T., & Westerink, J. J. (2008). Modeling the physics of storm surges. *Physics Today, 61*(9), 33–38. https://doi.org/10.1063/1.2982120.

Ribeiro, R. B., Leitão, J. C., Leitão, P. C., Puia, H. L., & Sampaio, A. F. P. (2016, October 16–21). Integration of high-resolution metocean forecast and observing systems at Port of Santos. In *IX international conference on coastal and port engineering in developing countries (PIANC-COPEDEC)*, Rio de Janeiro, Brazil. Available via http://www.gapcongressos.com.br/trabalhos/z0167/20151130341_full.pdf. Accessed 14 May 2018.

Ribeiro, R., Sampaio, A., Leitão, J., Leitão, P., & Rodrigues, J. (2017, September 13–15). Sistema de previsão da qualidade das águas balneares como ferramenta de gestão. In 13° Simpósio de Hidráulica e Recursos Hídricos dos Países de Língua Portuguesa, Porto, Portugal.

Roversi, F., Rosman, P. C. C., & Harari, J. (2016). Análise da renovação das águas do Sistema Estuarino de Santos usando modelagem computacional. *Revista Ambiente e Água, 11*(3), 566–585. https://doi.org/10.4136/ambi-agua.1770.

Sampaio, A. F. P., Mateus, M., Ribeiro, R. B., & Berzin, G. (2008). A modelling approach to the study of faecal pollution in the Santos Estuary. In R. Neves et al. (Eds.), *Perspectives on integrated coastal zone management in South America* (pp. 425–434). Lisbon: IST Press Available via https://www.unisanta.br/arquivos/CoastalZoneManagementFinal.pdf. Accessed 14 May 2018.

Tompkins, E. L. (2005). Planning for climate change in small islands: Insights from national hurricane preparedness in the Cayman Islands. *Global Environmental Change, 15*(2), 139–149. https://doi.org/10.1016/j.gloenvcha.2004.11.002.

United Nations International Strategy for Disaster Reduction (UNISDR). (2009). *Terminology on disaster risk reduction*. Geneva, Switzerland. Available via https://www.unisdr.org/files/7817_UNISDRTerminologyEnglish.pdf. Accessed 14 May 2018.

Valchev, N., Eftimova, P., & Andreeva, N. (2018). Implementation and validation of a multi-domain coastal hazard forecasting system in an open bay. *Coastal Engineering, 134*, 212–228. https://doi.org/10.1016/j.coastaleng.2017.08.008.

Verlaan, M., Zijderveld, A., de Vries, H., & Kroos, J. (2005). Operational storm surge forecasting in the Netherlands: Developments in the last decade. *Philosophical Transactions of the Royal Society A: Mathematical, Physical and Engineering Sciences, 363*(1831), 1441–1453. https://doi.org/10.1098/rsta.2005.1578.

Part III
Impacts

Chapter 8
Rainfall Episodes and Local Stability Thresholds in Santos

Beatriz Barbi de Oliveira Santos, Lucí Hidalgo Nunes, and Marcos Pellegrini Bandini

Abstract The temporal distribution of rain and its impacts on the municipality of Santos for the period 1980–2015 was evaluated, with emphasis on the correlation between rainfall events and mass movements, estimating the local stability thresholds for accumulated rain over 24, 48, and 72 h. Additionally, calculations at annual, seasonal and monthly level were carried out to evaluate the rainfall distribution. The rainfall distribution did not reveal a pattern of increase or decrease, and although higher amounts of precipitation have occurred below the average of the period since 2000, they tended to be more concentrated in recent years. Results also suggested that there is a probability of the occurrence of gravitational mass movements when the volume of rain reaches a value equal to or greater than 75.0 mm over a 2-day period. Besides to mass movements, flooding and property collapses were the recurring impacts recorded, which brought in turn a large number of displaced, homeless and deaths.

Keywords Rainfall episodes · Natural disaster · Stability threshold

8.1 Introduction

Rainfall is one of the main energy sources that induce instability in natural and anthropic environments, provoking such impacts as flooding and gravitation mass movements. In tropical regions—especially in naturally fragile environments such as coastal zones—the joint analysis of local physical characteristics and of spatial reproduction and land-use planning are essential for understanding the existing

B. B. de Oliveira Santos (✉)
Azul Brazilian Airline, São Paulo, Brazil

L. H. Nunes
Santos, São Paulo, Brazil
e-mail: luci@ige.unicamp.br

M. P. Bandini
Civil Defense of Santos, Santos, São Paulo, Brazil
e-mail: marcosbandini@santos.sp.gov.br

L. H. Nunes et al. (eds.), *Climate Change in Santos Brazil: Projections, Impacts and Adaptation Options*, https://doi.org/10.1007/978-3-319-96535-2_8

environmental dynamics of urban areas. This aspect is recognised by authors such as Nunes et al. (1989) and Larsen and Simon (1993), who view intense and concentrated precipitation as a key element in triggering landslides, but they highlight the role of environmental modification through anthropogenic actions as an important contributor to the advent of mass movements.

In cases of natural events that result in harm to the population, institutional fragility and economic inequality can reduce the capacity of the most vulnerable people to respond, heightening their exposure to risks and hazards (Silva 2013). In this perspective, natural disasters can be understood as a social construction arising from the way space is occupied: without consideration of the risks and existing vulnerabilities (Gonçalves 2003; Carmo and Anazawa 2014; Nunes 2015). Disasters derive from natural events and, as Nunes (2009a, b, 2015) and Valencio (2012) warn, result from a process that develops over time, not only at the moment of eruption.

In the face of numerous human and material loses caused by natural events, concern about risk reduction is growing, particularly since the 1990s, which the United Nations (UN) declared the International Decade for Natural Disaster Reduction. The initiative aimed to build capacity for disaster prevention and mitigation through risk reduction and the identification, evaluation and planning of structural and non-structural measures (Undro 1991 in Ferreira et al. 2009). Nonetheless, even though this initiative collaborated to create and/or strengthen ties between scientific, technological and political communities, the number of incidents and those affected did not decrease, but rather increased during the period, and the mortality trend persisted. Such facts confirm the ineffectuality of efforts to change the pattern of continuous increase in natural catastrophes (Bryant 1997; Tobin and Montz 1997; Kunkel et al. 1999; Pelling 2006). In accordance with the global perspective, Brazil initiated the National System of Civil Defence in 1988, aiming to integrate the various levels of government to consolidate and support strategies for disaster reduction, whereas the Civil Defence had previously operated with a focus on disaster response.

The 2nd World Conference on Disaster Reduction, organised in 2005 by the United Nations, was a global milestone in confronting natural disasters. At that conference, the Hyogo Framework of Action (HFA) was instituted for the period 2005–2015, focusing on increasing the resilience of nations and communities against disaster events. In the same year, the National System of Civil Defence was reorganised, influenced by the discussion that took place during the definition of the HFA.

The United Nations Conference on Sustainable Development in 2012 (Rio + 20) reaffirmed the importance of building community resilience. On 10 April 2012, the Brazilian government responded by enacting a law that established the National Policy on Civil Defence (PNPDEC) and created the National Council on Protection and Civil Defence (CONPDEC), which authorised the creation of a disaster information and monitoring system. This system prescribes plans and actions to reduce or prevent hazards, risks and vulnerability.

Advances in the political, juridical and social spheres occurred at various scales in Brazil, but the number of those affected by natural disasters continued to alarm—the country remained among the ten nations with the highest absolute number of victims of natural disasters between 1995 and 2015 (UNSDR 2016). According to the Brazilian Yearbook of Natural Disasters (CENAD 2014), the disasters that most affected the Brazilian population in 2013, touching more than 18 thousand people, were droughts (64.4%), heavy rainfall (9.4%) and floods (7.4%), though the greatest number of fatalities occurred due to landslides (22.4%), flash floods (20.7%) and inundations (19.6%). According to SEDEC (2014), the numbers of incidents and affected people are associated with areas of predominantly disorganised occupation and high susceptibility to gravitational mass movements, such as steep hillsides.

According to the classification presented by the Brazilian Yearbook of Natural Disasters, gravitational mass movements—a classification that includes landslides, earthflows, and embankment collapse, which we discuss in this chapter—can be defined as follows:

> the process by which rocky material moves under the force of gravity, necessarily resulting from collapsing soil and/or rock. This process includes landslides, debris flows, block slides, etc., which are classified according to the characteristics of the material, velocity and nature of the movement. (CENAD 2014, p. 51)

Due to their physical characteristics, coastal cities are already very susceptible to any type of atmospheric disturbance. The municipality of Santos, which is composed of mainland (85%) and insular (15%) areas, as shown in Fig. 8.1, experiences a high reoccurrence of this type of disturbance, according to the historical data of the Department of Civil Protection and Defence (DEDEC), the former Municipal Coordinator of Civil Protection and Defence of Santos (COMPDEC) and reports in the news media.

It should be noted that Santos is situated in a climatic transition zone, which features abrupt atmospheric variations such as warming and cooling within a short time interval (Tarifa and Armani 2000). Furthermore, part of the population lives on hillsides, where mass movements are natural and recurrent processes. In association with that characteristic, the urban spatial configuration and the socio-spatial inequalities – i.e., the deficiency and disparity of public infrastructure and services in the area – are important contributors to the growing physical susceptibility and vulnerability.

Singer (1977), in Silva (2013, p. 3), noted that the construction of Brazilian urban centres followed an economic policy whose objectives are profit and market interests. Feitosa et al. (2012) highlight that there was a tendency toward investment policies that favoured the construction of tourism facilities and their enhancement by strategic location (as in the case of the ports). Such policies have disadvantaged the excluded portion of the population that lives in the coastal municipalities of São Paulo state, thus undermined and occupying areas of risk, such as hillsides, and living in precarious housing.

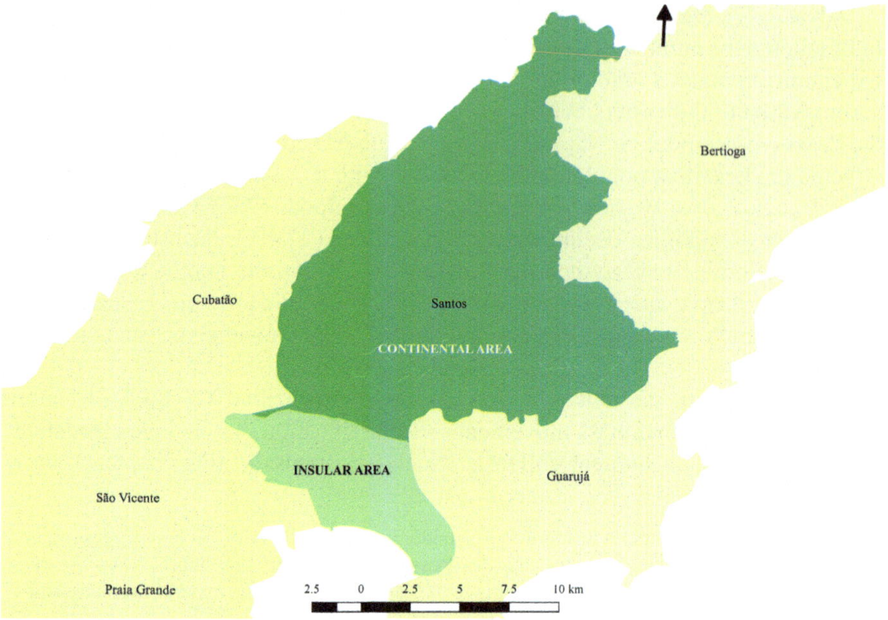

Fig. 8.1 Location of the insular and mainland areas of the municipality of Santos, SP

The municipality of Santos has a strategic geographic position due to its proximity to the capital city of São Paulo state. Interconnected with the most important centres of Brazil and neighbouring countries via main highways, the municipality of Santos acted as a coordinator in the formation of the metropolitan region to which it belongs: the Metropolitan Region of Baixada Santista (MRBS). In addition to the notable industrial activity in the MRBS, concentrated in the municipality of Cubatão, the port of Santos—the most important in Latin America—attracted urban functions and contributed to the early urbanisation of the region. Beyond the important port activities and their role in transforming the region, the recent discovery of petroleum in the pre-salt layer, and the recurrent investments in its exploration, has further enhanced the local dynamism. However, it has also contributed to an increase in the socioenvironmental impacts in this environment, which is already naturally fragile.

According to Zundt (2006), the rapid population growth and the expansion of the cities surrounding Santos occurred mainly due to the availability of the productive infrastructure and services that were established there. This author also states that Santos, São Vicente and Cubatão, which are the oldest municipalities on the coast of São Paulo state, have undergone intense urban densification through verticalisation and expansion of the urban area. This process brought about the invasion and illegal occupation of areas under environmental protection as well as areas that posed some risk to the population, such as hillsides. According to the information in the Carta Geotécnica dos Morros de Santos (IPT 1979a and b, in Figueira and

Bandini 2016), a pioneering work on the region, the hills were historically occupied by populations of low purchasing power, for whom finding housing in the high-cost residential areas of the lowland was impossible.

Although the study area's geographic characteristics are fundamental to an understanding of its environmental complexity, questions that address land-use planning and organisation, as well as public policies on risks management and prevention are also essential to an analysis of events that threaten social well-being. The transformations of the urban space of Santos brought benefits to the cities' economies and infrastructure; however, they also marginalised a segment of the population, which has not benefitted from the improvements. Furthermore, the rapid and intense transformations of the urban environment continue to affect the equilibrium of the natural environment, whether due to inappropriate land use and occupation, to the lack of public policies on land-use planning, or to non-compliance with existing standards—which restrict some uses and practices, but are not being respected—through degradation of the environment (pollution or deforestation, for example). Such practices drive the continuous increase of natural susceptibility and vulnerability to the occurrence of natural events in the region, and the severity of the impacts is not always proportional to the severity of the triggering phenomena—an indication of the great environmental disruption of the area (Araki and Nunes 2008).

In view of a scenario of environmental changes that occur at different scales, the natural susceptibility of the area, together with the socio-spatial configuration, would imply a process of worsening hazards and risks and greater vulnerability.

Considering records of natural disasters and rainfall data, the objective of this chapter is to investigate and analyse the temporal distribution of rain and its impacts on the municipality of Santos for the period 1980–2015. In this perspective, we discuss the correlation between rainfall events and mass movements, which frequently affect the insular part of Santos, and apply the envelopment technique, developed by Tatizana et al. (1987), to estimate the local stability thresholds for accumulated rain. This technique was adapted for the municipality of Santos and tested on three different criteria for rain accumulated over 24, 48, and 72 h.

Rainfall data were analysed at the Saboó station (Lat 23° 56′ 06″ S, Long 46° 20′ 22″ W; 60 m altitude), which has consistent daily data for the period between January 1980 and December 2015. The Saboó station is located on the insular part of Santos, close to the port and the commercial centre, and describes conditions that affect activites important to the municipality. As the mainland and insular sectors present different characteristics in terms of land use and occupation, they can respond differentially to the occurrence of instability hazards; therefore, this chapter focuses only on the insular area of the municipality.

The records of mass movements (landslides, earthflows and embankment collapse) were obtained from three databases: the natural disasters database SIMPAT (Sistema Integrado de Monitoramento, Previsão e Alerta de Tempestades para as Regiões Sul-Sudeste do Brasil), the database of the Laboratório de Estudos Climáticos at the Geosciences Institute of Unicamp (LECLIG), and the database of the Department of Civil Protection and Defence (DEDEC) of the municipality of

Santos. The data are based on information provided by the Civil Defence and the news media, as well as on information from the database maintained by the Institute for Technological Research (IPT). Although the databases store information on a variety of phenomena, the focus of this work is the mass movements that occurred from 1980 to 2015.

Among the scientific community, there is no consensus on the definition of *natural disaster*. The databases used in this study follow the conception held by the main source of the information, which in both cases is the Civil Defence of the state of São Paulo, which defines *disaster* thus:

> a result of adverse events, natural or provoked by man under a vulnerable scenario, causing grave disturbance to the functioning of a community or society, involving extensive human, material, economic and environmental losses and damages that exceed its capacity to cope with the problem using its own resources. (Instrução Normativa 2012, p. 30)

8.2 Natural Fragility and History of Impacts Associated with Mass Movements in Santos

The coast of São Paulo state, where the MRBS is situated, presents a complex relief, featuring changes in elevation, varying from 800 to 1200 m, between the coastal lowlands and the edges of the Atlantic Plateau, as well as floodplains composed of unconsolidated sediments. According to Gigliotti (2010), the cliffs of the Serra do Mar present embedded valleys, high declivity and widely variable elevation, resulting in the formation of slightly evolved soils and a high degree of instability; hence, the areas are susceptible to mass movements in the event of heavy or continuous rain. The humid tropical conditions favour chemical weathering, which makes the environment even more sensitive to rainfall episodes.

The coastal plains present low declivity, little variation in altitude, and unconsolidated soils (Ross and Moroz 1997). In addition to the natural susceptibility of the soil to flooding and land settlement, this area presents an extremely fragile ecosystem: the mangroves. Composed of deposits of fluvial and marine origins, the coastal plains present varied forms of relief such as marine terraces, dunes and restingas (Gigliotti 2010).

Associated with the region's morphological complexity and natural susceptibility to various processes, local climatic characteristics play a central role in the occurrence of gravitational mass movements. As noted by Futai (2014), water plays the role of both preparatory agent and trigger of landslides, as conditions of soil saturation constitute a central factor in the occurrence of mass movement processes.

At its latitudinal location on the western border of the south Atlantic, the MRBS presents constant high humidity due to the local conditions and to the action of mesoscale phenomena. Among these processes is the South Atlantic Convergence Zone (SACZ), which is fundamental to the transfer of heat and humidity from the Amazon to this region, especially in the summer, when it is associated with episodes of intense rainfall. Cold fronts are strongly associated with rains during the winter, though they can influence the area throughout the year.

Extratropical cyclones and complex mesoscale convective systems are processes that also act on the area, contributing to episodes of intense rainfall. In years of strong El Niño, part of the Brazilian Southeast region, including the coast of São Paulo, can experience higher volumes of rain during the winter (Nunes 1997). The Southeast region is also affected by upper tropospheric cyclonic vortices, which can be defined as closed synoptic-scale low-pressure systems that are formed in the upper troposphere (Gan 1982 in Malvestio 2013). Chapter 3 is dedicated to the analysis of extreme events.

At the local level, the considerable urbanisation favours the formation of heat islands, and the high verticalisation of buildings creates a barrier to local circulation and influences climatic conditions. Added to the complexity of the physical characteristics, the reproduction of urban space alters the dynamic of the natural and transformed system and initiates processes that reshape the landscape, increasing the dangers of instability (Sant'Anna Neto 1998).

Znamensky (2014) listed some physical characteristics of the Serra do Mar that favour the frequent occurrence of landslides and mud and debris flows: high volumes of rain in conjunction with residual soils and colluvial mantles that may be more than 10 m thick over permeable weathered rocks, and large differences in the natural ground level (1000 m on average). The hillside sector of Santos is no exception, having already experienced various calamitous incidents associated with gravitational mass movements, among them the tragedy that occurred on the northern slope of Monte Serrat in March 1928, which caused 80 deaths, affected many people, and destroyed part of the hospital facilities of Santa Casa de Misericórdia. In addition to this episode, other incidents of significant impact are notable in terms of the numbers affected, such as that in Morro Santa Terezinha on the 1st of March 1956, with 21 fatalities. On the 24th and 25th of March 1956, an event affected several hills in Marapé and Caneleira, with 43 fatalities (Pichler 1957). And on the 16th of December 1979, 12 people died in events in Morro da Nova Cintra and Morro do Pacheco.

Assessing the proposed period in this chapter (1980–2015), we identified 930 episodes of gravitational mass movement in Santos (Fig. 8.2), 68.9% of which occurred between December and March, the summer and rainy season in the region. These events have become more frequent in recent years and have affected more than 400 people.

Other types of impacts also happen in the municipality: more than 80 floods and over 100 incidents of falling trees have been triggered by rainfall events. In addition to the material damages, numerous people were affected (eight fatalities, as well as displacements and/or injuries). More than 90.0% of the reported impacts occurred from the year 2000 onward, with significant peaks in 2000, 2011 and 2015.

Although a search of the SIMPAT and LECLIG databases finds records from the 1980s onward, more information is available for after the year 2000. The data provided by the Civil Defence includes detailed records beginning in 2000, contributing to the large quantity of information after that year.

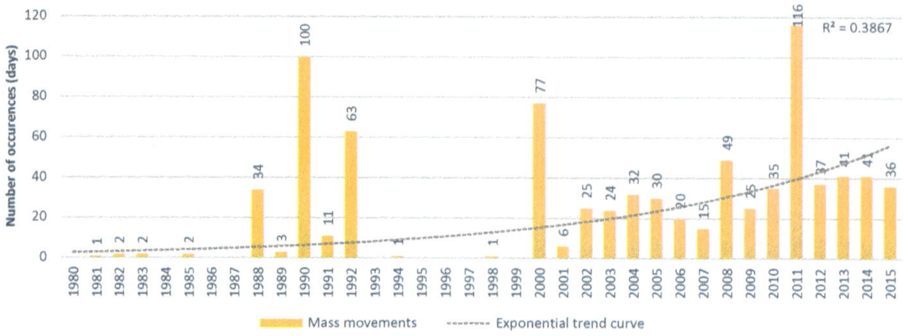

Fig. 8.2 Records of gravitational mass movements in Santos (1980–2015). (Source: Civil Defence, SIMPAT and LECLIG 2016)

No direct proportional relationship exists between the reported incidents of mass movements and the number of affected individuals, which indicates that other elements aside from the magnitude and intensity of natural phenomena determine the vulnerability of individuals. In the episode on the 21st of February 1988, for example, one mass movement event in Morro Marilú caused considerable property damage and affected 167 people (Jornal Folha de São Paulo, edition 22/02/1988).

8.3 Correlation Between Rainfall Events and Mass Movements: Estimate of the Local Stability Thresholds and Discussion of the Results

We analysed data on rainfall and its impacts for the municipality of Santos, for the 1980–2015 period, with a focus on the events related to mass movements. The maximum monthly volume of rain during this period (770.5 mm) occurred in March 1995, which was the rainiest year during the analysed period (3716.9 mm). The year 2001 was the least rainy of the period, with 1641.6 mm.

Considering that the average rainfall for the entire period was 2508.2 mm, 40.0% of the years presented above-average volumes of rain (1983, 1987, 1988, 1989, 1992, 1994, 1995, 1996, 2000, 2005, 2010, 2011, 2013 and 2015), while the other 60% were below the average (Fig. 8.3). Analysing the amounts of precipitation for the period does not reveal a pattern of increase or decrease in rainfall events in Santos; however, there were more incidents of rainfall below the period average since the year 2000, with peak highs in 2005, 2010, 2011, 2013 and 2015. That fact agrees with the greater number of records of mass movement in the same period.

With relation to the seasonal distribution of rainfall events over the entire period, the highest volumes are concentrated in the summer, here delimited as the period from December to March. The greatest amplitudes were identified in the months of December and March (Fig. 8.4). March presented the greatest amplitude (687.2 mm),

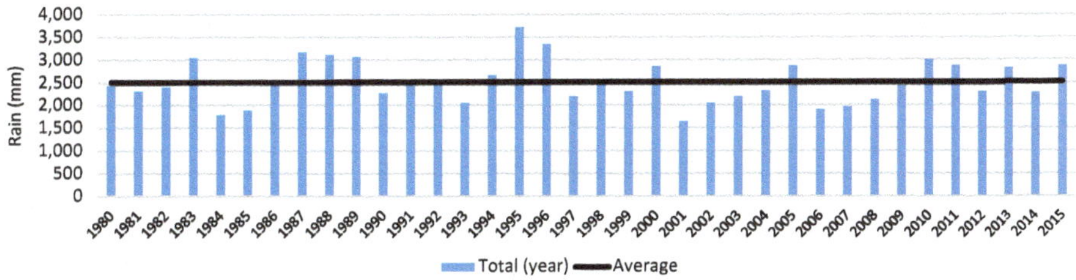

Fig. 8.3 Annual rainfall totals and average (mm) (1980–2015), Saboó station, Santos

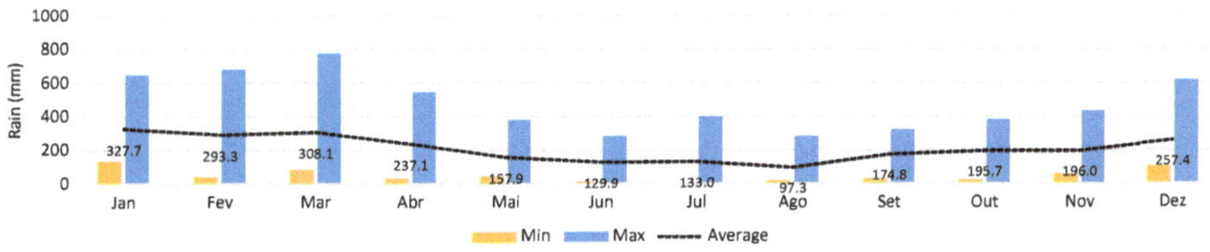

Fig. 8.4 Monthly distribution of rain (mm), considering the minimum, maximum and average volumes and the coefficient of variation (1980–2015). (Source: Saboó pluviometric station (P6 205), Santos, SP)

though it had an average similar to the other more rainy months (308.1 mm), demonstrating that some days of the month probably had a greater concentration of rain.

December presented a gradient of 514.0 mm, with more concentrated rains in the years 2000 (7% of the total), 1999 (6% of the total) and 1987 (5% of the total). In February, the amplitude was 634.3 mm, the predominant years being 1987 and 1996 (5% of the total for each year), 1988, 1989 and 1995 (6% of the total for each year).

A comparison of the representation of each month over the studied decades, shown in Table 8.1, reveals that January, November and December contributed increasing percentages of the total rain volume, while the contributions of other months oscillated inconsistently. In the first decade analysed, the November–March periods represented 53.4% of the total annual volume of rain in Santos, while from 2010 to 2015, the same months corresponded to 57.6% of the total annual volume rain, signalling a greater seasonal concentration of rainfall in the months that are historically the rainiest.

Evaluating the monthly distribution between 1980 and 2015, we identified a greater number of above-average rainfall episodes in May and June (17 each), though the most significant volumes of rain had occurred in other months (October–March). Of the 184 above-average monthly rainfall episodes, 27.7% occurred in the 1980s; 28.3% in the 1990s; and 44.0% since 2000, well distributed throughout the years, indicating that above-average rainfall events have occurred with greater frequency in recent years.

The establishment of correlation between precipitation and mass movements has already been pursued for some time by various authors, including Caines (1980) and Guzzetti et al. (2008). In Brazil, the pioneering contributions of Guidiccini and Iwasa (1977) and Tatizana et al. (1987) deserve mention.

Table 8.1 Comparison of monthly contribution per decade (1980–2015)

	Jan	Feb	Mar	Apr	May	Jun	Jul	Aug	Sep	Oct	Nov	Dec
1980–1990	12.3%	13.5%	12.1%	11.3%	7.0%	5.9%	5.6%	3.0%	6.2%	7.5%	6.1%	9.4%
1990–2000	12.2%	12.4%	14.6%	8.0%	6.0%	5.1%	4.1%	4.6%	8.7%	8.5%	7.3%	8.5%
2000–2010	14.0%	9.2%	10.6%	8.6%	6.2%	4.2%	6.1%	4.6%	6.6%	8.2%	9.6%	12.1%
2010–2015	14.5%	11.2%	11.1%	10.0%	5.8%	5.6%	5.8%	3.2%	5.3%	6.6%	8.8%	12.0%

Source: Saboó pluviometric station (P6 205), Santos, SP

To relate intensity of daily precipitation with the occurrence of mass movements, the present study uses the envelopment technique of Tatizana et al. (1987). This technique was initially applied in the evaluation of earthflows recorded in the region of Serra de Cubatão, the municipality adjacent to Santos, analysing 35 events and cumulative periods of precipitation of 2, 3, 4 and 8 days. In the analysis, the rain accumulated over 4 days was considered a determinant of earthflow.

The envelopment technique (Tatizana et al. 1987) is based on the volume of accumulated rain, over a certain period of days, that has the potential to trigger mass movements. Thus, the preceding rainfall is a preparatory element—infiltration causes an increase in the degree of saturation, or a decrease in the absorption, that, when associated with heavy rainfall, can trigger landslides (Futai 2014). In a study in Singapore, Raharjo et al. (2001, in Futai 2014) arrived at results similar to those in Tatizana et al. (1987), observing that the precipitation accumulated over 5 days acted as a preparatory element of the peak mass movement event that occurred there in February 1995.

In the present analysis, we applied the technique to the rain accumulated over 24, 48 and 72 h, establishing for each period the volume of rain above which rainfall would very possibly trigger a mass movement. Criterion "A" was the most conservative, i.e. having the lowest accumulation of rain; and criterion "T" was the most extreme, with the highest daily accumulation of rain.

To apply the technique, we first selected the impact incidents (426 records) for the purpose of relating those events to the accumulated rain, identifying what cumulative volumes in which time interval had the potential to trigger mass movement in the study area. For this, we empirically tested three criteria for the selection of incidents by searching for a correlation between rainfall and mass movements (Table 8.2). The criteria "A" and "B" were estimated based on the empirical classification of this research, and criterion "T" is the same as adopted by Tatizana et al. (1987).

Juxtaposing the rainfall data with the recorded incidents revealed that a greater number of mass movement events were associated with accumulation over 48 h (159 incidents). We selected 133 incidents of 24 h and 112 incidents of 72 h according to the test criteria. Twenty-two cases of mass movement were not associated with accumulated rain and were therefore not included in the analysis.

Table 8.2 Selection criteria for incidents of 24, 48 and 72 h

Criterion	24 h (mm)	48 h (mm)	72 h (mm)
A	50	75	120
B	75	120	150
T	100	150	200

Source: Empirical classification in the present study

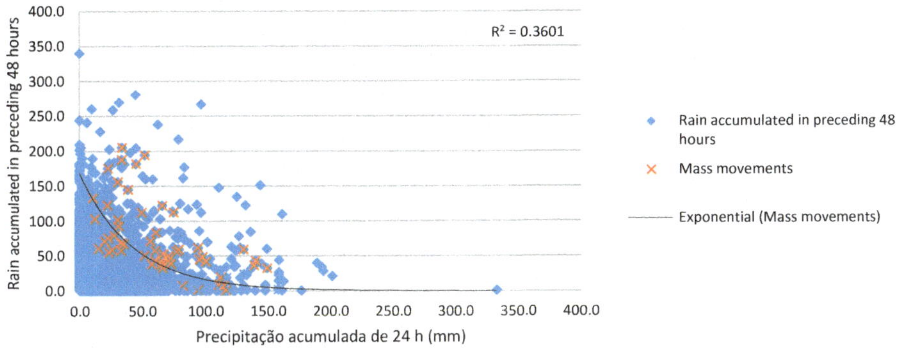

Fig. 8.5 Correlation between rain accumulated over 48 h with rain accumulated over 24 h (mm), using classification criterion "A"

When the criteria "A", "B", and "T" were tested for the accumulations over 24, 48 and 72 h, the strongest statistical correlation was found using the test criterion "A" for rain accumulated over 48 h (Fig. 8.5). After establishing the trend curve, based on the exponential model, the correlation was 0.36.

Evaluating the overall results identified criterion "A" as best fitting the model (Table 8.3). Of all the selected records, 228 were categorised under criterion "A", which is the most conservative of the tested criteria. This result is similar to that found in the work of Soares (2006) on the hills of the coastal municipality of Angra dos Reis, in the state of Rio de Janeiro.

The result obtained through application of the envelopment technique suggest that, according to the tested thresholds, there is a probability of the occurrence of gravitational mass movements when the volume of rain reaches a value equal to or greater than 75.0 mm over a 2-day period. For example: if 50 mm of rain accumulates over two days, an accumulation of approximately 25 mm in the following 24 h would necessary to trigger a mass movement.

Although criterion "A" presented the strongest correlation, Fig. 8.5 shows that there is a wide variation of rain volumes that trigger mass movements, and therefore, it was not possible to establish a clear relationship of cause and effect based on this threshold. However, this result suggests that mass movement events are also related to rain volumes classified as not extreme and possibly even below average, a fact that points to the environmental instability of the region.

Table 8.3 Result of the classification of incidents, per criterion

Criterion	24 h	48 h	72 h	Total/criterion
Criterion A	68	98	62	**228**
Criterion B	40	40	34	114
Criterion T	25	21	16	**62**
Total/accumulated	**133**	**159**	**112**	

Source: Empirical classification in the present study

Currently, the Civil Defence of Santos considers 80.0 mm of rain per day as a volume within the norm for a 24-h period, and declares a state of concern in the municipality when the precipitation volume reaches 100.0 mm in 72 h. Considering the present assessment, which suggests a more conservative approach, a new contemplation of these thresholds is needed.

The municipality of Santos is generally characterised as having high levels of wealth, life expectancy and education. According to the vulnerability index proposed by the government of the state of São Paulo (Índice Paulista de Vulnerabilidade Social – SEADE 2010), where Santos is located, most of the population lives in areas of extremely low, very low, or low vulnerability (11.1%, 69.8% and 2.7% respectively). In contrast, approximately 5.4% reside in areas of very high vulnerability, 2.3% are in areas of high vulnerability, and 8.7% are in areas of medium vulnerability, which corresponds to a total of 11% of the population living in the Metropolitan Region that includes the municipality of Santos.

Figure 8.6 shows the concentration of groups with some degree of vulnerability, from the central to northern parts of the insular sector of Santos, which corresponds to the areas with the most records of calamitous events. These areas are hilly and in a zone that comprises around 20 of the lowest-income neighbourhoods, including occupations that fall outside planning standards, known as the Northwest Zone—one of the sectors assessed in detail in the Metropole Project (Chap. 11). The locations with the best indices are located near the waterfront and the Port of Santos.

8.4 Final Considerations

The urban expansion dynamic of Santos intensified the process of peripheralisation of the low-income population to areas with deficient infrastructure. Added to the existing natural susceptibility, and intensified by the disorganised occupation of space (such as in the central area, which is composed of hills), this pattern contributes to an unequal territorial configuration and results in risks and hazards, manifesting in the form of material and social impacts on the most vulnerable population.

It should be noted that risk management is quite present in the municipality of Santos, particularly with respect to geotechnical risks, aligning with various laws and directives from other levels of government. To illustrate, we highlight the creation of the Civil Defence of Santos (1980) and the Grupo de Morros (1988),

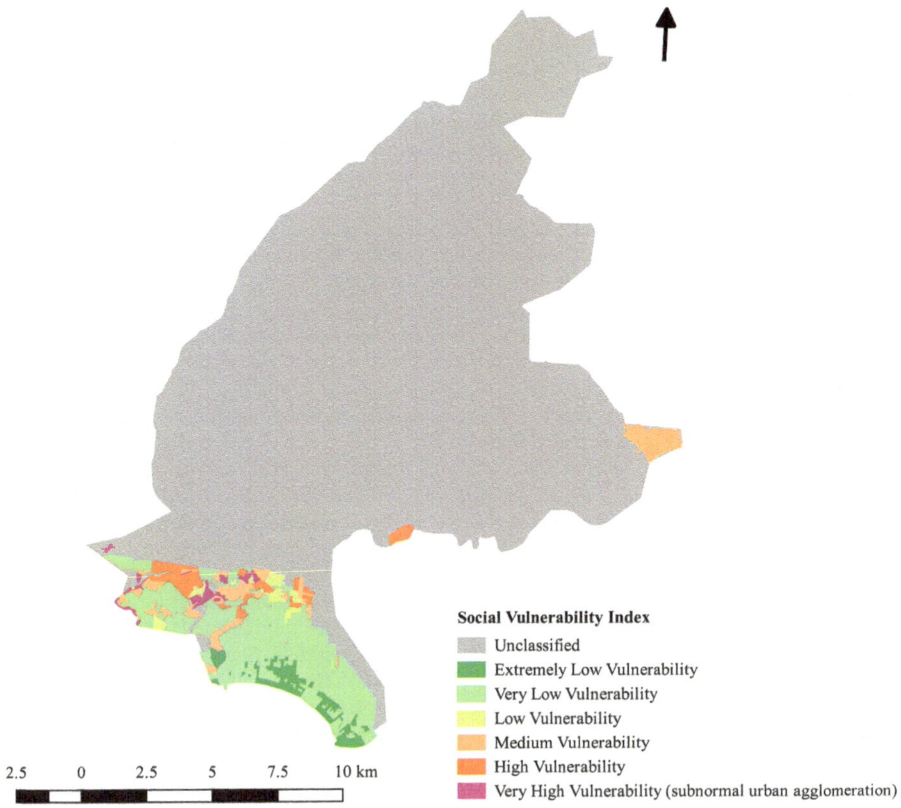

Fig. 8.6 Distribution of IPVS Santos per census sector (2010). (Source: Fundação Sistema Estadual de Análise de Dados – SEADE and Índice Paulista de Vulnerabilidade Social – SEADE 2010)

which expanded assistance services and action in relation to natural disasters. Recently, the Municipal Commission on Adaptation to Climate Change was created under the Decree n° 7.293 of 2015, Which is responsible for the elaboration and monitoring of the Municipal Plan for Adaptation to Climate Change in Santos (see Chap. 14).

References

Araki, R., & Nunes, L. H. (2008). Vulnerability associated with precipitation and anthropogenic factors on Guarujá City (São Paulo, Brazil) from 1965 to 2001. *Terrae- Geosciences, Geography and the Environment, 3*, 54–57.

Bryant, E. A. (1997). *Climate process and change*. Cambridge: Cambridge University Press.

Caine, N. (1980). The rainfall intensity: duration control of shallow landslides and debris flows. *Geografiska Annaler. Series A, Physical Geography, 62*, 23–27.

Carmo, R. L., & Anazawa, T. M. (2014). Mortalidade por desastres no Brasil: O que mostram os dados. *Ciência & Saúde Coletiva, 19*, 3669–3681.
CENAD. (2014). National center for risk and disaster management. Yearbook of natural disasters, 2013.
Feitosa, F. F., Maretto, R., Monteiro, A. M. V., et al. (2012). *Urbanização e vulnerabilidade social em Zonas Costeiras: a construção de um modelo de simulação das dinâmicas residenciais de Caraguatatuba, SP*. Presented at 18th Encontro Nacional de Estudos Populacionais. Águas de Lindóia.
Ferreira NJ, Ramírez, MV, Gan, MA (2009) Vórtices ciclônicos de altos níveis que atuam na vizinhança do nordeste do Brasil. Cavalcanti IFA, Ferreira, NJ, Justi da Silva, MGA et al. Tempo e clima no Brasil. São Paulo: Oficina de Textos, São Paulo, 43–60.
Figueira, R. M., & Bandini, M. P. (2016). *A incorporação das ações de proteção e defesa civil em âmbito municipal. Redução de Risco de Desastres: uma construção de resiliência local* (pp. 15–29). São Paulo: Coordenadoria Estadual de Proteção e Defesa Civil.
Futai MM (2014) Landslides in unsaturated conditions caused by rainfall infiltration. Lacerda WA, Palmeira EM, Coelho Neto AL et al. Extreme rainfall induced landslides – an international perspective. Oficina de Textos, São Paulo 41–69.
Gan, M. A. (1982). *Estudo observacional sobre as baixas frias da alta troposfera nas latitudes subtropicais do Atlântico Sul e leste do Brasil*. PhD. Thesis INPE, São José dos Campos.
Gigliotti, M. S. (2010). *Zoneamento geoambiental da região da Baixada Santista – SP como subsídio ao uso e ocupação da terra* (152 p). Dissertation, University of Campinas.
Gonçalves, N. M. S. (2003). Impactos pluviais e desorganização do espaço urbano em Salvador. In C. A. F. de Monteiro & F. A. de Mendonça (Eds.), *Clima Urbano*. São Paulo: Contexto.
Guidicini, G., & Iwasa, O. Y. (1977). Test of correlation between rainfall and landslides in a tropical humid environment. *Bulletin of Engineering Geology and the Environment, 16*(1), 13–20.
Guzzetti, F., Peruccacci, S., Rossi, M., & Stark, C. P. (2008). The rainfall intensity – duration control of shallow landslides and debris flows: an update. *Landslides, 5*, 3–17.
Instituto de Pesquisas Tecnológicas do estado de São Paulo. (1979a). *Carta Geotécnica dos Morros de Santos e São Vicente*. Technical Report 11559, São Paulo.
Instituto de Pesquisas Tecnológicas do Estado de São Paulo. (1979b). *Levantamento das condicionantes do meio físico e estabelecimento de critérios normativos para a ocupação urbana dos morros de Santos e São Vicente*. Technical Report 11599.
Instrução Normativa 1. (2012). Diário Oficial da União. 169.
Kunkel, K. E., Pielke, R. A., & Changnon, S. A. (1999). Temporal fluctuations in weather and climate extremes that cause economic and human health impacts: a review. *BAMS, 80*, 1077–1098.
Laboratório de Estudos Climáticos do Instituto de Geociências da Unicamp (LECLIG). (2016). Unpublished.
Larsen, M. C., & Simon, S. A. (1993). Rainfall intensity-duration threshold for landslides in a humid-tropical environment, Puerto Rico. *Geograsphica Annaler, 75*(1–2), 13–23.
Malvestio, L. M. (2013). *Variabilidade da precipitação pluviométrica da região Sudeste do Brasil no período chuvoso e suas consequências ambientais*. Dissertation, Universidade Estadual de Campinas, Campinas.
Nunes, L. H. (1997). *Distribuição espaço-temporal da pluviosidade no Estado de São Paulo: variabilidade, tendências, processos intervenientes*. PhD. Thesis Universidade de São Paulo, São Paulo.
Nunes, L. H. (2009a). Compreensões e ações frente aos padrões espaciais e temporais de riscos e desastres. *Revista Territorium, 16*, 179–189.
Nunes, L. H. (2009b). Mudanças climáticas, extremos atmosféricos e padrões de risco a desastres hidrometeorológicos. In D. J. Hogan, E. J. Marândola Jr (Eds.), *População e mudança climática: dimensões humanas das mudanças ambientais globais* (pp. 53–73). NEPO/UNFPA, Campinas.
Nunes, L. H. (2015). *Urbanização e desastres naturais na América do Sul: abrangência América do Sul*. São Paulo: Editora Oficina de Textos.

Nunes, L. H., Modesto, R. P., de Almeida, M. C. J., et al. (1989). *Estudo de episódios pluviais associados a escorregamentos – Município do Guarujá, SP*. Paper presented at the 2nd Encontro Nacional de Estudos sobre o Meio Ambiente, UFSC, Florianópolis, *1*(1), 402–408.

Pelling, M. (2006). Paradigms of risk. In M. Pelling (Ed.), *Natural disasters and development in a globalizing world* (pp. 3–16). London: Routledge.

Pichler, E. (1957). Aspectos geológicos dos escorregamentos de Santos. *Boletim da Sociedade Brasileira de Geologia, 6*(2), 68–77.

Ross, J. L. S., & Moroz, I. C. (1997). *Mapa geomorfológico do estado de São Paulo*. São Paulo: IPT/FAPESP.

SEADE. (2010). Fundação Sistema Estadual de Análise de Dados. Social Vulnerability Index of State of São Paulo (IPVS) http://www.seade.gov.br/. Accessed 5 Dec 2016.

SEDEC. (2014). National secretary for civil protection and defense. Yearbook of natural disasters, 2013.

Silva, R. B. (2013). *Urbanização e vulnerabilidade na Região Metropolitana da Baixada Santista, SP: Um olhar geográfico das desigualdades intraurbanas em bairros de Santos, Praia Grande e Peruíbe*. PhD. Thesis Universidade Estadual de Campinas, Campinas.

SIMPAT. (2016). *Sistema integrado de monitoramento, previsão e alerta de tempestades para as Regiões Sul-Sudeste do Brasil – SIMPAT*. http://www.ipmet.unesp.br/index2.php?menu_esq1=&abre=ipmet_html/defesa_civil/index. Accessed 25 Mar 2016.

Singer, P. (1977). *Economia política da urbanização*. São Paulo: Brasiliense.

Tarifa, J. R., & Armani, G. (2000). *Atlas geoambiental do município de São Paulo*. São Paulo. (Fase I – Diagnósticos e Bases para a Definição de Políticas Públicas para as Áreas Verdes no Município de São Paulo).

Tatizana, C., Ogura, A. T., Cerri, L. E. S., et al. (1987). *Análise de Correlação entre Chuvas e Deslizamentos – Serra do Mar, município de Cubatão*. Presented at 5th Congresso Brasileiro de Geologia e Engenharia. São Paulo.

Tobin, G. A., & Montz, B. E. (1997). *Natural hazards – explanation and integration*. New York: The Guilford Press.

UNSDR. (2016). United nations office for disastes risk reduction. The human cost of weather related disasters 1995–2015. https://www.unisdr.org/we/inform/publications/46796. Accessed 20 Oct 2017.

Valencio, N. (2012). *Para além do 'dia do desastre': o caso brasileiro*. Curitiba: Appris.

Znamensky, D. (2014). Debris and mudflow initiation processes in Brazilian tropical and subtropical humid and montainous environments. In W. A. Lacerda, E. M. Palmeira, A. L. Coelho Neto, et al. (Eds.), *Extreme rainfall induced landslides – an international perspective*. São Paulo: Oficina de Textos.

Zundt, C. (2006). Baixada Santista: uso, expansão e ocupação do solo, estruturação de rede urbana regional e metropolização. In J. M. da Cunha (Ed.), *Novas Metrópoles Paulistas: população, vulnerabilidade e segregação*. Campinas: NEPO.

Chapter 9
Land Use Change Dynamics in the Metropolitan Region of Baixada Santista MRBS (SP): Between Development and Environmental Impacts

Antonio Carlos da Silva Oscar Júnior, Beatriz Barbi de Oliveira Santos, Eduardo Kimoto Hosokawa, Pamela Pereira de Araújo, and José Marques Carriço

Abstract The chapter offers an analysis of the territorial organization of the Metropolitan Region of Baixada Santista (RMBS) under the perspective of global environmental changes, with emphasis on changes in land cover and use. The central argument is that the development ways traced and/or imposed by and for each metropolitan entity in the study area reverberates significantly in environmental quality, being a key context for the discussion of mitigation and adaptation to climate change. To achieve this, satellite images were used between 1996 and 2016, from which it was possible to detect changes in land cover, as well as projection of a trend scenario for 2021. Socioeconomic data and socio-territorial formation were incorporated, as well as carbon emission estimation. From this analysis it is possible to understand the process of organization of the metropolitan territory in a historical, political and socioeconomic perspective, identifying transforming agents, impacts and challenges. It is shown that the deleterious environmental impacts of urban expansion are explained by metropolitan land-use planning policies,

A. C. da Silva Oscar Júnior (✉)
Geography Institute, Rio de Janeiro State University (UERJ), Rio de Janeiro (RJ), Brazil
e-mail: antonio.junior@uerj.br

B. B. de Oliveira Santos (✉)
Azul Brazilian Airline, São Paulo, Brazil

E. K. Hosokawa
Municipal Government of Santos, Secretariat of Urban Development, Santos, SP, Brazil
e-mail: eduardohosokawa@santos.sp.gov.br

P. P. de Araújo
Municipality of Santos, Trainee in Municipal Secretary of Environment, Santos, SP, Brazil

J. M. Carriço
Municipality of Santos, Municipal Secretary of Government/Post-graduation Program in Law and the Architecture and Urbanism Course, Catholic University of Santos (UNISANTOS), Santos, SP, Brazil

L. H. Nunes et al. (eds.), *Climate Change in Santos Brazil: Projections, Impacts and Adaptation Options*, https://doi.org/10.1007/978-3-319-96535-2_9

controlled mainly by the real estate sector and central municipalities. Such a model contributes to greater potential for greenhouse gas emissions and limited opportunities for adaptation.

Keywords Adaptation · Mitigation · Socioterritorial formation · Metropolization

9.1 Introduction

There is a growing discussion about climate change around the world; after all, its impacts threaten living conditions and human subsistence. Environmental changes resulting from climate change have significantly affected the Earth's landscape, but given our level of concentration and technical development, these changes can now cause economic and production damage and affect mortality, morbidity and population mobility (Rose and Testa 2013).

According to the Intergovernmental Panel on Climate Change (IPCC 2007), 'climate change' refers to a statistically significant variation of average climate parameters, including its natural variability, persistence over an extended period of time. This change is due to both natural processes and the increasing interference of human activities in environmental systems, including climate. Human interference in the environmental system is so important that we are currently discussing the beginning of a new era for the recent history of Earth: the Anthropocene, with indelible marks of human activities (Crutzen 2006).

Although climate variations are cyclical, for some researchers, such as Dearing et al. (2006) and Nobre et al. (2011), significant impacts on the environmental systems, economy and society may be associated with the consequences of climate change. Since the discoveries of Tyndal and currently with the IPCC, studies have tried to demonstrate the association between carbon dioxide concentration in the atmosphere and temperature elevation, with physical explanation in the absorption capacity of this gas of long-wave radiation coming from the planet (Petit et al. 1999). According to the IPCC (2014), besides carbon dioxide (CO_2), other substances such as methane (CH_4), nitrous oxide (N_2), and ozone (O_3), have had a significant increase as a consequence of anthropic actions, particularly due to changes in land cover and land use, as this is the case in Brazil, with suppressed native vegetation for industrialization purposes and the consequent implementation of cities, and forfarming purposes (CETESB 2011). This aspect is represented in Fig. 9.1, taken from the First Inventory of Greenhouse Gas of the State of São Paulo, which shows that 60.6% of the emissions of these gases come from UTMUTF (land use, land use change and forests).

According to the IPCC reports, since the Industrial Revolution a significant increase in the concentration of greenhouse gases (GHGs) in the atmosphere has been observed. An increase of 280 parts per million (ppm) from 1000 to 1750–368 ppm in 2000 is estimated, reflecting a mean increase of the Earth's temperature of ±0.2 to 0.6 °C. More recently, at the Conference of the Parties (COP21), the Met Office (British meteorological service) observed that, for the first time since 1850,

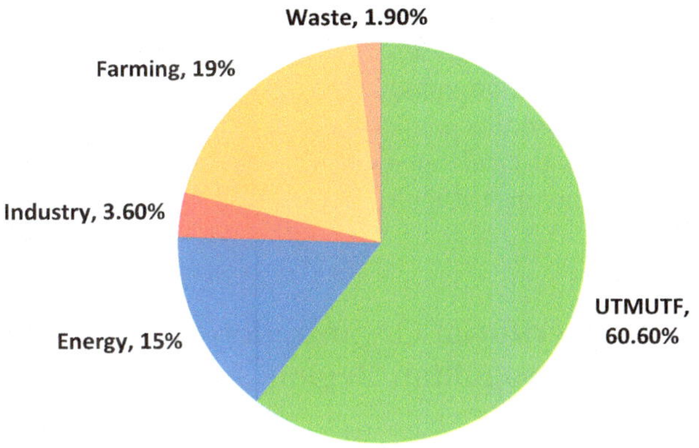

Fig. 9.1 GHG emission in Brazil (1990–2008). (Source: First inventory of greenhouse gas of the State of São Paulo 2011)

between January and September 2015, the Earth's average temperature increased 1.02 °C, which was expected at this point (MetOffice 2016).

Although the impacts of these changes are still uncertain, some regions of the world are already having climate-related calamities such as Tuvalu Island, located in the southern Pacific Ocean, with more tropical cyclones and higher sea level in the last decade, events influenced by the increased temperature of the surface waters of the ocean and its interference in the incidence of storms (Carmo and Silva 2009).

In this context of changes and impacts, the urban areas and its social, economic, demographic and environmental dynamics play an important role in the discussion of mitigation and adaptation to climate change. However, the spatial organization enters a new stage. The urban phenomenon gradually embraces the phenomenon of metropolization, promoting significant changes in spatial processes (Ferreira et al. 2014), and in territorial governance (Dallabrida and Becker 2003), ensuring a new scale to the analysis of this issue.

Considering this scale of analysis of climate change, this chapter offers an analysis of the Metropolitan Region of Baixada Santista (MRBS), which covers nine municipalities, including Santos, in the perspective of global environmental changes, with emphasis on changes in land cover and use. The central argument is that the development systems adopted and/or imposed by and for each metropolitan entity in the study area has an effect on the atmospheric emissions, and it is, therefore, a key context for the discussion of mitigation and adaptation to climate change. This trend has been largely debated in recent years, so it is not new in MRBS: Cubatão's petrochemical and steel complex, installed in the area in the mid-twentieth century, has turnedCubatãointo the most polluted municipality in the world in the 1980s, according to the United Nations United Nations.

In order to identify and analyse these aspects and considering the importance of the impact of land use and cover changes on atmospheric emissions, taking Landsat satellite images from 2005, 2011, to 2016 and using detected changes in land cover to estimate a trend for 2021. And to analyse the link between this process and the occupation morphology of the region, images from the same satellite from 1979, 1980, 1991, 1992, to 2011 were used, obtained from Instituto Pólis (2013).

This study was also based on socioeconomic data from IBGE (Brazilian Institute of Geography and Statistics), municipal and metropolitan bodies, and carbon emission estimates provided by CETESB (Environmental Company of the State of São Paulo). From this analysis, it is possible to identify the main transforming agents, impacts and challenges of the territorial management of the metropolitan region considering the difficult scenario of global environmental changes.

9.2 Territorial Organization of Baixada Santista Metropolitan Region

In terms of global logic, the Metropolitan Region of Baixada Santista has challenges and seeks solutions in a scenario of growing and increasingly dynamic environmental changes. This metropolitan region, as well as the São Paulo coast in general, has been through intense transformation in the last decades, especially due to the strong relationship with the Metropolitan Region of São Paulo (MRSP) (Carmo et al. 2012).

The RMBS was a pioneer in territorial organization after the 1988 Federal Constitution (Zundt 2006), although it was officially created after the State Complementary Law 815 of July 30, 1996. The MRBS comprises continental and insular areas and covers nine municipalities: Bertioga, Cubatão, Guarujá, Itanhaém, Mongaguá, Peruíbe, Praia Grande, Santos, and São Vicente.

According to Jakob (2003), the MRBS territory presented significant growth between 1940 and 2000 (7.5 times) due to the Port of Santos, the petrochemical complex of Cubatão, and improvements in transport infrastructure that favoured the connection with the Metropolitan Region of São Paulo (MRSP), which intensified the tourism activity and increased the population flow resulting from the labour relations with the MRSP (Table 9.1). These phenomena resulted in early urbanization of the MRBS when compared to the other Metropolitan Regions of the state of São Paulo – in 2000, the MRBS presented 99.6% urbanization against 95.8% of the RMSP.

These data show the gradual stagnation of Santos' growth, especially after the 1970s, when the island of São Vicente[1] presented a saturation of proper areas for occupation. In the beginning of this period, the population of Santos was 76.0% of the region's population, but in that decade the region's population became twice the population of Santos, showing the urban spreading process caused by industrialization in the region.

After World War II, according to Carriço (2006), the national economy *"had a deep transformation, with the breakdown of agro-exporting hegemony and the beginning of an extensive stage of accumulation, in which the industry of the Southeast, especially from São Paulo, led the process."* In this phase of

[1]The headquarters of the municipalities of Santos and São Vicente are in this insular area, which concentrates most of the regional jobs, generating serious problems of urban mobility.

Table 9.1 Progress of Brazil, State of São Paulo, MRBS, and Santos populations (1950–2010)

Location	1950	1960	1970	1980	1991	1996	2000	2010
Brazil	51,944,397	70,070,457	93,139,037	119,002,706	146,825,475	157,079,573	169,799,170	190,755,799
São Paulo	9,134,423	12,809,231	17,771,948	25,040,712	31,588,925	34,120,886	37,032,403	41,252,160
MRBS	267,387	416,963	653,441	961,249	1,220,249	1,309,263	1,476,820	1,664,136
Santos	203,582	262,997	342,055	412,448	417,450	412,243	417,983	419,400

Source: IBGE censuses (1950–2010)

industrialization oriented to the domestic market, the heavy industry was essential, complementing the industrialization in the capital of São Paulo. That was how the Cubatão industry emerged,[2] resulting from investments in infrastructure.

In 1947, the first lane of the first highway was inaugurated: Via Anchieta. In 1953, the second lane was inaugurated, leading to a reconfiguration of the political geography of the region; after all, it resulting in the foundation of the municipality of Cubatão, separated from the territory of Santos in 1948, and the consolidation of the municipality of Guarujá, also separated from the territory of Santos.

Due to the short distance from and easy connection with the capital of São Paulo and the Port of Santos, Cubatão attracted the interest of the industrial sector, for example, PresidenteBernardesCubatão – RPBC (1955) and CompanhiaSiderúrgicaPaulista – COSIPA[3] (1953), which set their facilities in the area, recruiting workers from Baixada Santista and migrants from various regions of Brazil, especially from the Northeast. Until the 1940s, Santos accounted for 80% of the industrial production, but in the 1950s, it was surpassed by Cubatão, which already corresponded to 72.5% of the region (Santos1992 *apud* Jakob 2003).

With the policy of economic development of Juscelino Kubitscheck, who assumed as the President of Brazil in 1956, automotive plants were set up in the ABC Paulista region, promoting the implementation of the petrochemical and steel complex in Cubatão and attracting a large number of workers to civil construction in the 1950s and 1960s.

Due to the expansion of industrial, commerce and service activities, and the population concentration in urban areas, Silva (2013) points out that in the 1950s the region began to experience environmental problems. However, Coelho (2009) highlights that it is not the population growth itself that causes environmental problems, but the lack of public policies and proper land use planning. In this context, the municipality of Cubatão is a good example of the development model planned at higher scales and imposed to the site, causing the decay of natural heritage and environmental quality, after all industrialization at all costs, combined with the geographical characteristics of the area, were decisive to generate countless environmental and social damages (Hogan 1990).

The intense transformation and increased dynamism in the region contributed to the generation of well-developed urban and administrative centres (Zundt 2006), leading to the creation of new municipalities, such as Guarujá in 1934, Mongaguá in 1959, and Praia Grande in 1966.[4]

Of the municipalities that today comprise the MRBS, three were created after they separated from Santos: Guarujá, Cubatão, and Bertioga; one separated from São Vicente: Praia Grande; and two from Itanhaém: Peruíbe and Mongaguá.

---

[2] In this context, the abundant supply of water at the base of Serra do Mar mountain range, in Cubatão, was essential for the industry.

[3] Later becameUsiminas.

[4] This municipality was already independent, according to a state law of 1964, which was revoked by the Justice and later revalidated.

With this scenario of the MRBS territorial organization, Santos and São Vicente appear as the central areas, driving the development of the metropolitan region and highlighting their influence and importance for the metropolitan policies. Such spreading and conurbation with neighbouring cities is a process explained by Zundt (2006) associated with the "real estate boom generated by the summer and the implementation of great infrastructures and industries in the region".

The development model in force encouraged the migration and economic growth of the Baixada Santista region, based on intense land use and occupation, and socio-spatial segregation. This model, combined with the ineffectiveness of regional plans developed so far[5]promotedthe occupation of peripheral areas, contributing to the emergence of urban environmental problems, aligned to a large extent with the highly exclusionary and unequal pattern of the urban environment (Faria 1978).

The transformations were so intense that in 1940 the region already presented an urbanization rate of 91%, while the state average was 40.0%. Silva (2013) points out that the 1950s were characterized by the arrival of migrants who started working in the civil construction sector. The 1960s and 1970s were characterized by the consolidation of the Cubatão industrial complex, foreign trade growth, transportation infrastructure improvements (construction of the Immigrants highway in 1976) and a population growth rate of 4.6% of the municipalities of Peruíbe and Mongaguá. Between 1970 and 1980, Peruíbe and Praia Grande presented the highest geometric rates of population growth: 10.2% and 12.8%, respectively.

When analysing the most recent data on population growth and population density, it is clear that the municipalities of Santos, São Vicente, Guarujá, and Praia Grande concentrate most population of the metropolitan region and present the highest demographic density rates. In 1980, the population of Santos accounted for around 43.0% of the total region, while in 2015 it fell to 24.0%. [6] The most expressive municipality in this comparison was Praia Grande, which in 1980 accounted for 7.0% of the total population, and in 2015 it reached 17.0%, indicating a type of migration to the municipalities outside the central axis of the Baixada Santista (Silva 2013).

The highest population density is in the areas closest to the coast favoured by the geomorphological characteristics of the region, with a narrow flat strip of intense conurbation between the municipalities, which is limited by Serra do Mar (mountain range) and the coast that are important geographical obstacles (Zundt 2006).

As illustrated in Fig. 9.2, other municipalities presented significant population growth rates in recent times (1990–2010): Bertioga, Mongaguá, Praia Grande, Itanhaém, and Peruíbe. Bertioga, Praia Grande, and Mongaguá have presented high population growth rates in the last two decades, indicating intense transformation of their urban space. Santos has presented rates close to 0% starting that same year,

[5]In 1971, the CEMBS (Metropolization Study Committee of Baixada Santista), created by Municipal Decree 3.894, of August 4, 1971, submitted its general report to a federal influencer recommending measures related to the "Master Plan of Baixada Santista Integrated Development" (CEMBS 1971).

[6]This reduction was particularly due to the migration to neighbouring municipalities.

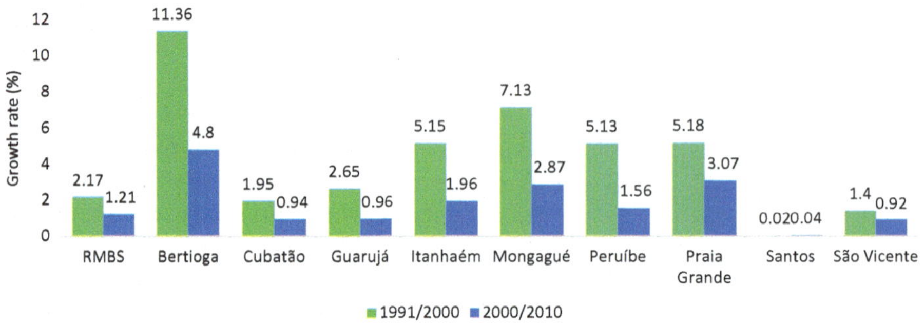

Fig. 9.2 Annual population growth rate of the MRBS municipalities (1991–2015). (Source: SEADE, organized by the authors)

followed by São Vicente, Cubatão and Guarujá. These data reinforce the continuous process of population spreading towards the neighbouring municipalities of Santos and São Sebastião.

As illustrated in Fig. 9.3, in late 1970s, the urban area in the region had already expanded to the south through a range of construction developments along the coast from the tip of Itaipu in Praia Grande to the present central area of Peruíbe, with some discontinued portions. In this period, the implementation of the PriestManoel da Nóbrega highway (SP 55), linking the central area of Baixada Santista to Vale do Ribeira, to the south coast of the state of São Paulo, was determinant to allow a spreading and extremely deleteriousurbanization model, and in environmental terms, based on the production of summer facilities, through the production of horizontal construction units.

The occupation between the central area of the region and its southern end was characterized by fragmentation and the predominance of households occupied by floating population. While the regulated urban lots were destined to households of occasional use, the worst locations, mainly to the north of the highway, were greatly occupied by resident mostly low-income population.

This socially segregated occupation model, widespread and discontinuous, led to suppressed vegetation in sensitive ecosystems, such as mangroves, restingas,[7] and beaches, demanding high investments in urban infrastructure. In addition, the widespread urban area induces the intense use of motor vehicles, resulting in more greenhouse gas emissions. This factor is aggravated by the region's long-line configuration and its two typical use patterns: port/logistic/industrial and residential activities, both highly relying on automobiles, lightweight commercial vehicles, urban buses, heavy and semi-heavy trucks, which according to CETESB, are the main cause of atmospheric emissions of pollutants in the state of São Paulo (CETESB 2011).

---

[7]A type of vegetation of tropical and subtropical coastal areas with individuals present in herbaceous, shrubby and arboreous layers, that grow on sandy plains.

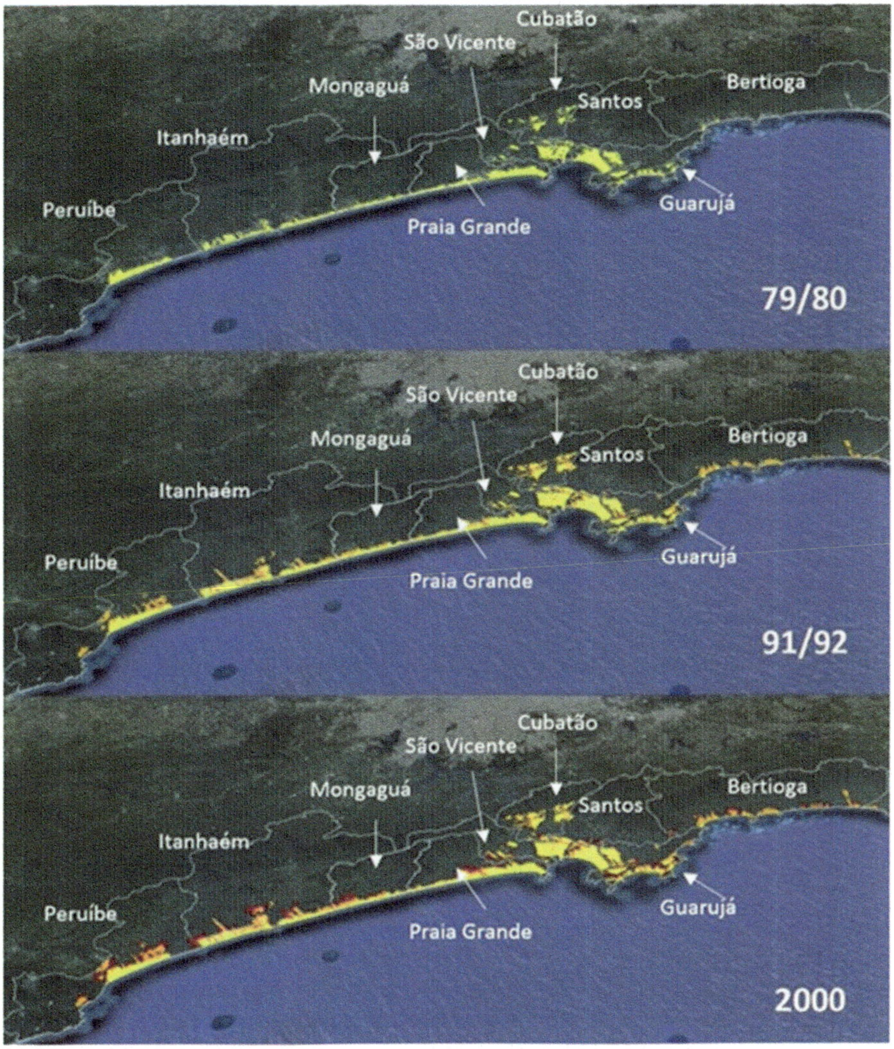

Fig. 9.3 Progress of urban areas in Baixada Santista (1979–2000). (Source: Images of Landsat 1979–2000, *apud* Instituto Pólis (2013), adapted by the authors. The yellow area refers to the existing urban area and the red area refers to the progress compared to previous period)

As Seabra pointed out in 1979, the São Vicente Island shorereceived vertical constructions mainly due to beach tourism. In this area, seafront apartment buildings of the 1940s and later were first occupied by floating population. With this more compact pattern of occupation, tertiary sub-centres soon emerged along the seafront in São Vicente and later in Guarujá.

After this period, the floating population gradually changed to permanent residents in these vertical homes, increasing the beach-downtown commuting in

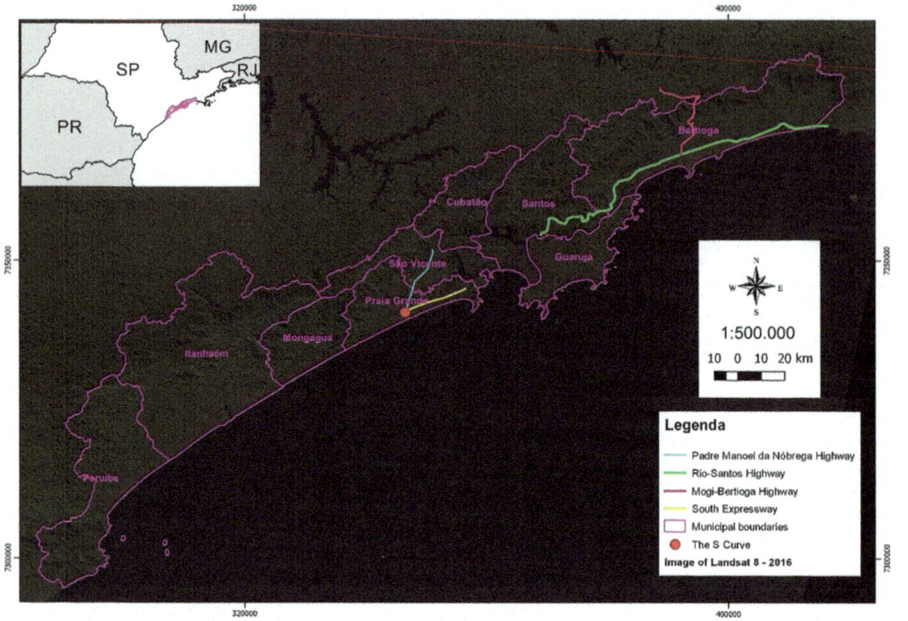

Fig. 9.4 Location of some territorial elements. (Source: developed by the authors (2017))

Santos.[8] Therefore, despite a more compact urban form and the creation of an economic base near the shore, such displacement remains to date, based on motorized commuting, as indicated in the Baixada Santista Source-Destination Surveys conducted in 2007 and 2012 (São Paulo 2008), resulting in increased greenhouse gas emissions.

From the results of satellite images (Fig. 9.3) and the location map of the images (Fig. 9.4), the occupation to the northeast of the São Vicente and Santo Amaro Islands were intensified in the 1980s, with the construction of the Manoel Hyppolito do Rego Highway, known as Rio-Santos (SP 55), and with the implementation of the Mogi-Bertioga Highway (SP 98), which created a new connection between the coast and the country side. The urban form of this occupation to the north of the region is very similar to that on the south coast, but even more fragmented. And like the south, homes of occasional use prevail, with the same deleterious effects highlighted above.

Between the Serra do Mar and Santos and São Vicente Estuaries, Cubatão presents territorial occupation characterized by discontinuous even more fragmented occupation, with poor settlements on the hillsides[9] of the Serra do Mar and close to the highways, railways, and waterways.

[8] This phenomenon did not happen in São Vicente in the same period, as the central area of this municipality is located near the shore, between Gonzaguinha and Itararé Beaches.

[9] This occupation began due to construction sites for the construction of Via Anchieta, generating

Figure 9.4 shows that in the 1980s there was a significant increase in occupation in Bertioga, but even more disconnected and fragmented than in the southern coast in the previous decade. In the municipalities of Itanhaém and Peruíbe, a great expansion of the urban area was seen in this period to the north of the highway, indicating expansion of occupation to areas of poor infrastructure. In Praia Grande, the occupation expanded to the area known as "the S Curve," at the confluence of the Padre Manoel da Nóbrega highway and the current South Expressway. In almost all of these cases, the process was characterized by the appearance or expansion of low-income settlements in irregular lots. This process demanded an even greater effort of the state government to provide these municipalities with basic sanitation (Instituto Pólis 2013).

In the central area of the region, in a decade of economic stagnation due to the failure to implement the Second National Development Plan, poor settlements occupied hillsides unsuitable for occupation and mangroves. During this period, a well-defined pattern of sociospatial segregation in the periphery consolidated in the municipalities of Santos, São Vicente, and Guarujá. In the latter, the verticalization process of the shore extends to neighbourhoods farther from the central area. However, the urban planning legislation of Guarujá, based on more sustainable terms that allow the adoption of acceptable standards of natural ventilation and light, had some changes in relation to the verticalization process in central beaches from the previous period.

In Cubatão, a high number of workers attracted by the industrial sector in the previous decade, were unemployment and further expanded the occupation without access to proper infrastructure in the Serra do Mar and lowlands. In São Vicente, the occupation also increased in the continental area and later consolidated as the preferential destination of low-income families with labour and income connection with the tertiary insular areas of Santos and São Vicente.

Figure 9.3 shows that the 1990s were characterized by expanded occupation to the north end of SP-55 highway, to the north of settlements, in the four municipalities of the south coast and in Bertioga. This pattern of horizontal expansion is associated with summer attractiveness, in terms of workforce of low qualification and remuneration. Thus, new irregular occupations emerged in locations farther and farther from the seafront, affecting deeply the environmental and social conditions.

Another characteristic of this process was the increased commuting, since these areas are predominantly monofunctional residences, locally known as dormitoryareas. Therefore, the evident impact from suppressed mangrove and restinga vegetation and the intense commutingresulted in higher production of greenhouse gas.

The expansion of irregular occupations at the edge of existing irregular occupation was also observed in the central area of the MRBS in the period mentioned above. It happened mainly in Guarujá, Cubatão and in the continental areas of São Vicente and Santos. What this image does not show is that, based on census data of

---

neighbourhoods whose residents were initially the highway workers. In the 1970s, this phenomenon happened again, now even denser in these neighbourhoods, with the construction of the Rodovia dos Imigrantes.

the period, the poor settlements became denser in this central area, in hillside areas or lowlands, that is, areas of susceptibility. The beginning of a verticalization process is observed in certain central areas, intensified in the subsequent decade.

In 2000, the same figure shows decelerated expansion of horizontal occupation at the edge to the north of the SP-55 highway, which was accelerated in the 1990s. These areas are mainly in Peruíbe, Itanhaém, and Bertioga. Studies such as that of Rifino et al. (2015) show that the development of housing programs of social character like*Minha Casa Minha Vida* (my house my life) contributed to the peripheral horizontal housing expansion in theMRBS. Due to the high cost of land in the central area of the region, this program concentrated most of its production in the municipalities located in the south of the Baixada Santista. And currently, expansions are observed in this fabric through irregular occupations.

In the 2000s and the first half of the current decade, the main Brazilian metropolises also had intense housing production for high-income population. In the MRBS, Santos and Praia Grande were the municipalities where most of these high-income areas were produced, with a high supply of predominantly residential tower-type buildings, withmore commercial and corporate buildings in Santos, due to the expectations from the exploration of the pre-salt layer in the Santos Basin.

The real estate sector, as in the first decades of verticalization in the central area of the region, concentrated most of the developments near the seafront, as reported by Carriço and Barros (2015). As in this period most lots on the blocks closest to the sea were already vertical, the sector began to incorporate lots between these blocks and the old Sorocabana railroad, producing significant changes in the urban design of the neighbourhoods to the north of this boundary. In general, although the occupation remained at the edge, almost all neighbourhoods in the east side of the city had real estate launches in a much more intense verticalization pattern than in previous decades.

The economic growth in the last decade and the pre-salt layer expectations, combined with changes in the national regulation of the sector and the revision of the 1998 urban legislation, which among other things eliminated the limit height of buildings, were responsible for a vigorous verticalization process in the area mentioned above. This situation caused serious effects in terms of urban form and landscape and stronger environmental impact due to the architectural style adopted, based on the huge availability of common areas and car spaces on the lower floors of the towers, a fact encouraged by municipal legislations on land use and occupation.

This building model, as reported by Carriço (2011), is characterized by the construction of towers of max. 40 floors on a base made up of basement floors and above-ground floors, which are the floors between the ground floor and the standard floors, whose insertion in the lot greatly affects the adjacent buildings, because the sideset-backs are usually reduced and the height can reach about 20 m, affecting the natural ventilation and light around, a model that produces a less energy efficient city, with more greenhouse gas production due to the high number of car spaces available.

According to a study by the Núcleo de Estudos da População (Nucleus of Population Studies) of UNICAMP (NEPO 2013), in the MRBS there is a greater volume of commuting between the city of Santos and the "dormitory cities", such as São Vicente, which represents 38,0% of the total working age population (WAP). This study shows that among all studied regions, in 2000 the MRBS presented the highest percentage of commuting WAP. These data confirm the role, especially of Santos, as the gravitational centre of this metropolitan region, directly or indirectly imposing a dynamic of territorial organization in the metropolitan region and guiding the urban and environmental agenda. During the last decade, this process became worse: in 2010, the region had 15.51% of the commuting WAP against 14.8% of all other regions. According to this study, in 2010 more than 200,000 people worked or studied in another municipality, regularly commuting every day in the week prior to the census, that is a growth of over 57,0% in 10 years. In addition, less populous municipalities presented the greatest variation in relation to total commuting, with increases exceeding 150,0% between 2000 and 2010 in cities such as Itanhém, Mongaguá, Peruíbe and Bertioga (NEPO 2013).

These data are related to the peripheral horizontal expansion dynamic presented above. Although Praia Grande was the municipality that received the most intraregional migrants in the last decade, it was the only municipality of the MRBS that reduced its proportion of commuters in relation to the WAP in 2000, with more than 35,000 commuters, so we can say that it happened due to the relative growth of the municipality's economic base, which helped reduce the commuting growth rate. Therefore, the expansion of the economic base in the municipalities located farther from the central area and the settling of low- and middle-income population in the central area of the region seem to be important strategies to reduce migration and commuting and, consequently, reduce their harmful effects on the environment and social reproduction.

In other areas with strong tourism, in high season (school holidays and holidays throughout the year), the MRBS faces challenges to accommodate the visiting population that, according to Silva (2013), corresponds to more than twice the resident population. It creates an exponential increase in demand for transportation infrastructure, sanitation, water availability and services, and consequent pressure on the natural environment.

From the economic point of view, the region's GDP is basically from Santos, Cubatão and Guarujá, although all municipalities showed growth in 2014 (Fig. 9.5). Santos' GDP accounted for 38% of the total GDP forthe region in 2014, a stable share since 2002 (max. 1%).

Regarding the GDP progress since 2002, Bertioga showed the highest growth (more than 4 times) from 2002 to 2010, followed by Praia Grande (2.8 times) and Mongaguá (2.4 times). Silva (2013) reported this scenario in his study on the Baixada Santista, and according to him, it is due to the increased tourism activities that are strong in these municipalities.

According to Marandola et al. (2013), the city growth is usually associated with an idea of positive development, but the environmental opposition to this idea is growing stronger, bringing to light the negative effects of uneven and unlimited growth. The fact that natural disasters are increasing and causing great human, material and economic losses, especially in urban centres, highlights the importance

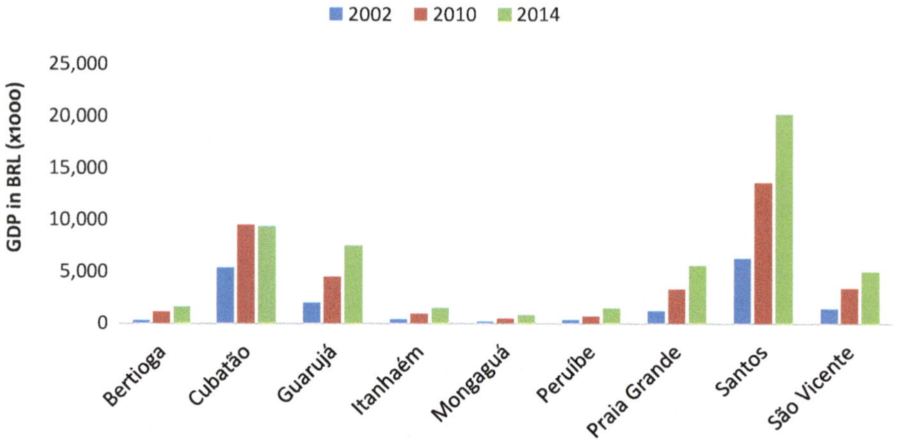

Fig. 9.5 GDP process (R$) (×1000), 2002–2014. (Source: SEADE, organized by the authors)

of discussing the challenges involved in city development, whether due to the growth and population density, or the way in which urban occupation has occurred (Marandola et al. 2013).

Changes of the urban area of the MRBS have brought benefits to the economy and infrastructure of the cities, although they have created a marginal population segment, and a clear sociospatial segregation pattern, with the low-income population placed in areas where access to goods, services and urban infrastructure is precarious, intensifying the social exclusion of these population segments (Silva and Nunes 2016). In this sense, it is necessary to understand that the natural susceptibility of the studied area, besides the sociospatial configuration of the territory, involve aggravated risks that, when combined with its characteristic fragmentation and marginalization, leading to increased vulnerability. Finally, the territorial organization model of the metropolitan region and the economic and urban development options reinforce the complex situation of the MRBS, with a significant increase in the atmospheric emissions of greenhouse gases, triggering physical and chemical changes in the atmosphere that result in low environmental quality and contribute to changes in the climate system.

9.3 Changes in Land Cover and Use and Atmospheric Emissions

According to the Greenhouse Gas Emission Calculation System, changes in land use have emitted 30.20 billion tons of carbon equivalent (t CO_2e), that is, 62.1% of the total national emissions between 1990 and 2013. These data reflect the dynamic and potential transformation of land cover and use in Brazil. However, this offer is still focal, especially in the northern region.

According to the Greenhouse Gas Emission Report issued in 2011 for the state of São Paulo, between 1994 and 2002, 45.4% of CO_2 emissions from changes in land use came from unmanaged forest converted into reservoirs, 28.6% from unmanaged forest that became urban area, and 16.1% of unmanaged forest that became agricultural area. From 2002 to 2005, the main cause of emission was the change from agricultural areas to urban areas (68.4%), while the conversion of unmanaged forest to agricultural area accounted for 30.7% of CO_2 emission in the period. Finally, between 2005 and 2008, 60.3% of the emission came from reforestation areas converted into urban areas; 17.3% of unmanaged forest that became agricultural area, and 10.6% of agricultural area that became urban area.

The results of the inventory indicate that land use category with the greatest CO_2 removal contribution came from the maintenance of Mantajada Forest, followed by the planted pasture converted into reforestation and unmanaged forest converted into managed forest. In this sense, environmental protection systems have been implemented in Brazil through the forest code and the National System of Conservation Units.

Satellite images from the thematic mapper (TM) sensor installed in Landsat-5 and Landsat-7 satellites, obtained with the Image Generation Division (DGI) of the National Institute of Space Research (INPE) were used in the analysis of the studied area. This sensor operates in seven bands, each band representing a range of the electromagnetic spectrum.

Reference images from 2005, 2011 to 2016 (Table 9.2) were selected, with a cloud cover factor equal to zero, or as close to zero as possible. The images from 2005 to 2011 were obtained from Landsat-5 and the image from 2016 was obtained from Landsat-7. Four scenes of the satellite images were required to represent the nine municipalities of Baixada Santista.

Image pre-processing and processing were performed in SPRING, version 5.4.3, with records in the WGS84 Geodetic Reference System of 20 reference points. Resampling, which consists of correction of image coordinates when they do not match their raw state, was performed with the allocation interpolator, commonly applied in the absence of heterogeneousgray values of the image (INPE 1997). Histogram equalization was the contrast enhancement technique used to obtain the maximum variance of an image histogram. An operation is performed to approximate the histogram of the original image to a uniform histogram, which calculates its accumulated histogram and uses it as an intensity function, thus producing an image of better contrast (Câmara et al. 1996).

Image post-processing (cutout image of the Baixada Santista Region and delimitations of the municipalities) was performed in QGIS, version 2.18, Las Palmas. The shapefile file with the delimitation of Baixada Santista municipalities was

Table 9.2 Acquisition dates of Baixada Santista images

Area	Scene	2005	2011	2016
MRBS	219/76, 219/77, 218/76 and 218/77	11/09	23/05	13/06

Source: Developed by the authors (2017)

obtained free of charge from the IBGE website. Image georeferencing was performed using the WGS 84 Geodetic Reference System and the Mercator UTM Universal Transverse Geographic Coordinate System.

The satellite images are not orthorectified and generate shading, especially in hills and mountain ranges, where the hillside presents such shades, giving the idea of depth. For the study, because of the 30-m spatial resolution, the following classes were used: water, beach, exposed soil, urban area, vegetation, and hillside, with vegetation and hillside areas included in the same group.

From the data presented in Table 9.3 and illustrated in Fig. 9.6, we can observe the existence of three groups of municipalities: (1) municipalities with land cover changes, urban area expansion, and suppressed vegetation: Guarujá, Itanhaém, Mongaguá, and São Vicente; (2) municipalities with a decelerated pace of land cover changes, decelerated urban conversion, increase in green areas: Cubatão and Peruíbe; and (3) municipality with stable land cover changes: Santos.

In theory, from land cover and use data, Praia Grande and Bertioga would fit the last group; however, demographic data point to the persistence of significant population growth. Although the population is not the determining factor, or the only one, it is important to point out that many people live in Praia Grande and work in Santos, São Vicente, Cubatão and in the MRSP. Praia Grande is a dormitory city for the low cost of living in relation to those more central municipalities, and for this reason it should have a more conservative interpretation of its stability in terms of land use changes.

As for Bertioga, the tourism activity should be highlighted. It is a municipality that receives tourists of higher economic profile, that is, different from those of Praia Grande, Itanhaém and Mongaguá. As Bertioga's population growth has been concentrated in the last decades, its separation from Santos is relatively recent, there is a potential high trend towards land cover and use changes due to the pressure from the real estate sector, for example, highlighting the fragmented and discontinuous pattern of this process.

Table 9.4 helps identify the municipalities that stood out in terms of demographic growth, ratifying the classification mentioned above.

Corroborating the analysis of the demographic dynamic, the growth vectors of the MRBS illustrated in Fig. 9.7 indicate a reorientation of this vector in the last three decades, especially in Bertioga.

With IDRISI Taiga software, it was possible to interpret and generate a trend for land cover and use change. The years 2011 and 2016 were used for the calculation, as they express the most current social, economic, political and demographic dynamic. Figure 9.8 shows reduced vegetation and urban area between 2011 and 2016. In this sense, it is important to interpret this result considering the three classes of municipalities mentioned above, in which this reduced urban area is observed in the municipalities of the third group: Bertioga, Praia Grande and Santos. On the other hand, vegetation losses are mainly in the first group: Guarujá, Itanhaém, Mongaguá and São Vicente.

With the LCM (Land Change Modeler) application, an IDRISI module (Geographic Information System), that evaluate and project land cover changes,

Table 9.3 Percentage of land use and cover (2005–2016)

	2005				2011				2016			
	Urban		Exposed	Vegetation	Urban		Exposed	Vegetation	Urban		Exposed	Vegetation
Municipalities	area	Beach	soil	Hillside	area	Beach	soil	Hillside	area	Beach	soil	Hillside
Bertioga	16.3	1.9	0.0	81.8	17.4	2.9	0.0	79.7	16.8	1.4	0.0	81.8
Cubatão	17.8	0.0	0.0	82.2	19.0	0.0	0.4	80.6	14.1	0.0	0.2	85.7
Guarujá	23.3	0.7	0.0	76.0	25.0	0.6	0.1	74.4	29.8	0.9	0.1	69.3
Itanhaém	4.1	0.2	0.0	95.8	6.5	0.1	0.0	93.3	7.0	0.2	0.0	92.8
Mongaguá	0.5	0.3	0.0	99.2	10.9	0.4	0.0	88.7	12.7	0.5	0.0	86.9
Peruíbe	17.0	1.2	0.0	81.8	17.6	1.0	0.1	81.3	19.1	1.2	0.1	79.5
Praia Grande	22.5	0.7	0.0	76.8	14.2	0.5	0.0	85.3	14.4	0.6	0.0	85.0
Santos	7.1	0.1	0.0	92.8	7.1	0.1	0.1	92.7	7.8	0.1	0.0	92.1
São Vicente	6.7	0.1	0.0	93.2	7.4	0.1	0.1	92.4	8.9	0.1	0.1	91.0

Source: Developed by the authors based on Landsat images (2005–2016)

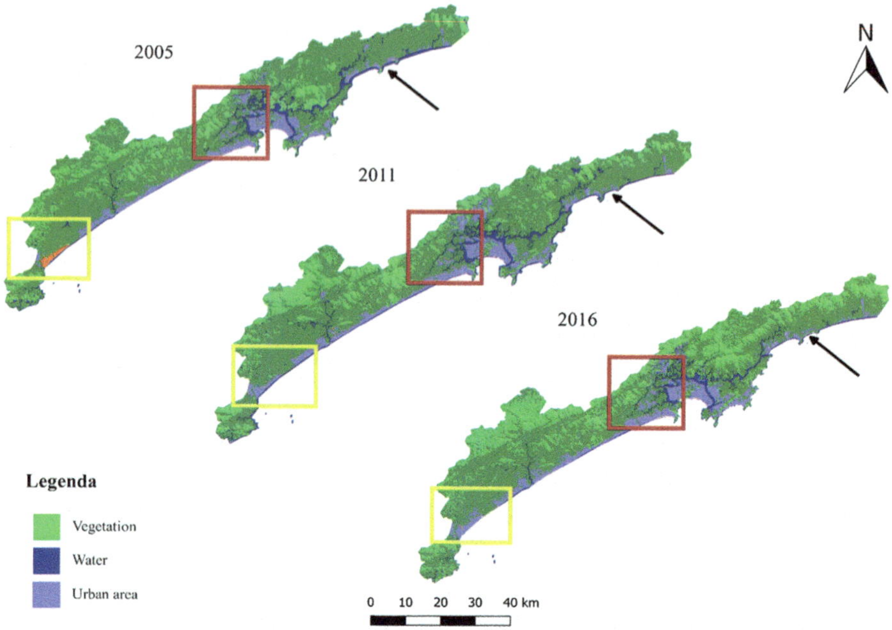

Fig. 9.6 Classification and use and cover in the MRBS (2005–2016). (Source: Developed by the authors based on Landsat images of 2005, 2011, and 2016)

calculating transition potentials, simulating future changes in the landscape, identifying trends and class transitions, predicting future scenarios, as well as planned land use interventions and impacts on biodiversity (Eastman 2009), was possible understanding the tendentialland cover changes to MRBS.

The results show the dynamic of vegetation suppression for other uses that may occur peripherally by 2021, reinforcing the historical pattern of MRBS territorial organization presented in the previous section, mainly driven by the still intense peripheral urban expansion, started from Santos.

By 2021 there is a 98.0% probability of vegetation conversion into urban area, notably in the periphery, while the probability of urban area converted into vegetation is only 0.11%, foreseen for the central area of the MRBS.

Thus, in addition to the removal of CO_2 sinks, increased emissions from land use changes, the possibility of continuous increase of motorized commuting in the region will significantly increase these greenhouse gas emissions. However, it should be noted that changes can be observed in this scenario, mainly due to the creation of more effective public policies that address climate change, especially in urban and regional planning. Otherwise, an unsustainable pattern of marginalized and exclusionary urban development will be reinforced, exposing the metropolitan periphery to greater risks of all kinds, including climate risks.

Table 9.4 Demographic progress of Baixada Santista by municipalities

Period/municipality	Bertioga	Cubatão	Guarujá	Itanhaém	Mongaguá	Peruíbe	Praia Grande	Santos	São Vicente
1980–1990	–	13%	26%	37%	45%	41%	44%	3%	26%
1990–2000	–	17%	23%	39%	49%	39%	40%	-2%	14%
2000–2010	37%	9%	9%	18%	24%	14%	26%	0%	9%
2010–2015	15%	4%	4%	6%	9%	5%	10%	1%	4%

Source: Fundação SEADE, organized by the authors

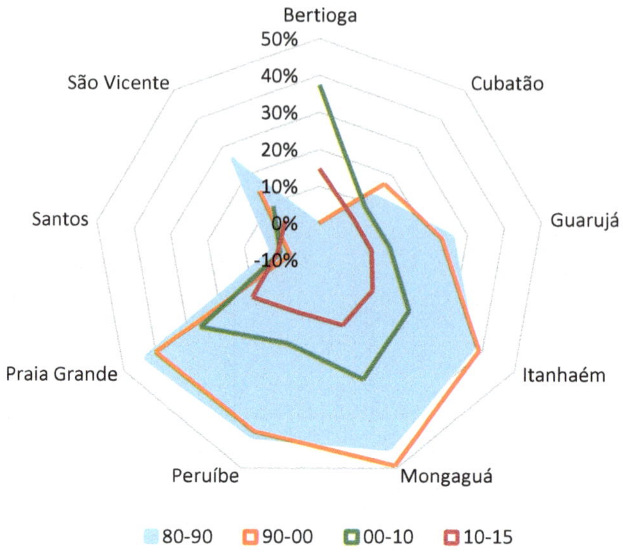

Fig. 9.7 Growth vectors in the municipalities of the MRBS (1980–2015). (Source: SEADE, organized by the authors)

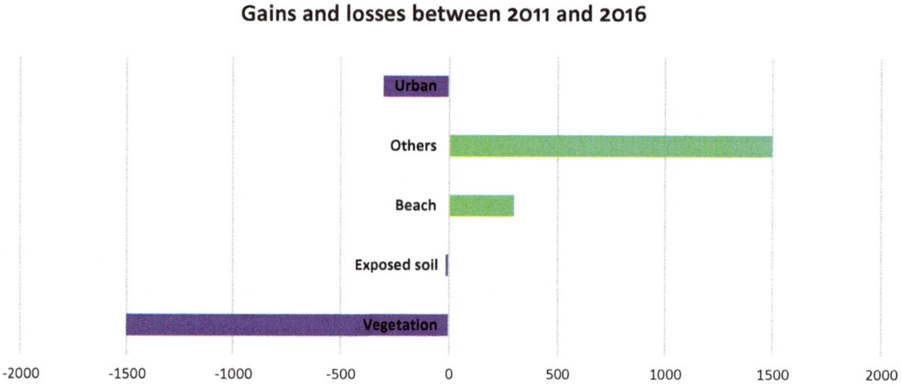

Fig. 9.8 Progress of land cover in Baixada Santista between 2011 and 2016. (Source: Developed by the authors (2017))

9.4 Final Considerations

The results of this study provide general information about the dynamic of the municipality of Santos and its metropolitan region.

The MRBS territorial organization was presented, including its urban development and changes in land use and cover. This study attempted to demonstrate that the deleterious environmental impacts of the peripheral horizontal expansion can be explained by the migration process and the development model of the municipalities that are far from the central region, based on second home tourism. It is a result of allotments, with strong unbalance in the urban quality of the areas and higher recurrence of households for occasional use, which present better infrastructure and public facilities, contrasting with the greater urban deficiencies observed in areas predominantly occupied by the resident population. The real estate dynamic in the central areas is also highlighted, which prevents population settling in locations of better infrastructure and contributes to spreading the urban area. With no interference and reorientation of urban and regional planning policies focused on land use planning, this pattern will intensify in 2021, reflecting greater potential for greenhouse gas emissions, by removing natural sinks, converting land use, or due to intensified commuting that is mainly through automotive vehicles.

Cubatão, despite its still intense industrial activity, is an excellent example for the MRBS, since it significantly improved its poor environmental quality indexes and its green area. However, environmental depreciation is observed in its peripheral area, where the land price is cheaper, attracting poor families, while in the central areas, such as Santos, the environmental changes and quality indexes are better.

From the results of this uneven development or the sociospatial fragmentation of the metropolitan fabric, the unequal distribution of environmental (but also social or technological) risks is discussed. In addition, considering the panorama of environmental risks related to climate change, other issues are highlighted: the adaptive and mitigation ability either by the direct effect on risk intensification or consequences of uneven development.

In this context, it is necessary to consider the metropolitan management model and the distinct abilities of metropolitan entities to understand urban environmental problems. In the MRBS context, these central municipalities directly or indirectly impose an urban logic and metropolitan planning to peripheral areas that is closely related to the conversion of land cover, urban mobility, greenhouse gas emission, susceptibility of the environment, vulnerability of social organizations and groups, and finally exposure to risks.

Also, the urban model of metropolitan centres based on verticalization should be noted, as it has negative consequences in terms of energy efficiency and greenhouse gas production. A more compact and sustainable development model is required, using proper architectural typologies, accessible to low-income families, combined with the expansion of the economic base in the dormitory areas as a strategy to reduce commuting.

Acknowledgement The authors would like to thank to the translation provided by the Espaço da Escrita, UNICAMP and Lucí Hidalgo Nunes by the review and enriching contributions.

References

Camara, G., Souza, R. C. M., Freitas, U. M., & Garrido, J. (1996). Spring: Integrating remote sensing and GIS by object-oriented data modelling. *Computers & Graphics, 20*, 395–403.

Carmo, R. L., & Silva, C. A. M. (2009). População e mudanças climáticas no contexto litorâneo: uma análise da Região Metropolitana da Baixada Santista. *Revista VeraCidade, 4*, 1–15.

Carmo R L, Marques C, & Miranda Z I (2012) *Dinâmica demográfica, economia e ambiente na zona costeira de São Paulo*. Textos NEPO 63. Available via http://www.nepo.unicamp.br/publicacoes/scripts/textos_nepo_63.php. Accessed 14 June 2018.

Carriço, J. M. (2006). *Baixada Santista: transformações produtivas e sócio-espaciais na crise do capitalismo após a década de 1980*. Thesis, Universityof São Paulo.

Carriço, J. M. (2011). Produção do espaço urbano voltada às elites: doze anos de aplicação da Lei de Ordenamento do Uso e Ocupação do Solo na área insular de Santos. In D. A. Vazquez (Ed.), *A questão urbana na Baixada Santista: Políticas, vulnerabilidades e desafios para o desenvolvimento* (pp. 145–176). Santos: Editora Universitária Leopoldianum.

Carriço J M, Barros MF (2015) *Crise de mobilidade urbana em Santos/SP: produção imobiliária, segregação socioespacial e desenraizamento*. Paper presented at 16th Encontro Nacional da Associação Nacional de Pós-graduação e Pesquisa em Planejamento Urbano e Regional, ANPUR, Belo Horizonte, 2015.

CEMBS. (1971). *Study commission on the metropolization of the Baixada Santista – RelatórioGeral*. Santos: Prefeitura Municipal de Santos.

CETESB (São Paulo). (2011) *Environmental Company of the State of São Paulo*. Inventario de emissões antrópicas de gases de efeito estufa diretos e indiretos do Estado de São Paulo: comunicação estadual/CETESB. Alves J. W. Set al (ed), 2.ed. CETESB, São Paulo. Available via: http://www.cetesb.sp.gov.br. Accessed 30 May 2016.

Coelho, M. C. N. (2009). Impactos ambientais em áreas urbanas: teorias, conceitos e métodos de pesquisa. In A. J. T. Guerra & S. B. Cunha (Eds.), *Impactos ambientais urbanos no Brasil* (5th ed., pp. 19–46). Rio de Janeiro: Bertrand Brasil.

Crutzen, P. J. (2006). The anthropocene. In E. Ehlers & T. Krafft (Eds.), *Earth system science in the anthropocene* (pp. 13–18). Heidelberg: Springer.

Dallabrida, V. R., & Becker, D. F. (2003). Governança Territorial Um primeiro passo na construção de uma proposta teórico-metodológica. *Desenvolvimento em Questão*, 73–98.

De Rose, A., & Testa, M. R. (2013). *Climate change and reproductive intentions in Europe*. Austria: Vienna Institute of Demography.

Dearing, J. A., Battarbee, R. W., Dikau, R., Larocque, I., & Oldfield, F. (2006). Human–environment interactions: Learning from the past. *Reginal Environmental Change, 6*, 1–16.

Eastman, J. R. (2009). *IDRISI Taiga: Guide to GIS and image processing*. Worcester: ClarkLabs. Available via: http://web.pdx.edu/~nauna/resources/TaigaManual.pdf. Accessed 19 May 2016.

Faria, V. (1978). *O processo de urbanização no Brasil: algumas notas para seu estudo e interpretação*. Paper presemted at Encontro Nacional de Estudos Populacionais, Campos do Jordão, SP, 1978.

Ferreira, A., Rua, J., & Mattos, R. C. (2014). Metropolização do espaço, e gestão territorial. *GEOUERJ, 4*, 277–504.

Hogan, D. J. (1990). *Quem paga o preço da poluição? Uma análise de residentes e migrantes pendulares em Cubatão*. Paper presented at 7th Encontro Nacional da ABEP, Caxambu, 1990.

INPE – National Institute of Space Research. (1997). *Manual do Spring: Noções de Geoprocessamento*. São José dos Campos: National Institute of Space Research.

INSTITUTO PÓLIS. (2013). *Diagnóstico Urbano Socioambiental Regional Preliminar. Litoral Sustentável*. São Paulo: Instituto Pólis. Available via: http://litoralsustentavel.org.br/diagnosticos/diagnostico-urbano-socioambiental-regional/. Accessed 25 May 2017.

IPCC-Intergovernmental Panel on Climate Change. (2007). Climate change 2007: Synthesis report. Reino unido: Cambridge University Press, 2007. Available at: https://www.ipcc.ch/publications_and_data/publications_ipcc_fourth_assessment_report_synthesis_report.htm Accessed 7 Jan 2017.

IPCC-Intergovernmental Panel on Climate Change. (2014). Climate change 2014-synthesis report. A contribution of working groups I, II, and III to the fifth Assessment Report (AR5) of the intergovernmental panel on climate change. Watson, R.T. and the Core Writing Team (Eds.). Cambridge: Cambridge University Press: 398 pp.

Jakob, A. A. E. (2003). *Análise sócio-demográfica da constituição do espaço urbano da Região Metropolitana da Baixada Santista no período 1960–2000*. Thesis, University of Campinas.

Marandola, J. R. E., Marques, C., De Paula, L. T., & Cassaneli, L. B. (2013). Crescimento urbano e áreas de risco no litoral norte de São Paulo. *Revista Brasileira de Estudos de População, 30*, 35–56.

METOFFICE – UK Weather Service. (2016). *Global Weather extremes and climate changes*. Boletim 16/0844. Available via: http://www.metoffice.gov.uk/binaries/content/assets/mohippo/pdf/climate/cop22/theme_3-weather-extremes.pdf. Accessed 23 June 2017.

NEPO – NucleusofPopulationStudies. (2013). *O fenômeno da mobilidade pendular na macrometrópole do Estado de São Paulo: uma visão a partir das quatro regiões metropolitanas oficiais*. São Paulo: EMPLASA.

Nobre, C. A., Young, A. F., Orsini, J. A. M., Salvida, P. H. N., Nobre, J. A., Ogura, A. T., Thomaz, O., Valverde, M., Párraga, G. O. O., Silva, G. C. M., Silveira, A. C., & Rodrigues, G. O. (2011). Vulnerabilidade das Megacidades brasileiras às mudanças Climáticas: Região metropolitana de São Paulo. In R. S. Motta, J. Hargrave, G. Luedemann, & M. B. S. Gutierrez (Eds.), *Mudança do Clima no Brasil: Aspectos econômicos, sociais e regulatórios* (pp. 233–257). Brasília: IPEA.

Petit, J. R., Jouzel, J., Raynaud, D., Barkov, N. I., Barnola, N. I., Basile, I., Bender, M., Chappellaz, J., Davis, M., Delaygue, G., Delmotte, M., Kotlayakov, V. M., Legrand, M., Lipenkov, V. Y., Lorius, C., Pépin, C., Saltzman, E., & Stievenard, M. (1999). Climate and atmospheric history of the past 420,000 years from the Vostok ice core, Antarctica. *Nature, 399*, 429–436.

Rifino, M. B. C., Klintowitz, D. C., Menegon, N. M., Uemura, M. M., Ferreira, A. C., Frignani, C., & Barreto, F. (2015). A produção do Programa PMCMV na Baixada Santista: habitação de interesse social ou negócioimobiliário? In C. S. Amore, L. Z. Suimbo, & M. B. C. Rifino (Eds.), *Minha casa... e a cidade?avaliação do programa minha casa minha vida em seis estados brasileiros*. Rio de Janeiro: Letra Capital.

Silva, R. B. (2013). *Urbanização e vulnerabilidade na Região Metropolitana da Baixada Santista, SP: um olhar geográfico das desigualdades intraurbanas em bairros de Santos, Praia Grande e Peruíbe*. Thesis, University of Campinas.

Silva, R., & Nunes, L. H. (2016). *Vulnerability profiles in a metropolitan area of Brazil*. LAP Lambert AcademicPublishing.

Zundt, C. (2006). Baixada Santista: uso, expansão e ocupação do solo, estruturação de rede urbana regional e metropolização. In J. M. P. Cunha (Ed.), *Novas Metrópoles Paulistas: População, vulnerabilidade e segregação*. Campinas: Núcleo de Estudos de População – NEPO. Unicamp.

Chapter 10
Emerging Impacts of Climate Change on Human-Health in Santos Municipality in the Context of São Paulo State

Luiz E. O. C. Aragão and Fernanda R. F. Carvalho

Abstract In Brazil, around 12.5 million cases of Dengue fever were recorded from 1990 to 2016. The Southeast is the most affected region with just over 6 million cases registered during this period with linear increase of 47,000 cases per year. In this chapter we present an overview of environmental risk management, associating the concepts with case studies of how Dengue fever outbreaks respond to climate variability. We also bring a discussion on how some factors can influence future patterns of Dengue fever in São Paulo State. Dengue fever incidence clearly responds to temperature increases that are predicted to increase in the future. Without any prophylactic measures public treatment costs are likely to increase in the future. These costs are likely to be much higher if we consider the other diseases associated with the same vector such as Zikka and Chicungunha, plus emergency actions to mitigate the problem during outbreaks. Investments must be allocated for prevention focusing on long-term monitoring of temperature and El Niño data. Authorities must be alerted when mean temperature surpasses 24 °C, because of the measured increase of dengue cases during the El Niño years. Finally, prophylactic measures must be intensified between March and May, when the disease peaks in the region.

Keywords Dengue · Climate change · Coastal cities · Risk reduction

10.1 Introduction

Dengue fever is one of the most important threats to human-health in the tropics. Over 2.5 billion people are at risk from Dengue fever, with an estimate between 50 and 100 million infections worldwide every year (Pessanha 2012). Dengue virus has been quickly spread across tropical countries since the 1970s (Fig. 10.1), as a

L. E. O. C. Aragão (✉) · F. R. F. Carvalho
Remote Sensing Division, National Institute for Space Research (INPE),
São José dos Campos, SP, Brazil
e-mail: luiz.aragao@inpe.br

L. H. Nunes et al. (eds.), *Climate Change in Santos Brazil: Projections, Impacts and Adaptation Options*, https://doi.org/10.1007/978-3-319-96535-2_10

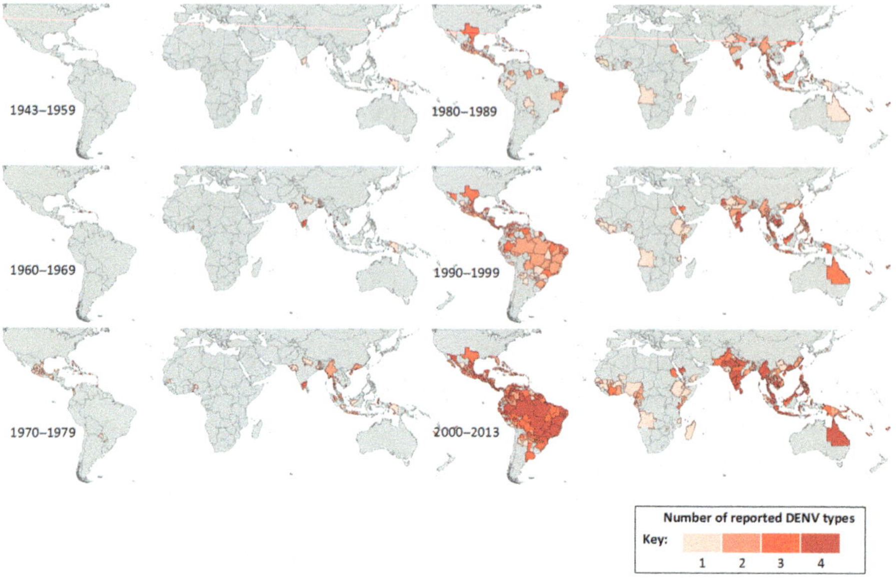

Fig. 10.1 Global distribution of cumulative number of Dengue fever cases since 1943. Red colors indicates viruses types ranging from 1 to 4 for cases confirmed in the given decade. (Figure from Messina et al. 2014)

consequence of global urbanization and intensification of international travel (Messina et al. 2014).

In Brazil, around 12.5 million cases were recorded from 1990 to 2016 (Fig. 10.2). The Southeast is the most affected region with just over six million cases registered during this period, followed by the Northeast, Centre-east, North and South regions. Global trends observed by Messina et al. (2014) are corroborated by data from the Brazilian Ministry of Health, which show an average linear increase of 47,000 cases per year in Brazil.

The control of Dengue fever vector, the *Aedes aegypti* mosquito, is largely done by education of citizens aiming for larvae habitat elimination and reduction (Degallier et al. 2000) and space spraying (Horstick et al. 2010). However, mitigation of the epidemic process has been shown to be a function of herd immunity and depletion of susceptible individuals rather than efficient vector control (Medronho 2006). So, control requires integrated policies between various sectors of society to improve the effectiveness of vector control and to develop an efficient vaccine to immunize the population against Dengue fever serotypes. The first step to alleviate this problem, therefore, is to implement an active disease surveillance system as proposed by the World Health Organization Global Strategy for Prevention and Control of Dengue Fever and Dengue Hemorrhagic Fever (WHO 1995).

Despite causes of spread of Dengue virus has been attributed to urbanization and international travels (Messina et al. 2014), mathematical model results indicate that climate change is also likely to affect the Dengue fever viruses and vector

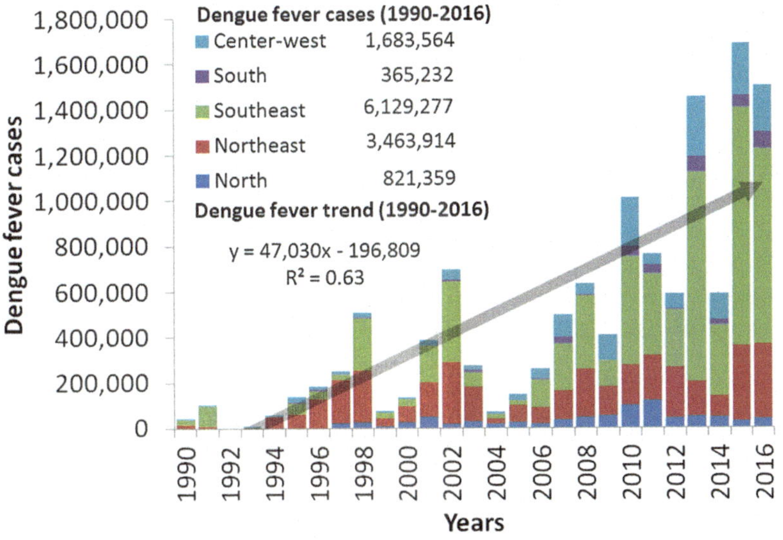

Fig. 10.2 Dengue fever cases in Brazil from 1990 to 2016. The linear trend shows a mean increase of 47,030 cases per year during this period. (Data are from the Brazilian Ministry of Health – SISNAN program. Available at: http://portalarquivos.saude.gov.br/images/pdf/2017/fevereiro/10/Dengue-classica-ate-2016.pdf)

populations (Morin et al. 2013). Both virus and vector are sensitive to climate changes. Rainfall directly influences habitat availability for the Dengue fever mosquito *Aedes aegypti* (Morin et al. 2013) and temperature influences vector development rates, mortality, and behaviour, as well as controls viral replication within the mosquito (Hales et al. 2002). Climate change combined with a suite of physical, biological, economic, social, cultural and psychological factors should be considered as risk factors for Dengue fever outbreaks.

A recent study in the Brazilian Amazon (Aragão et al. 2016) depicted a coherent seasonal pattern in Dengue fever cases, showing that the majority of cases occurred in the wet season. This result follows similar pattern in other tropical countries (Pham et al. 2011). The exacerbation of Dengue fever cases in the wet season is expected as rainfall is the main variable that influences habitat availability for the Dengue fever mosquito *Aedes aegypti* (Morin et al. 2013). Other studies have demonstrated the impact of temperature upon Dengue fever cases in Brazil. Horta et al. (2014) showed that a rise in minimal temperature causes an increase of 37% in the risk of Dengue fever. Moreover, Aragão et al. (2016) found a significant covariation between trends in minimal temperature and Dengue fever occurrence. By analyzing the outputs of global climate models from the Intergovernmental Panel on Climate Change CMIP5 dataset, Aragão et al. (2016) concluded that the combination of rainfall variability and temperature indicates that if wet season rainfall is able to provide enough habitat area for the development of mosquito larvae, the increased trend in temperature tend to favor *Aedes aegypti* and Dengue fever virus development, escalating the proportion of vulnerable populations. Therefore, positive trends in temperature, as a consequence of climate change, are likely to lead to an increase of Dengue fever incidence in Brazil.

In this chapter, therefore, we aim to present an overview of environmental risk management, associating the concepts with case studies of how Dengue fever outbreaks respond to climate variability. We also discuss the main factors that can influence future patterns of Dengue fever in São Paulo State and the macro region of Baixada Santista and the City of Santos. Study cases are focused on the period from 2008 to 2015 to show how these patterns are associated to weather fluctuations. Based on the examples depicted in the chapter we propose recommendations for minimizing the risk of this disease in the future.

10.2 A Perspective on Environmental Risk Management

Integration of the scientific knowledge with social, economic and governance processes has, in the last decade, been recognized as an important step towards a sustainable society (Costanza 2003). The Sendai Framework for Disaster Risk Reduction 2015–2030 (UNISDR 2015), explicitly recognizes the need of the risk-informed decision making based on the open exchange of data and information. Moreover, both the United Nations, under the Hyogo Framework for Action 2005–2015 (UNISDR 2007) and the Sendai Framework for Disaster Risk Reduction (UNISDR 2015), recognized the need for an agreement that enforces and shares responsibility to reduce the risk of natural disasters by states, local governments, private sector and other stakeholders. Scientists across the globe are developing a greater understanding of the underpinning causes and associated impacts of climate change in tropical countries. A major emergent challenge in this context is to encourage the continuous scientific development with close engagement with the national and international stakeholders, policymakers and local communities. The consolidated dialog among scientists, civil society and decision makers aiming to develop robust mitigation and adaptation options for the consequences of extreme climate fluctuations is not trivial (Aragão et al. 2016). One critical constraint for policymakers to expand regional perceptive for developing mitigation and adaptation policies and cope with present and future impacts of climate changes is the complexity of climate and environmental datasets (Marengo et al. 2016).

For allowing the development of efficient strategies for reducing risk and enhancing adaptation capacity of cities, there is a need for putting in place a new paradigm where solutions for the challenges imposed by global environmental changes are designed and conducted by a partnership between scientists and society. This partnership is critical for producing essential information for societal transformations towards sustainability, in line with international demands (Mauser et al. 2013; Meadow et al. 2015; Weaver et al. 2014). One pivotal actor in the conception and establishment of this new paradigm has been the Future Earth program that is supported by the 'Science and Technology Alliance for Global Sustainability' (Mauser et al. 2013). The Sendai Framework (UNISDR 2015) is also a keystone on setting the boundaries for managing climate risk. Four priorities are clearly identified in this framework: (i) Understanding disaster risk; (ii) strengthening disaster risk governance to manage disaster risk; (iii) investing in disaster risk reduction for resilience; and (iv) enhancing disaster preparedness for effective response and to "Build Back Better" in recovery, rehabilitation and reconstruction (UNISDR 2015).

With the understanding of science and policy requirements Aragão et al. (2016) proposed a conceptual overarching framework for delivering the information from scientists to policymakers in order to make effective decisions for increasing resilience. This framework is based on six steps that align scientific information with policy needs (Fig. 10.3). The understanding of past and future climate patterns is mostly connected to academic research (Fig. 10.3a). However, the realization of the consequences is also perceived by the society (Fig. 10.3b), which is a key partner to elaborate possible solutions. Therefore, establishing the patterns and possible consequences of environmental impacts on human health and ecosystems is the primordial step to identify stakeholder's needs and its role in the co-design and co-production process (Fig. 10.3c) of an analytical and diagnosing framework (Mauser et al. 2013). The result of this synergism is a comprehensive routine for monitoring, analysing, understanding and propose pathways for building institutional resilience (Fig. 10.3d). Building capacity within the institutions to analyse results provided is required for allowing decision makers to plan ahead (Fig. 10.3e). It is also essential to quantify and interpret impacts (Fig. 10.3f) that is aimed to ultimately decrease vulnerability and increase the efficacy of decisions made.

However, for efficiently apply this concept for reducing vulnerability and consequently the risk of catastrophic Dengue fever epidemics, actors involved in knowledge production and decision making must clearly understand the processes that

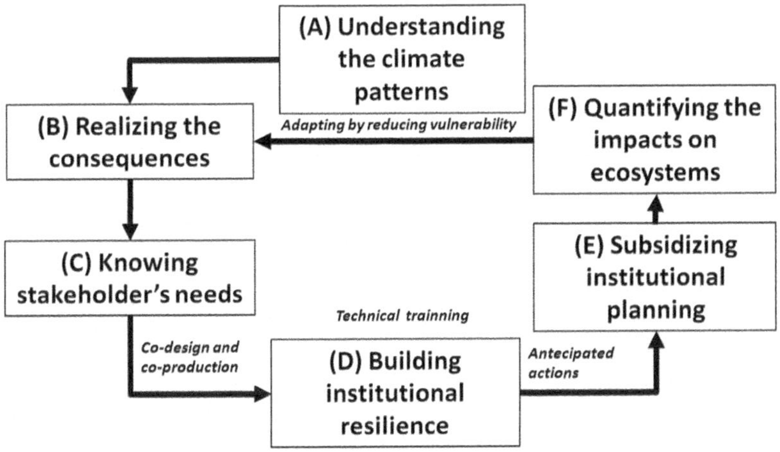

Fig. 10.3 Architecture of a conceptual model to promote coherent knowledge exchange between science and policy for increasing institutional resilience and reducing vulnerability of cities, states and countries to climate and environmental change impacts. (Aragão et al. 2016)

lead to the materialization of the risk. Cardona et al. (2012) understand that the risk, which is the possibility of adverse effects in the future, is a consequence of the interaction of social and environmental processes. This interaction emerges from the combination of hazards and vulnerabilities of exposed groups (Cardona et al. 2012). Risk is, hence, defined by three main pillars: (1) hazard, which refers to the envisaged occurrence of natural or human-induced events (e.g. climate extremes, diseases, landslides, floods, fires) with adverse effect on the society or environment; (2) exposure, which is related to target elements (e.g. populations or infrastructure) that exist in areas that may be hit by an specific hazard; and (3) vulnerability, which refers to the predisposition of target elements to experience the effects when impacted by a hazard (Fig. 10.4). Vulnerability is a function of several components (Burkett et al. 2014). Exposure and sensitivity of the target element to a hazard are two critical components that define the potential impact. Another key component is the adaptation capacity that acts in an opposite direction, minimizing the impact. The combination of all these will define the vulnerability of a society or environment (Fig. 10.4).

Following these concepts, in the next sections of the chapter we provide a detailed assessment of Dengue fever, a vector borne disease in which cases have recently escalated in Brazil. We focus on our analysis at State and Municipal levels to elaborate recommendations for informing decision makers about potential hazards and their feedbacks as well as vulnerabilities of exposed populations.

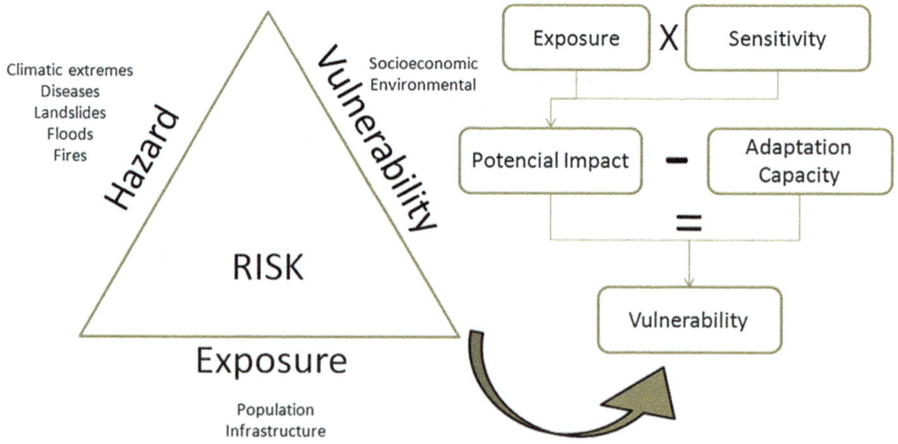

Fig. 10.4 Concept of risk and their multiple components based on definitions from Cardona et al. (2012)

10.3 Climate Patterns and Dengue Fever Outbreaks

10.3.1 Study Region and Datasets

The analysis presented in this chapter contemplates the State of São Paulo, the region of Baixada Santista and the city of Santos (Fig. 10.5). This study region is of upmost importance as São Paulo state contributes 32.2% of Brazil's gross domestic product and hosts the largest population in the country (45 million inhabitants) (IBGE 2017). São Paulo is also inserted in the Southeast region of Brazil, which is the most affected by Dengue outbreaks since the 1990s (Fig. 10.2). The coastal zone of the Baixada Santista, moreover, has a population of around 1.8 million people exposed to several potential hazards including sea level rising, vector-borne diseases, landslides and floodings. This is critical not only because the presence of vulnerable populations, but also because of the massive infrastructure present at the city of Santos for supporting the largest port in Brazil (ANTAQ 2017).

For analysing the relationship between climate and Dengue fever incidence we focused on a new database containing spatio-temporal information on human health, climate model outputs, sea level rise data, developed by CEPAL-Universidad de Cantabria (Spain) for sea level rise for the Americas, population density and topography (Table 10.1). Human health and demographic data are publicly available from the Brazilian Health System (SUS) and the Brazilian Institute for Geography and Statistics (IBGE), respectively.

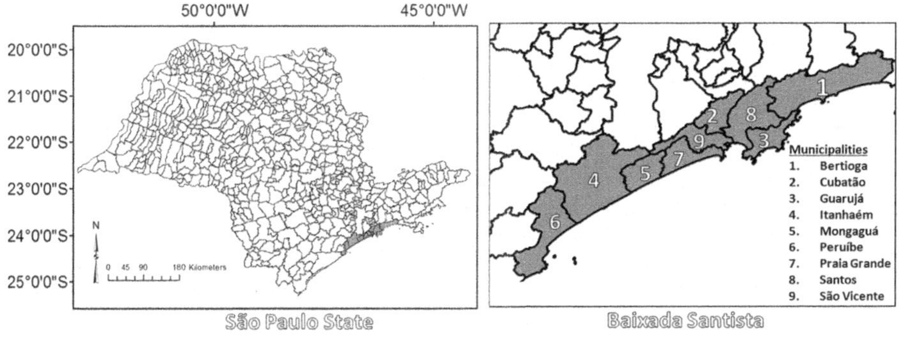

Fig. 10.5 Location of the study region. On the left, a map of São Paulo State with the limits of all Municipalities. Grey shade represent the Municipalities from the Baixada Santista region that can be seen in detail on the right panel

Table 10.1 Health, environmental and demographic variables used in the analysis

Dataset	Variables	Coverage	Spatial resolution	Source
Human-health	Dengue fever	São Paulo State	Municipality	DATASUS (http://www2.datasus.gov.br/DATASUS/index.php?area=02)
Topography	Elevation and slope	São Paulo State	30 × 30 m	http://www.dsr.inpe.br/topodata/acesso.php
Sea level rise and impacts	Coastal dynamics, storm surge, coastal flooding	Brazilian coast	Point-based data	CEPAL-Universidad de Cantabria
Climate observations	temperature and precipitation anomalies	National level maps of, at	1 × 1° spatial resolution data from 1929 to 2012	Climatic Research Unit (CRU TS 3.21)
Climate projections from global climate models	Decadal average maps for temperature, precipitation anomalies	National level maps of, at	1 × 1° spatial resolution data from the 2010s to the 2090s in relation to the 1961–1990 baseline period	CMIP5 archive from IPCC 5th Assessment

10.3.2 Temporal Patterns of Dengue Fever

The data on the cases of hospitalization for Dengue fever in the State of São Paulo show a striking feature during the year 2015. There was a massive increase in the number of cases in relation to previous years. To understand these notable changes in the Dengue fever pattern we considered the mean observed temperature data and total accumulated precipitation for the State of São Paulo, from 2008 to 2015. Moreover, we also considered data on the El Niño Southern Oscillation or ENSO, which is a phenomenon of atmosphere-ocean interaction, associated with changes in the normal patterns of sea surface temperature (SST), changing significantly tropical temperature and precipitation pattern. To evaluate the distribution of cases of hospitalization for Dengue or DHF, we normalized this data by population data from the 2010 Census and Population estimates of the State of São Paulo, released by IBGE. So, the incidence rate of Dengue fever was calculated following the Eq. 10.1 below:

$$\text{incidence} = \frac{\text{number of new cases}}{\text{total of population}} \times 100.000 \qquad (10.1)$$

The results show that during the period analysed (2008–2015), the State of Sao Paulo recorded 36,134 hospitalizations for Dengue fever, being the year of 2015 the most representative, encompassing 40.5% of those cases. The monthly distribution of Dengue fever cases in the State followed the climate seasonality. Our observations showed increase in the number of cases of Dengue fever from January to May, corresponding to the rainy season, and decrease of the cases between June and November. It was possible to demonstrate a close relationship between high temperatures and high rainfall with the significant increase in cases of Dengue fever.

Interestingly, the occurrence of the El Niño phenomenon seems to be aligned with the incidence of Dengue fever in 2010 and 2015. This association is clearer for 2015, in which the cases have reached worrying levels, with nearly 4000 confirmed cases in the State, only in the rainy months (Fig. 10.6).

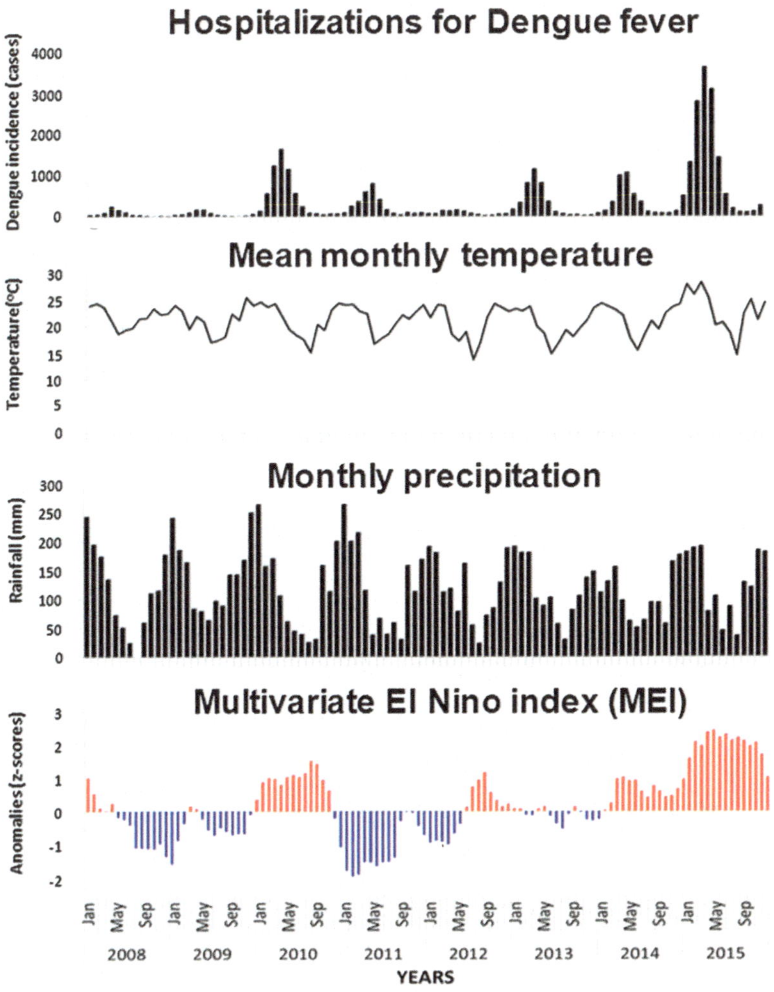

Fig. 10.6 Temporal patterns of Dengue incidence for the State of São Paulo, in comparison to mean monthly temperature (°C), rainfall (mm month^{-1}) and ENSO anomalies

To complement this analysis we evaluated the anomaly of the historical series for the cases of hospitalization data for dengue fever, temperature and precipitation in the State of São Paulo in Brazil. The anomaly was calculated from the relationship between data average subtracted by total value by the standard deviation, as the equation:

$$\text{Anomaly} = \frac{\text{number of cases} - \text{monthly cases average}}{\text{standart deviation of monthly cases}}$$

This next analysis confirmed a larger number of Dengue cases in 2010 and 2015, overcoming the long-term mean value of dengue outbreaks (Fig. 10.7). However, the only clear climate pattern is related to temperature anomalies. Rainfall anomalies do not present a clear pattern during the years of maximum Dengue fever outbreaks. This leads to the hypothesis that Dengue fever outbreaks are associated with

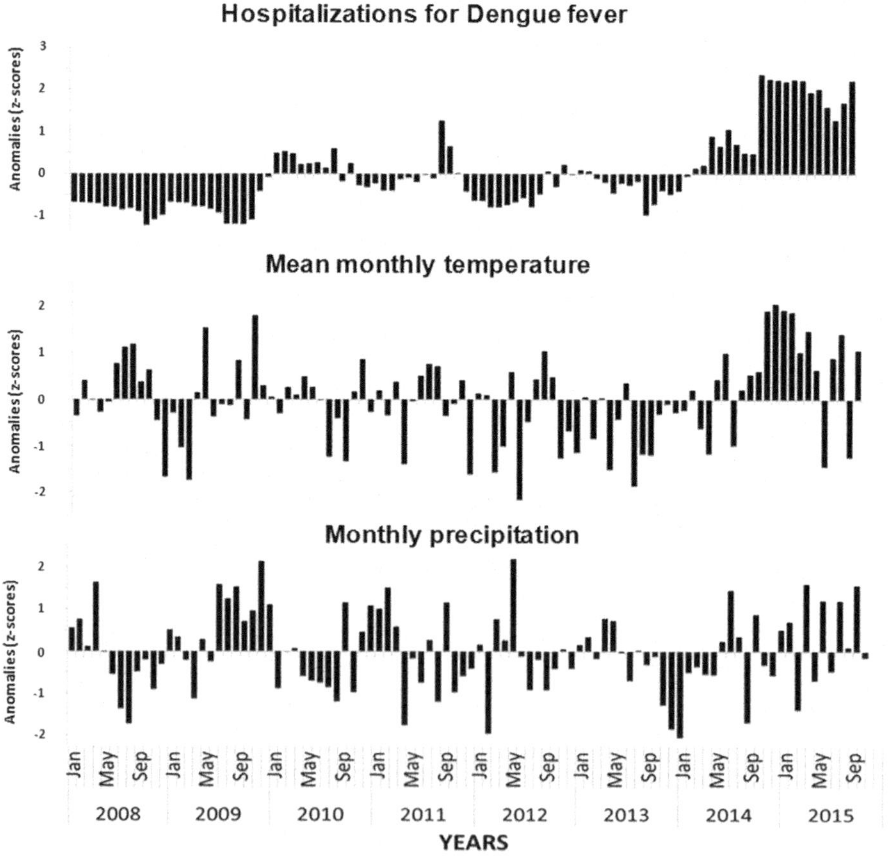

Fig. 10.7 Anomaly analysis for cases of Dengue fever hospitalization and monthly temperature and precipitation anomalies for the State of São Paulo

hotter climate, as rainfall background is always high during summer time in São Paulo state.

To test the hypothesis we perform a regression analysis between cases of hospitalization for dengue and average temperature (a), between the cases of hospitalization for dengue and precipitation (b) and the cases of hospitalization by dengue fever and the Multivariate Enso Index-MEI- (c), calculated for the four quarters of the year 2015 (Fig. 10.8).

The data indicate a significant correlation between the hot months and the highest incidence of cases of hospitalization for dengue, especially when there is occurrence of the El Niño phenomenon, which increases summer temperatures.

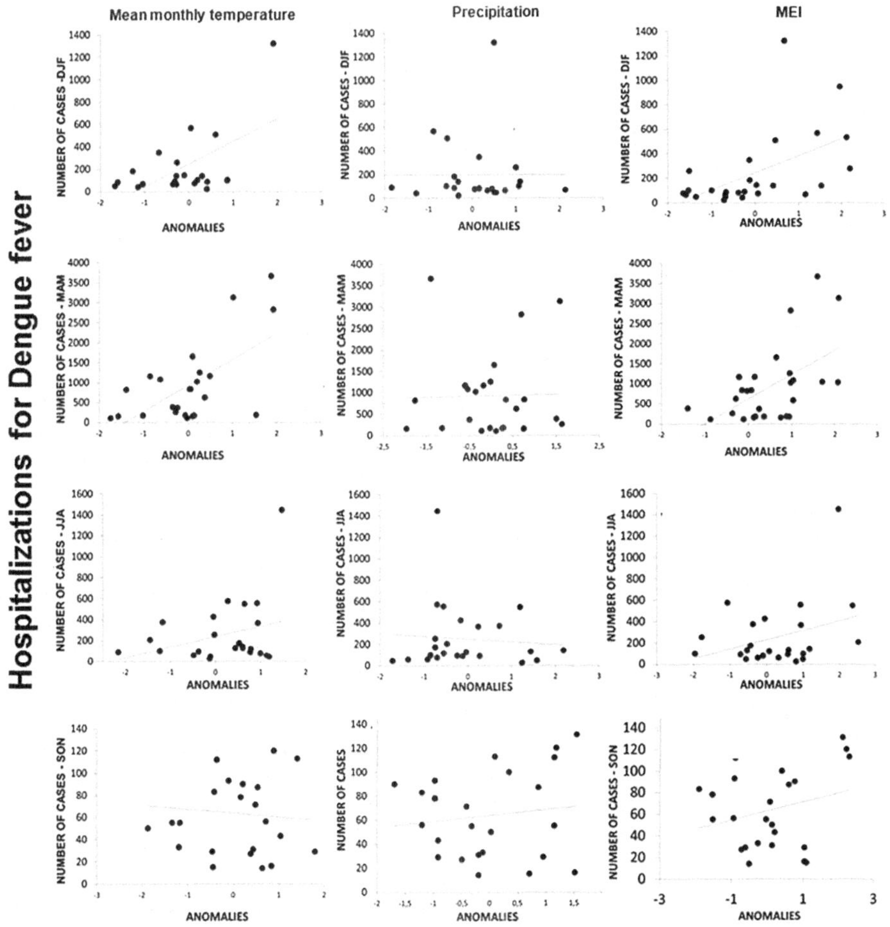

Fig. 10.8 Regression analysis between Dengue fever cases and mean temperature (first column), Dengue fever cases and precipitation (second column) and Dengue fever cases and MEI (Last column). Each line corresponds to the cases aggregated by trimesters. The first line corresponds to December-January-February (DJF) and the bottom line to September-October-November (SON)

Table 10.2 Total annual hospitalizations for diseases from 2008 to 2015 in the Baixada Santista

	2008	2009	2010	2011	2012	2013	2014	2015
Cólera	2	4	0	0	2	1	2	1
Dengue	66	52	2022	68	94	375	71	481
Diarreia	150	192	175	165	158	107	177	80
Leptospirose	38	28	52	34	27	44	18	40

Surprisingly, no relationship was found between Dengue fever cases and rainfall (Fig. 10.8). However, it is known that the proliferation of mosquitoes' transmitters of dengue can be higher in years when the temperature is higher, without the respective increase of precipitation. Thus, when considering the temperature recorded in summer months the year 2015 and the cases of hospitalization for dengue (Fig. 10.6), the hypothesis is validated. Another factor considered in recent research refers to the increased minimum temperature, especially in years with epidemic Dengue fever. These studies suggest that the minimum temperature is an important atmospheric component to be considered in epidemic years, being this a limiting factor as the maturation of eggs of *Aedes aegypti* (Horta et al. 2014).

Comparing Dengue fever with other vector-borne diseases at the Baixada Santista, it is clear that Dengue fever is the most important hazard, in which population are exposed to. Dengue fever incidence seems to be aligned with the patterns found for São Paulo state, with higher number during 2010 and 2015 (Table 10.2).

Despite high dengue incidence in the Baixada Santista in 2010 and 2015, for Santos and São Vicente, we observed higher incidence in the year 2010 in comparison to 2015. This pattern is different from the observed for São Paulo state, where the number of cases in 2015 was higher than that recorded in 2010 (Fig. 10.9).

Dengue cases are also associated with high temperature observations, during periods of recent El Niño. However, the unexpected reduction from 2010 to 2015 is likely to be related to prevention actions in this region.

10.3.3 Spatial Patterns of Dengue Fever

Analysing the spatial dataset we show that on average cases of hospitalization for Dengue fever for the whole period of analysis occurs during the months of March, April and May. The incidence is widespread in São Paulo State, with most of the region with low to medium incidence. In the case of the Baixada Santista and Santos, incidence is on average low (Fig. 10.10). It is clear, however, that the geographical area of activity of the vector *Aedes aegypti* is increasing annually, with peaks in 2010, 2013 and 2015. While most of the areas of high incidence in the state of Sao Paulo were observed in 2015, we show that Baixada Santista turned from a low incidence area to a medium incidence area during 2010, and some municipalities of this region also in 2015 (Fig. 10.11).

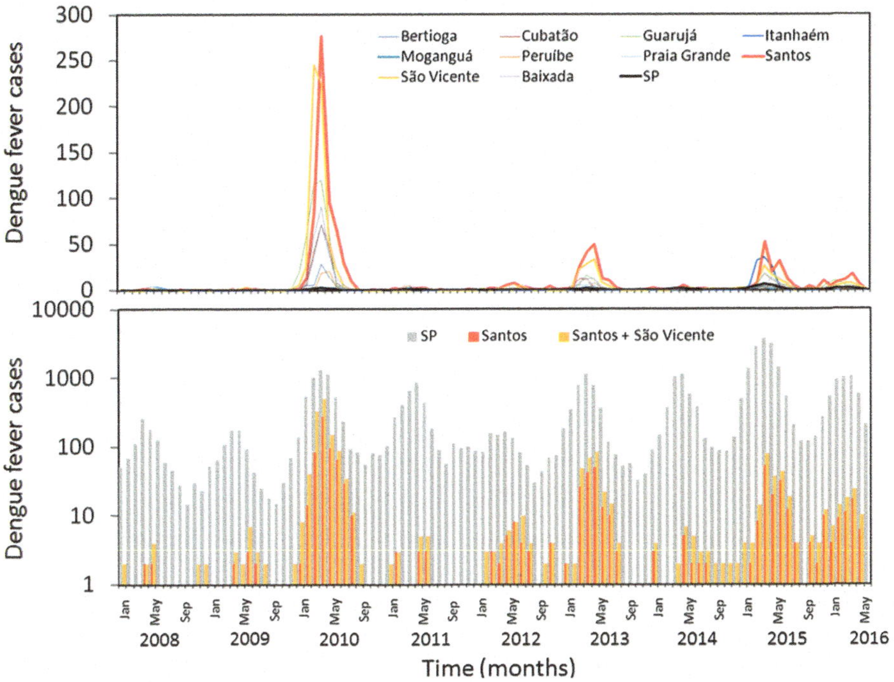

Fig. 10.9 Incidence of dengue cases in the municipalities of the Baixada Santista from 2008 to 2016 (top) and the comparison in log scale between incidence in Santos, Santos plus São Vicente and São Paulo

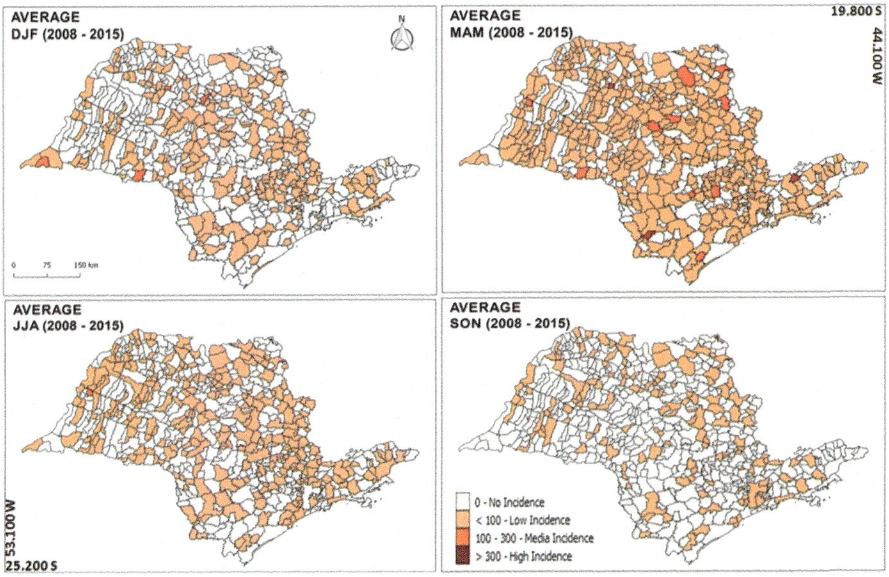

Fig. 10.10 Quarterly average of Dengue incidence in São Paulo state

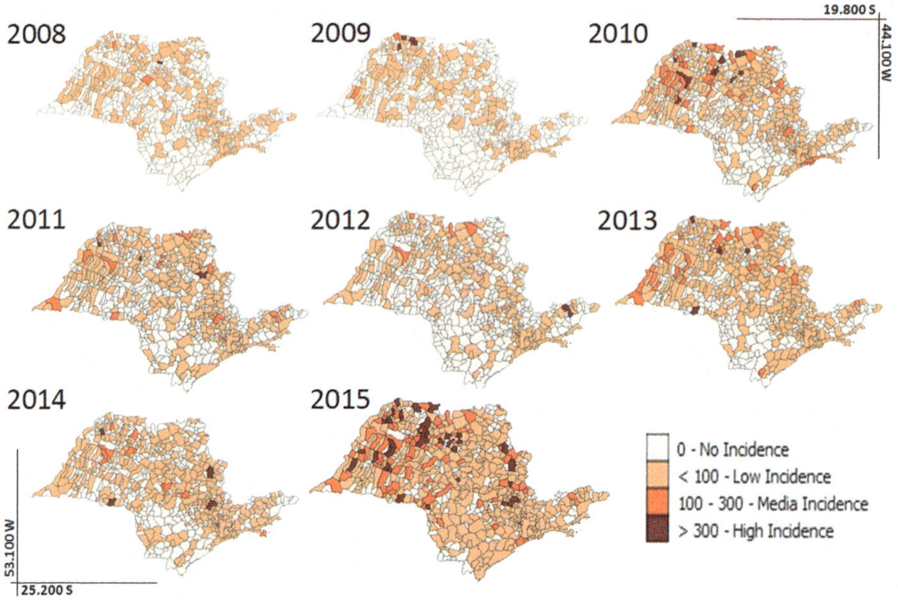

Fig. 10.11 Annual incidence of Dengue in São Paulo state

These spatially explicit data are critical for planning state level investments for prevention actions. Moreover, they are essential for developing scenarios of future vulnerabilities and mitigation strategies. In this sense, we investigate in the next section some potential factors that may intensify the risk of dengue in the region.

The Baixada Santista is one of the main economic regions of São Paulo state, with a population of about 1.8 million inhabitants. This turns this region into an area of high exposure to the disease, which in turn increases the risk of outbreaks.

Among the nine cities that are part of the Baixada Santista, Santos is the most populated with 430.000 inhabitants, with the neighbouring city of São Vicente the second in the list with about 350.000 inhabitants (Table 10.3). These two cities are, therefore, the most exposed to Dengue fever outbreaks, in the region. We hence performed an analysis to evaluate the behaviour of Dengue fever in this region, in comparison to the patterns observed in São Paulo state.

10.4 Factors Increasing Vulnerability of Santos to Dengue Fever Outbreaks

By analyzing the long-term series of temperature anomalies (Fig. 10.12), our data show that there is a clear trend of increased temperature anomalies in the state of São Paulo. Temperature anomalies from 1990 onwards tend to be higher than the long-term mean for most of the years observed.

Table 10.3 Population density from the 2010 IBGE census and estimated for 2016 for cities in the Baixada Santista

Municipality	Population 2010	Population 2016
Bertioga	47.645	57.942
Cubatão	118.720	127.887
Guarujá	290.752	313.421
Itanhaém	87.057	97.439
Mongaguá	46.293	53.384
Peruíbe	59.773	65.907
Praia Grande	262.051	304.705
Santos	419.400	434.359
São Vicente	332.445	357.989
Total	**1.664.136**	**1.813.033**

It is worth highlighting that by evaluating the spatial distribution of historical mean temperature anomalies from CRU for the state of São Paulo for the year 2012 (Fig. 10.13), 83% of the months in 2012 have widespread occurrence of positive temperature anomalies. This indicates that temperatures in the months analysed are above the long-term mean temperature value (1961–1990). It is interesting to note that the months with the highest anomalies are September, October and December, corresponding to spring time in the south hemisphere, when incidence of vector-borne diseases such as Dengue fever is higher.

Based on our population density data we observe that São Paulo state has three centres of high population density (people per km^2), one area in Campinas region, São Paulo Capital and Baixada Santista region (Fig. 10.14). In the Baixada Santista, Santos, Guarujá and São Vicente are the districts with higher population density.

It is clear, by analysing the topographic information, that the Baixada Santista area is dominated by elevations varying from 0 to 10 m above the sea level. Bertioga, Santos, São Vicente and Praia Grande, have a large fraction of their territory covered by low elevations (Fig. 10.15). The dominance of areas with low elevation is a critical factor for the incidence of Dengue in the future. The data from CEPAL, it is very likely that sea level rise will create more areas suitable for the proliferation of Dengue fever vector, favouring the increased development of the disease in this region.

The dataset from CEPAL-Universidad de Cantabria shows that trends in sea level rise from 2010 to 2040 in the Baixada Santista region correspond to around 3 mm/year. Despite the small tendency of sea level rise in the next 30 years, the vulnerability of this region is likely to increase. Because of the low elevation of a large fraction of the studied area, the combination with high tides, coastal storms and riverine floods due to above the average extreme rainfall events, this region will be even more exposed to natural disasters. Based on the stratification of elevation between 0 and 10 m adopted here to provide information in conformity with the dataset of urban area affected by sea level rise available from CEPAL-Universidad de Cantabria, near Santos and Guarujá, a total of 28,763,100 m^2 of urban area is

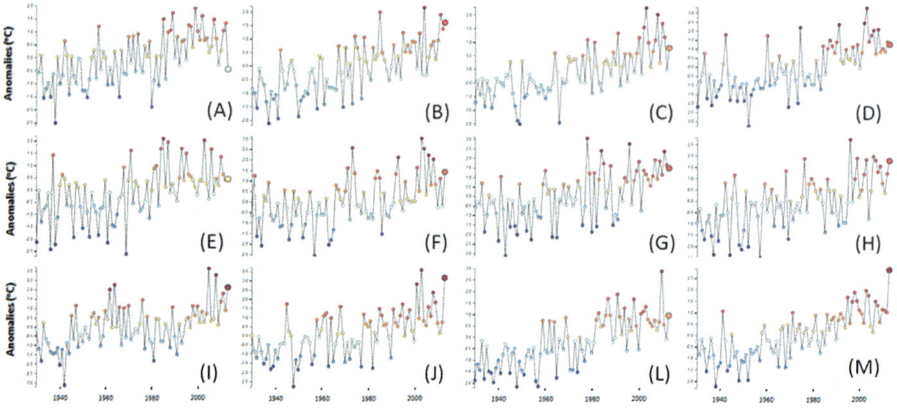

Fig. 10.12 Historical time-series of mean temperature anomalies from CRU dataset for the state of São Paulo. Panels (**a**–**m**) indicate the 2012 temperatures from January to December, respectively

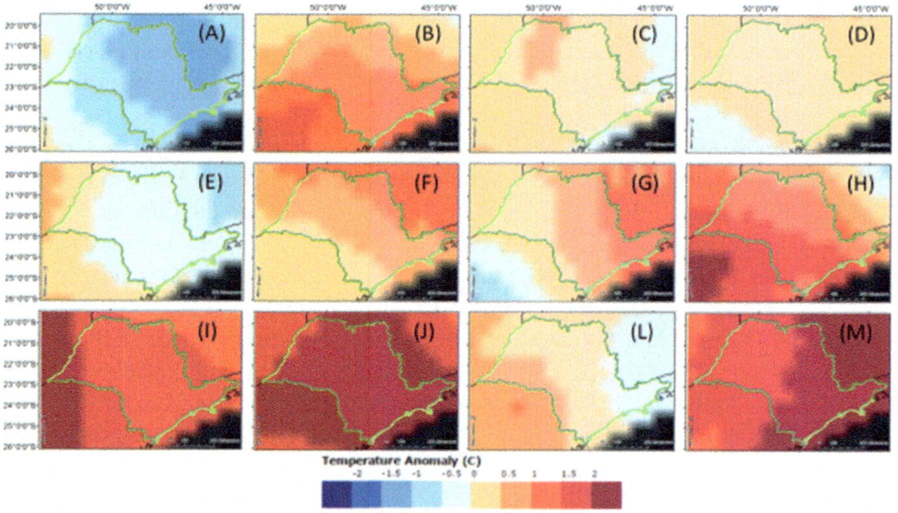

Fig. 10.13 Mean temperature anomalies from CRU dataset. Green polygon highlights the state of São Paulo. Panels (**a**–**m**) indicate the 2012 temperatures from January to December, respectively

established under the 10 m elevation threshold. Moreover, an additonal 264,900 m^2 of urban area, further south, in the Praia Grande and Mongaguá region is also exposed to sea level rise. This dataset is essential for for helping timely policy decisions in terms of developing effective adaptation policies. More importantly is that the potential for increasing the exposure of the Baixada Santista to natural disasters, especially related to floodings, is likely to directly influence the incidence of mosquito transmitted diseases such as dengue.

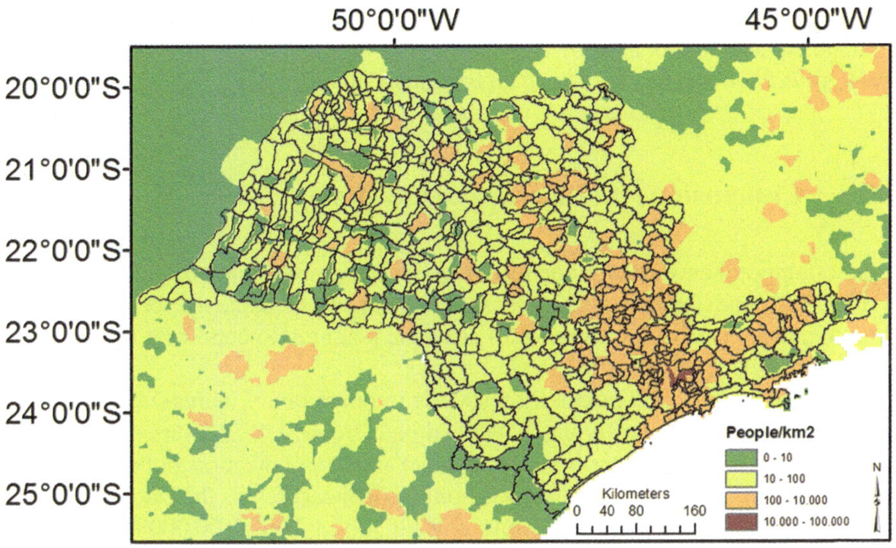

Fig. 10.14 Population density (people km^{-2}) in São Paulo state

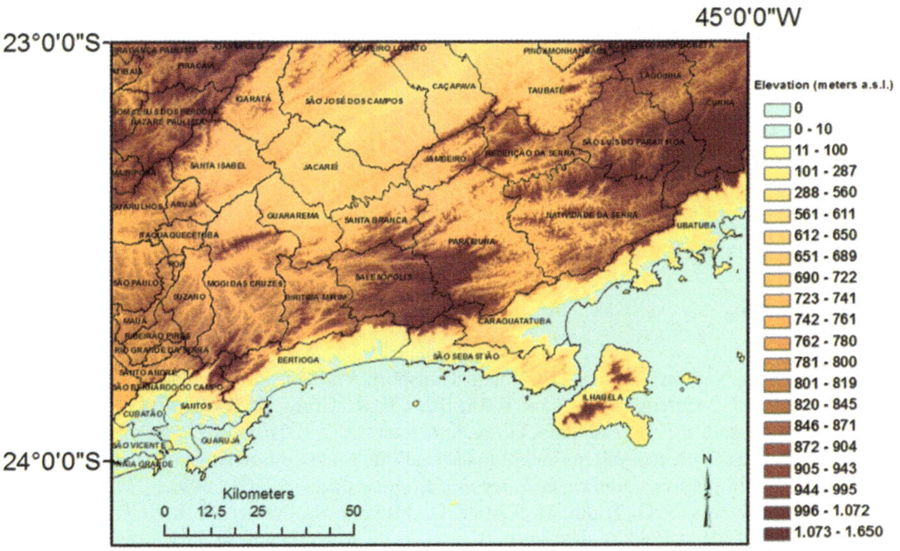

Fig. 10.15 Topographic map of the Baixada Santista showing in blue tones areas with elevation lower than 10 m above the sea level

In this chapter we provide resources for assisting policy-makers and the civil society to understand and vizualize the impacts of climate on human-health. Understanding these patterns allow for better planning resonable solutions for the changes predicted ahead. This information is the basis for evaluating risks related to climate change and its impacts on society, specially in Santos. This provides a key background for coastal cities to adapt to climate-induced risks.

10.5 Conclusions and Recommendations

Spatio-temporal datasets provide resources for integrative climate change risk analysis. This understanding increases the adaptation capacity of States and Municipalities by the development of strategic planning to cope with the predicted changes.

Dengue fever incidence clearly responds to temperature increases that are predicted to increase in the future. If we estimate from SUS an average cost of R$ 600.00 per patient, considering only treatment and hospitalization, outbreaks related to temperature increase, without any prophylactic measures, as the case of 2010, are very likely to increase costs in the future. With 300 cases per 100,000 inhabitants in 2010 and assuming a population of 400,000 inhabitants, we estimate a cost of R$ 720,000 in 2010 with dengue treatment. These costs are likely to be much higher if we consider the other diseases associated with the same vector such as Zikka and Chicungunha, plus emergency actions to mitigate the problem during outbreaks.

Therefore, the investments must be allocated for prevention that needs a long-term monitoring of temperature and El Niño index data that must start in November and December of each year. Authorities must be alerted when mean temperature surpasses 24 °C, because of the measured increase of dengue cases during the El Niño years. Finally prophylactic measures must be intensified between March and May, when the disease peaks in the region.

References

ANTAQ. (2017). National Agency for Aquatic Transport. Available at: http://portal.antaq.gov.br/

Aragão, L. E. O. C., Marengo, J. A., Cox, P. M., Betts, R. A., Costa, D., Kaye, N., Alves, L., Smith, L. T., Cavalcanti, I. F. A., Sampaio, G., & Anderson, L. O. (2016). Assessing the influence of climate extremes on ecosystems and human health in Southwestern Amazon supported by the PULSE-Brazil platform. *American Journal of Climate Change, 5*(03), 399.

Burkett, V. R., Suarez, A. G., Bindi, M., Conde, C., Mukerji, R., Prather, M. J., St. Clair, A. L., & Yohe, G. W. (2014). Point of departure. In C. B. Field, V. R. Barros, D. J. Dokken, K. J. Mach, M. D. Mastrandrea, T. E. Bilir, M. Chatterjee, K. L. Ebi, Y. O. Estrada, R. C. Genova, B. Girma, E. S. Kissel, A. N. Levy, S. MacCracken, P. R. Mastrandrea, & L. L. White (Eds.), *Climate change 2014: Impacts, adaptation, and vulnerability. Part A: Global and sectoral aspects. Contribution of working group II to the fifth assessment report of the intergovernmental panel on climate change* (pp. 169–194). Cambridge/New York: Cambridge University Press.

Cardona, O. D., van Aalst, M. K., Birkmann, J., Fordham, M., McGregor, G., Perez, R., Pulwarty, R. S., Schipper, E. L. F., & Sinh, B. T. (2012). Determinants of risk: Exposure and vulnerability. In C. B. Field, V. Barros, T. F. Stocker, D. Qin, D. J. Dokken, K. L. Ebi, M. D. Mastrandrea,

K. J. Mach, G.-K. Plattner, S. K. Allen, M. Tignor, & P. M. Midgley (Eds.), *Managing the risks of extreme events and disasters to advance climate change adaptation, A Special Report of Working Groups I and II of the Intergovernmental Panel on Climate Change (IPCC)* (pp. 65–108). Cambridge/New York: Cambridge University Press.

Costanza, R. (2003). A vision of the future of science: Reintegrating the study of humans and the rest of nature. *Futures, 35*(6), 651–671.

Degallier, N. I. C. O. L. A. S., Vilarinhos, P. D. T. R., de Carvalho, M. L., Knox, M. B., & Caetano, J., Jr. (2000). People's knowledge and practice about Dengue fever, its vectors, and control means in Brasilia (DF), Brazil: Its relevance with entomological factors. *Journal of the American Mosquito Control Association, 16*(2), 114–123.

Departamento de Informática do Sistema Único de Saúde (DATASUS). (n.d.). Dados do Sistema de Informações Hospitalares (SIH). Disponível em: http://www2.datasus.gov.br/DATASUS/index.php?area=0901&item=1&acao=25. Acesso 30 June 2015.

Hales, S., De Wet, N., Maindonald, J., & Woodward, A. (2002). Potential effect of population and climate changes on global distribution of Dengue fever: An empirical model. *The Lancet, 360*(9336), 830–834.

Horstick, O., Runge-Ranzinger, S., Nathan, M. B., & Kroeger, A. (2010). Dengue vector-control services: How do they work? A systematic literature review and country case studies. *Transactions of the Royal Society of Tropical Medicine and Hygiene, 104*(6), 379–386.

Horta, M. A., Fonseca, P. A., Costa, D., Barcellos, C., & Hacon, S. (2014). Temporal relationship between climatic factors and the occurrence of Dengue fever in an Amazonian Urban Center, Brazil. *International Journal of Epidemiology & Infection, 2*(4), 80–87.

IBGE. (2017). Brazilian Institute for geography and statistics. Demographic density census. Available at: https://censo2010.ibge.gov.br/sinopse/index.php?dados=10&uf=00

Marengo, J. A., Aragão, L. E. O. C., Cox, P. M., Betts, R., Costa, D., Kaye, N., Smith, L. T., Alves, L. M., & Reis, V. (2016). Impacts of climate extremes in Brazil: The development of a web platform for understanding long-term sustainability of ecosystems and human health in Amazonia (PULSE-Brazil). *Bulletin of the American Meteorological Society, e-View*. https://doi.org/10.1175/BAMS-D-14-00177.1.

Mauser, W., Klepper, G., Rice, M., Schmalzbauer, B. S., Hackmann, H., Leemans, R., & Moore, H. (2013). Transdisciplinary global change research: The co-creation of knowledge for sustainability. *Current Opinion in Environmental Sustainability, 5*(3), 420–431.

Meadow, A. M., Ferguson, D. B., Guido, Z., Horangic, A., Owen, G., & Wall, T. (2015). Moving toward the deliberate co-production of climate science knowledge. *Weather, Climate, and Society, 7*, 179–191.

Medronho, R. A. (2006). Dengue e o ambiente urbano. *Revista Brasileira de Epidemiologia, 9*, 159–161.

Messina, J. P., Brady, O. J., Scott, T. W., Zou, C., Pigott, D. M., Duda, K. A., Bhatt, S., Katzelnick, L., Howes, R. E., Battle, K. E., & Simmons, C. P. (2014). Global spread of dengue virus types: Mapping the 70 year history. *Trends in Microbiology, 22*(3), 138–146.

Morin, C. W., Comrie, A. C., & Ernst, K. (2013). Climate and Dengue transmission: Evidence and implications. *Environmental Health Perspectives, 121*(11–12), 1264–1272.

Pessanha, J. E. M. (2012). Risk assessment and risk maps using a simple Dengue fever model. *Dengue Bulletin, 36*, 73–86.

Pham, H. V., Doan, H. T. M., Phan, Y. Y. Y., & Minh, N. N. T. (2011). Ecological factors associated with Dengue fever in a central highlands Province, Vietnam. *BMC Infectious Diseases, 11*(1), 172.

Ripsa Rede Interagencial de Informações para a Saúde. *acesso em 21 fevereiro 2016*. Disponível em: http://www.ripsa.org.br/fichasIDB/record.php?node=D.2.3&lang=pt&version=ed3

UNISDR (United Nations International Strategy for Disaster Reduction). (2007). *Hyogo framework for action 2005–2015: Building the resilience of nations and communities to disasters*. Extract from the final report of the World Conference on Disaster Reduction (A/CONF.206/6). Geneva.

UNISDR (United Nations International Strategy for Disaster Reduction). (2015). *Sendai framework for disaster risk reduction 2015–2030*. Geneva: UNISDR.

Weaver, C. P., Mooney, S., Allen, D., Beller-Simms, N., Fish, T., Grambsch, A. E., & Winthrop, R. (2014). From global change science to action with social sciences. *Nature Climate Change, 4*(8), 656–659.

WHO. (1995). *Global strategy for prevention and control of Dengue fever and Dengue haemorrhagic fever*: Report of the consultation on key issues in Dengue vector control towards the operationalization of a global strategy, CTD/FIL(DEN)/IC/96. Geneva: World Health Organization.

Chapter 11
Land Reclamation Effects on Santos Estuary Inlet Stability

Thiago Bezerra Corrêa, Tiago Zenker Gireli, and Vinícius de Carvalho Neiva Pinheiro

Abstract During the past two decades, governmental incentives on port expansion have led Santos to be a hotspot of infrastructure investments. Nevertheless, land reclamation, mainly by port terminal construction in estuarine areas, has affected Santos estuary tidal inlet stability, increasing siltation along the estuarine channel. Herein, two scenarios were compared to assess the effects of land reclamation on Santos estuary tidal inlet stability. The baseline scenario (2006) is prior to the construction of two port terminals (EMBRAPORT and BTP) and deepening dredging in Port of Santos, and the foreseen scenario (2014) is posterior to these interventions. The assessment of Santos estuary tidal inlet stability consisted of a new method derived from the classical Area-Prism relationship. The Slice Method enhanced the cross-sectional area estimative accuracy and enabled the evaluation of tidal inlet cross-sectional profile changes. During these 8 years, Santos estuary tidal prism decreased from 55.1×10^6 to 53.6×10^6 m^3, tidal inlet got shallower and the cross-sectional area reduced 5.5%. The study confirms that land reclamation reduced Santos estuary tidal prism. Therefore, cumulative effects of land reclamation on tidal inlet stability must be studied to define mitigation strategies for channel siltation.

Keywords Land reclamation · Tidal inlet · Tidal prism · Cross-sectional stability · Area-prism relationship · Port of Santos

T. B. Corrêa (✉)
Ramboll Environment & Health, São Paulo, SP, Brazil
e-mail: tcorrea@ramboll.com

T. Z. Gireli · V. d. C. N. Pinheiro
School of Civil Engineering, Architecture and Urban Design, University of Campinas (Unicamp), Campinas, SP, Brazil
e-mail: zenker@fec.unicamp.br; pinheiro@fec.unicamp.br

L. H. Nunes et al. (eds.), *Climate Change in Santos Brazil: Projections, Impacts and Adaptation Options*, https://doi.org/10.1007/978-3-319-96535-2_11

11.1 Introduction

Santos estuary shelters the most relevant Brazilian port (Fig. 11.1). Port of Santos importance in Brazilian history began in the late nineteenth century, when Coffee Cycle in São Paulo State pushed the port economically (Scazufca 2012). In 2014, it was responsible for one quarter of Brazilian balance of trade, including exports and imports of industrial and agricultural goods (CODESP 2015).

Despite those benefits for Santos, major environmental impacts derive from the port. For instance, the port expansion, which mainly concerns on land reclamation, leads to basin reduction in bays or estuarine areas. This land reclamation decreases Tidal Prism (Feng et al. 2015; van de Kreeke 2004), which is the amount of water entering and leaving the inlet, so it works as a flow that controls sand deposit on bars (Bruun 1978).

In addition, siltation increases when inlet deviates from its equilibrium. As the channel gets deeper, it is harder to maintain depth, because the relationship of depth maintenance with siltation is non-linear, following exponential or potential rate (Gireli and Vendrame 2012). For that reasons, studies regarding land reclamation impacts on channel siltation are crucial to understand these processes and to mitigate this negative effect.

In the past two decades, Port of Santos has received a great deal of investments for deepening dredging, and for construction and expansion of terminals (Brasil 2007, 2013). Considering this scenario of recent anthropogenic intervention in Santos estuary, this study evaluates how the construction of terminals between 2006 and 2014 along Port of Santos navigation Channel affected the stability of Santos estuary tidal inlet.

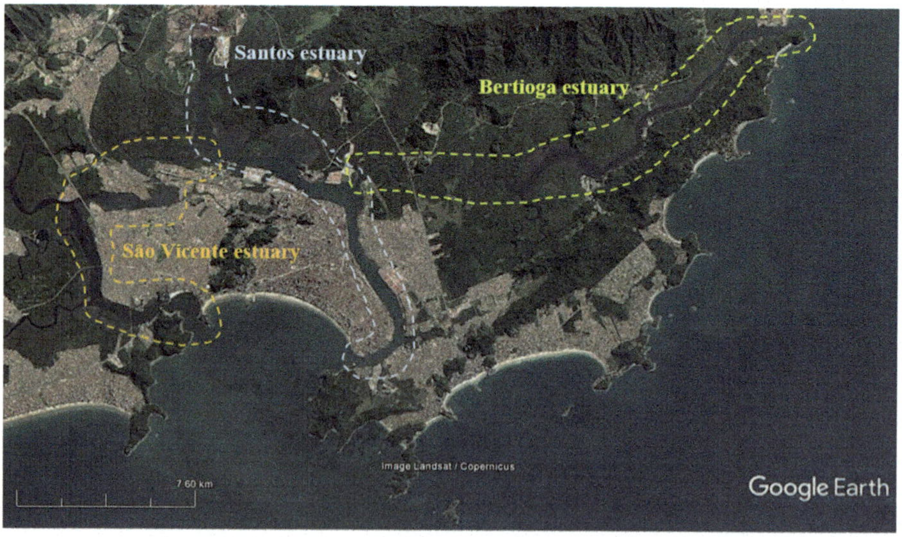

Fig. 11.1 Santos Estuary System composed by Santos estuary, São Vicente and Bertioga estuarine channels. (Source: Google Earth)

11.2 Santos Estuary

11.2.1 Hydraulic Characterization

The hydraulics in the Santos Estuary System depends mainly on tidal propagation, and salinity induces an internal circulation that changes vertical velocity profile, but does not change discharges (Sondotécnica 1977a, b). Three estuarine channels, São Vicente estuary, Santos estuary (Port of Santos) and Bertioga estuary, form a complex estuary system (Fig. 11.1), where meanders, channel ramifications and mangrove inundation areas influence the hydraulics in the estuary.

Considering the main objective of this study – the evaluation of Santos estuary inlet stability due to land reclamation along the estuary – the most important forcing in the region are tidal dynamics. In addition, estuary geometry also influences hydraulics (Sondotécnica 1977a, b; Leitão et al. 2008; Roversi et al. 2016), so that changes in estuary bathymetry due to deepening dredging and land reclamation are important factors.

11.2.2 Land Reclamation

During the last decade, Port of Santos deepened its channel from 13 to 15 m, and allowed the construction of two terminals.

EMBRAPORT (Brazilian Company of Port Terminals) is a private port terminal in the left margin of Port of Santos, which began its operations in July 2013 (EMBRAPORT 2013). During the terminal construction, 600,000 m^3 of contaminated sediments were dredged from berth and berth access, contained in 208 geotubes (2300 m^3 each), and dewatered inside three impermeable cells (WEDA 2013). Therefore, EMBRAPORT construction consisted of stepped land reclamation in the Santos estuary (Fig. 11.2).

BTP (Brazil Port Terminal) dealt with contaminated sediments too. The terminal building area used to be contaminated due to the Alemoa Dumping Ground. The construction is partially concluded (only phase 1 is achieved) and its operations began in August 2013. According to MKR (2003), the major land reclamation will occur when phase 2 finishes (Fig.11.3).

11.3 Channel Siltation Diagnostic

According to INPH (2007), Port of Santos deepening dredging allowed the traffic of larger vessels, and the channel widening allowed transit of two vessels side by side, depending on their beam. Figures 11.4 and 11.5 shows the channel bathymetries in 2006 and 2014, respectively, and confirms this observation.

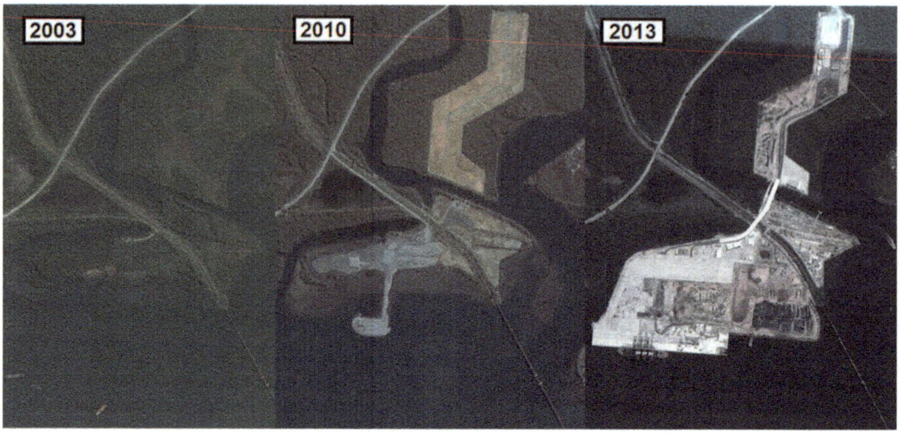

Fig. 11.2 Land reclamation in Santos estuary due to EMBRAPORT construction. (Source: adapted from Google Earth)

Fig. 11.3 Planned land reclamation in Santos estuary due to BTP construction, only phase 1 is finished. (Source: adapted from MKR 2003)

Nevertheless, in 2006, before deepening dredging, Santos estuary channel gorge, region with depth maintained only by tidal currents scour, was over 26 m deep (Fig. 11.4). Eight years later, in 2014, the gorge got shallower and its maximum depth was under 24 m deep (Fig. 11.5).

Curiously, observations from other coastal areas (Oliveira et al. 2006; Malhadas et al. 2009; Picado et al. 2010) indicate that deepening dredging may increase tidal prism.

Still, Santos estuary tidal inlet got shallower (Fig. 11.6), its cross-sectional area reduced 5.5% in 8 years, from 6105 m^2 in 2006 to 5769 m^2 in 2014. Considering that

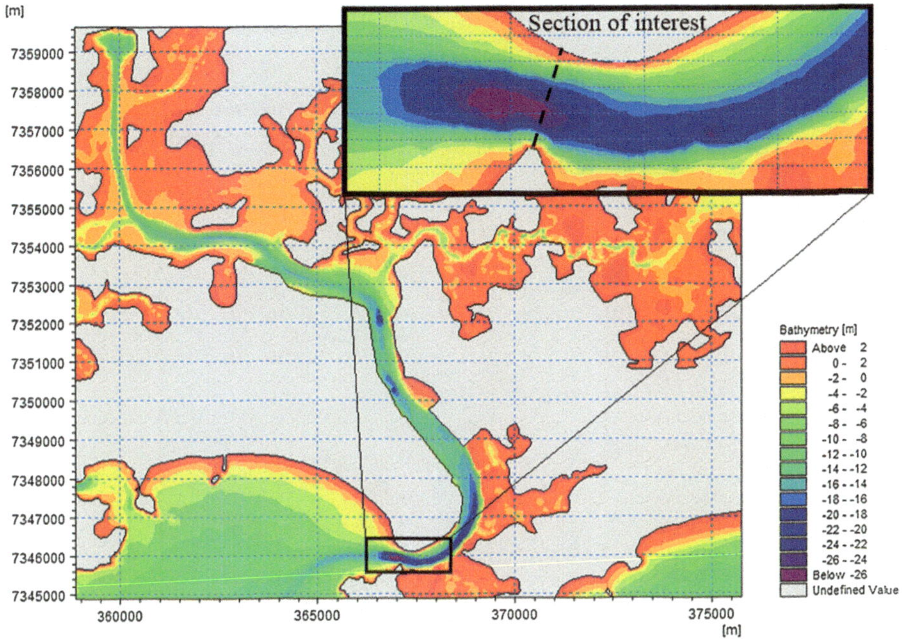

Fig. 11.4 Port of Santos channel bathymetry in the Santos estuary. (Data from 2006 INPH 2007)

land reclamation decreases tidal prism (van de Kreeke 2004; Feng et al. 2015), understanding tidal prism variation from 2006 to 2014 and the Area-Prism relationship for Santos estuary tidal inlet is crucial to determine the synergistic effect of several estuary occupations in tidal inlet stability.

11.4 Area-Prism Relationship

It was observed worldwide that inlets with large cross-sectional area were associated to large basins, such lagoons, estuaries, coastal embayment etc. (Jarrett 1976; O'Brien 1969; Hume and Herdendorf 1993; Townend 2005; Powell et al. 2006). The combination of basin surface area with tidal range lead to a volume of water in the basin stored during flood or released during ebb.

Thus, the Area-Prism relationship (AP relationship) derives from that empirical observation. Equation 11.1 shows the AP relationship, where A (m^2) is the tidal inlet equilibrium cross-sectional area, P (m^3) is the spring tidal prism, and C and q are coefficients of adjustment.

$$A = CP^q \tag{11.1}$$

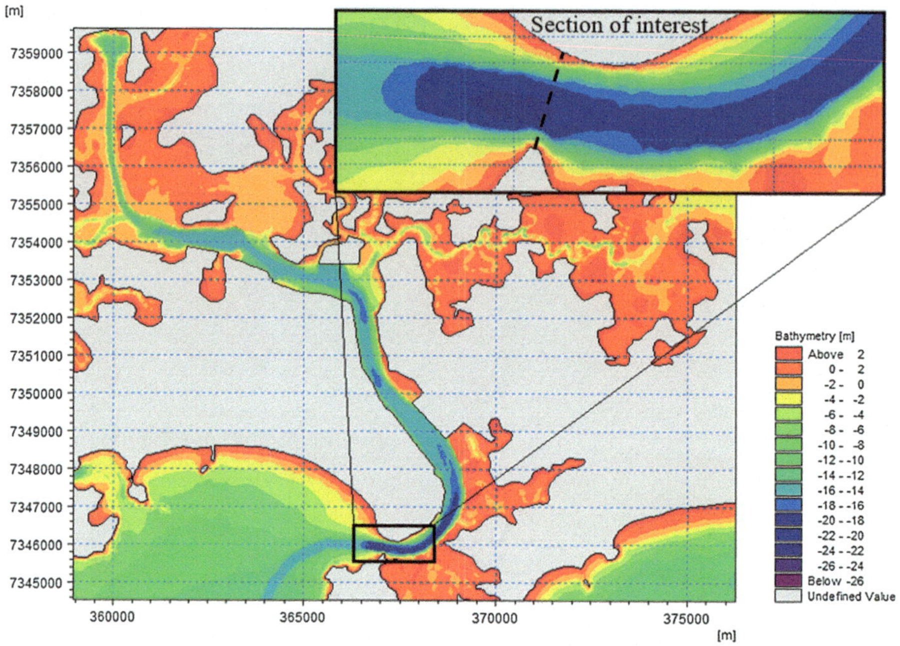

Fig. 11.5 Port of Santos channel bathymetry in the Santos estuary. (Data from 2014 conceded by CODESP)

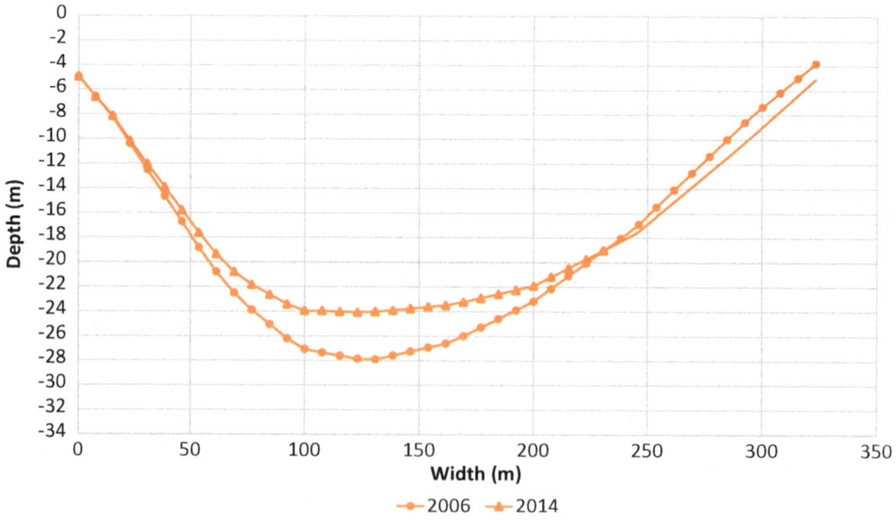

Fig. 11.6 Comparison of Santos estuary tidal inlet cross-sectional area in 2006 and 2014

Therefore, tidal prism has been acknowledged as an indicator of tidal inlet stability. According to Mehta and Ozsoy (1978), tidal prism is the volume stored during flood or released during ebb. So, tidal prism (P) is the integral of water discharge ($Q(t)$) through the tidal inlet during ebb or flood period (T) (Eq. 11.2).

$$P = \int_{T}^{0} Q(t)\,dt \tag{11.2}$$

For several tidal inlets, investigators measured cross-sectional area, computed tidal prism, and plotted those data to determine coefficients that best fit the equation for scatter data.

First, tidal inlets were grouped according to its location, and/or whether it had jetties (O'Brien 1969; Jarrett 1976). Later, Stive et al. (2010) re-scrutinized data from Jarrett (1976) and Powell et al. (2006), categorizing those inlets according to their mean grain size, tidal range, hydraulic radius, and littoral transport. In general, these categories lead to better correlation when compared to the complete dataset.

Kraus (1998) proposed the first theoretical approach for AP relationship, which accounts the dynamic balance between inlet ebb tidal transport and longshore sand transport. However, this method leads to larger predicted cross-section area when the littoral transport is low.

Hughes (2002) developed a theoretical AP relationship (Eq. 11.3), which coefficient C depends on tidal period (T), median grain size (d_e), channel width (W), gravity (g), sediment specific gravity (S), and a coefficient related to the effects of non-sinusoidal tides (k_a). This relationship may be applied for any tidal inlet, as long as the required information is available.

$$A = 0.65\,k_a \left[\frac{W^{\frac{1}{9}}}{\left[g(S-1)\right]^{\frac{4}{9}} d_e^{\frac{1}{3}} T^{\frac{8}{9}}} \right] P^{8/9} \tag{11.3}$$

11.4.1 The Slice Method

Tidal prism is the volume of water stored during flood or released during ebb computed by Eq. 11.2. Then, the equilibrium cross-sectional area may be predicted using any AP relationship (Eq. 11.1). However this approach does not provide the changes in cross-sectional profile. Furthermore, one-fit-to-all equation defines the coefficients C and q from different tidal inlets, which may not have similar characteristics.

Therefore, the Slice Method seeks to determine the AP relationship for a specific location and the cross-sectional profile of the tidal inlet of interest (Fig. 11.7). In order to achieve that, some abstractions are required.

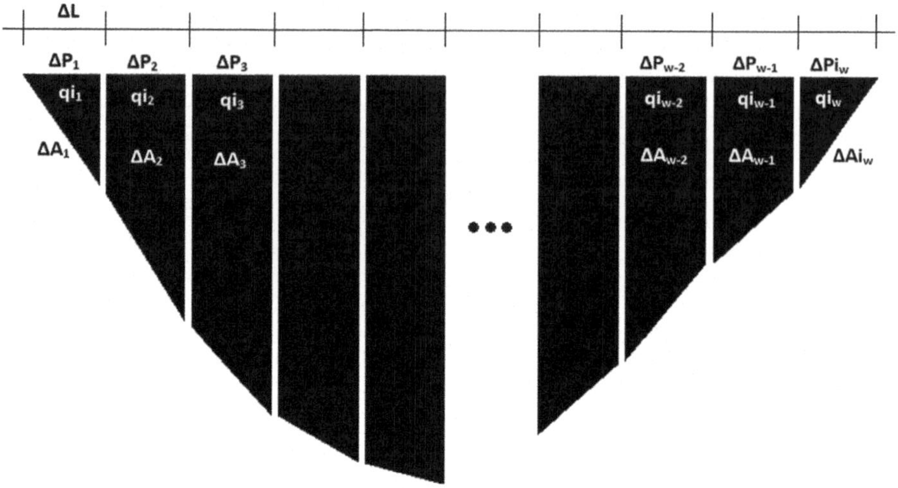

Fig. 11.7 Generic tidal inlet divided into slices

The analytical discharge integration to calculate tidal prism (Eq. 11.1) can be written in discrete form (Eq. 11.4), where *Δt* is the length of time step, *i* is the time step, *t* is the number time steps, and Q_i is the discharge in a given time step.

$$P = \sum_{t}^{i=1} Q_i \Delta t \tag{11.4}$$

$$Q_i = \sum_{w}^{j=1} q_{ij} \, \Delta l \tag{11.5}$$

Combining Eqs. 11.4, 11.5, and 11.6 represents tidal prism discretized in time and space (graphic representation of spatial dicretization in Fig. 11.7).

$$P = \sum_{t}^{i=1} \sum_{w}^{j=1} q_{ij} \, \Delta l \;\; \Delta t \tag{11.6}$$

Considering only one slice *j*, the integration of its specific discharge (q_{ij}) in time provides a volume of partial tidal prism, hereafter called *slice tidal prism* (*ΔP*) (Eq. 11.7). The sum of all *w* slice tidal prisms (ΔP_j) results in the tidal prism (*P*) (Eq. 11.8).

$$\Delta P_j = \sum_{t}^{i=1} q_{ij} \, \Delta l \;\; \Delta t \tag{11.7}$$

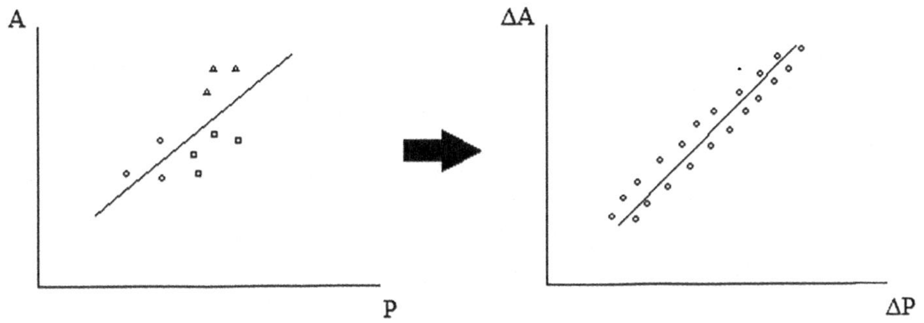

Fig. 11.8 Representation of scatter data from several tidal inlets ($A \times P$) and from one specific tidal inlet using the slice method ($\Delta A \times \Delta P$)

$$P = \sum_{w}^{j=1} \Delta P_j \tag{11.8}$$

Instead of defining the coefficients C and q from scatter data of area (A) versus tidal prism (P) from several inlets, the Slice Method uses each slice area (ΔA) and slice tidal prism (ΔP) from a unique tidal inlet (Fig. 11.8). Besides, when the slice width (Δl) is small enough, the slice area (ΔA) divided by the slice width (Δl) may be a fair approximation for the depth of each point along the cross-section. Then, the Slice Method may provide the tidal inlet cross-sectional profile.

The only issue for this method is the difficulty to measure the specific discharge for several slices along tidal inlet cross-section on the field. Depending on the inlet width and/or on the traffic through the inlet, the measurements may be unfeasible. Nevertheless, the time series for specific discharge along tidal inlet can be retrieved from a hydrodynamic numerical model calibrated and validated for tides and currents.

11.5 Hydrodynamic Numerical Modelling

The current study applies a 2D hydrodynamic model with flexible mesh (Mike 21 Flow Model FM) to retrieve velocities and currents from the study area. As the area of interest is mainly estuarine, the main hydrodynamic forcing is the sea level variation. Thus, for the proposed study, a simple model with continuity and momentum equations are enough. Figure 11.9 shows the bathymetry from 2006 interpolated with Mike Mesh Generator.

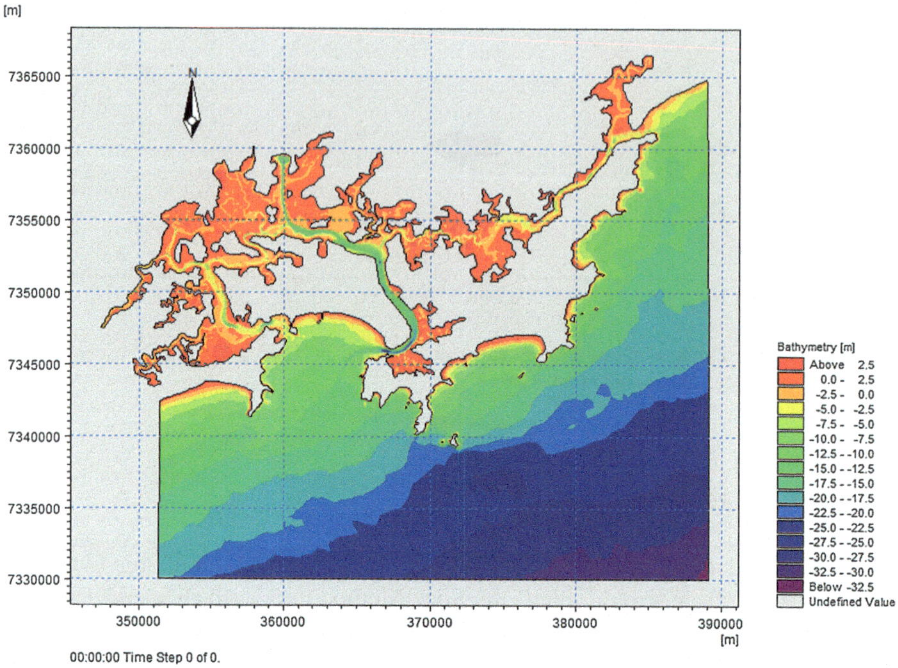

Fig. 11.9 Model of Santos Estuary System and nearshore region 2006 bathymetry interpolated using Mike Mesh Generator

11.5.1 Data Set

Since the baseline of this study is set for 2006, the Port of Santos Access and Navigation Channel (located in Santos estuary) are retrieved from bathymetric data of this year (INPH 2007). Foreseen scenario (from 2014) accounts Port of Santos navigation channel deepening dredging (from 13 to 15 m) and the construction of two terminals (EMBRAPORT and BTP).

Due to lack of current bathymetric data, the estuary consists of a mosaic of older surveys (MARMIL 2015; Garcia et al. 2002), and this mosaic is the same for the baseline and for the foreseen scenario. In both scenarios, mesh enhancement was performed by the module Mike Mesh Generator, which constructs an unstructured mesh with triangular elements.

The major model forcing is tidal elevation, so the sea boundary consists of nine nodes and eight segments (Fig. 11.10). The tidal forcing for each node were computed using the nine most energetic tidal constituents in Santos region, each node has amplitude adjustment and phase lag, based on tidal maps for these nine constituents (Harari and Camargo 1994). The hydrodynamic model has nine points of river discharge along the estuary (Fig. 11.10 and Table 11.1). The study considers long period river discharges retrieved from Roversi et al. (2016).

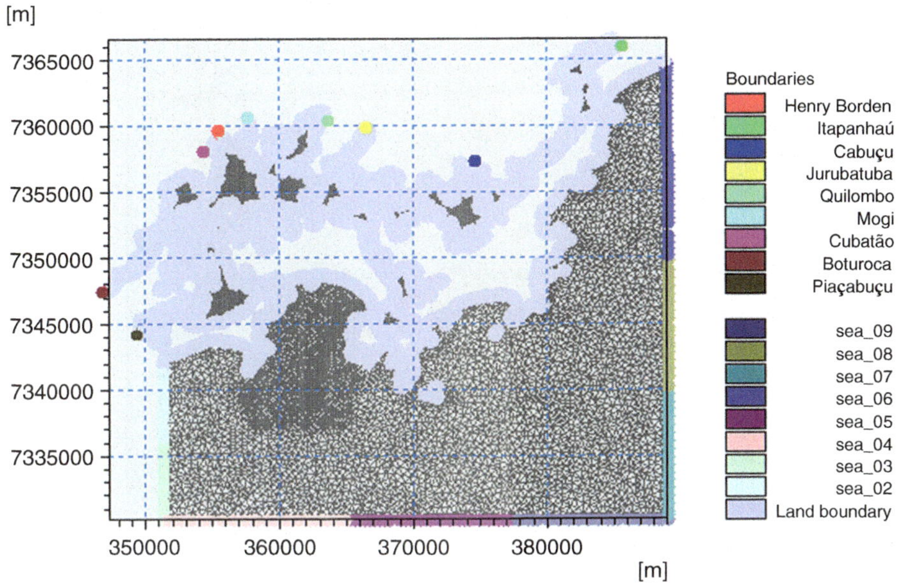

Fig. 11.10 Sea and riverine boundaries of the hydrodynamic model

Table 11.1 Rivers long period discharges in the Santos Estuary System

River	Discharge (m^3/s)
Boturoca	7.18
Cabuçu	3.43
Cubatão	8.09
Itapanhaú and Ibitinga	20.28
Jurubatuba	3.91
Mogi	3.58
Piaçabuçu	2.27
Quilombo	4.55
Henry Borden	6.00

Adapted from Roversi et al. (2016)

11.5.2 Model Calibration and Validation

The model period of simulation covers 13 days, from 4th to 17th March of 2006. The calibration consisted on adjusting the bed roughness to minimize the errors in five tide gauge stations (Fig. 11.11), and validation consisted of comparing current measurements in eight sections (INPH 2007) along the estuary (Fig. 11.12) with current velocities retrieved from model simulation.

Calibration showed good agreement with field data. The comparisons between harmonic analysis and simulated results are shown in Fig. 11.13 and the Root Mean Square Error (RMSE) is acceptable (Table 11.2).

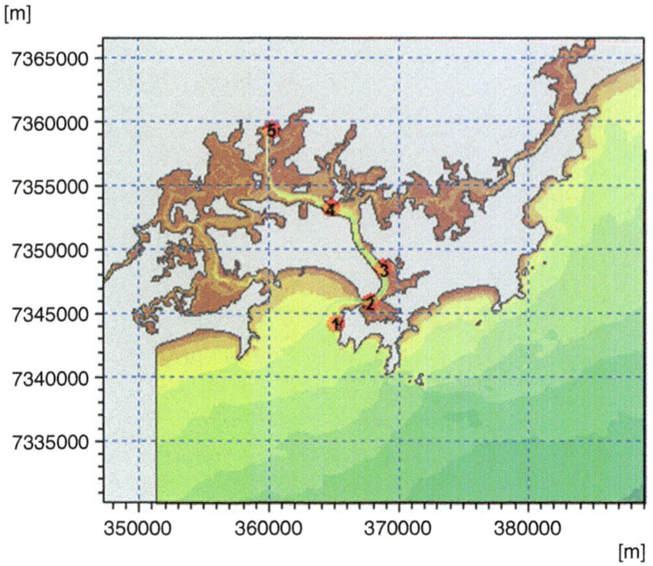

Fig. 11.11 Tide gauge stations along Santos estuary used to calibrate the hydrodynamic model. 1-Ilha das Palmas, 2-Praticagem, 3-Conceiçãozinha, 4-Ilha Barnabé and 5-Cosipa

Fig. 11.12 Eight flow stations (S04, S05, S06, S07, S08, S09, S10, S11) along Santos estuary used to validate the hydrodynamic model for currents

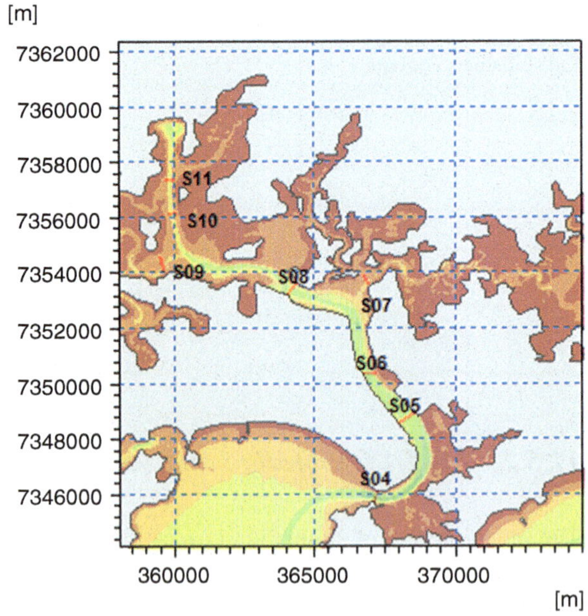

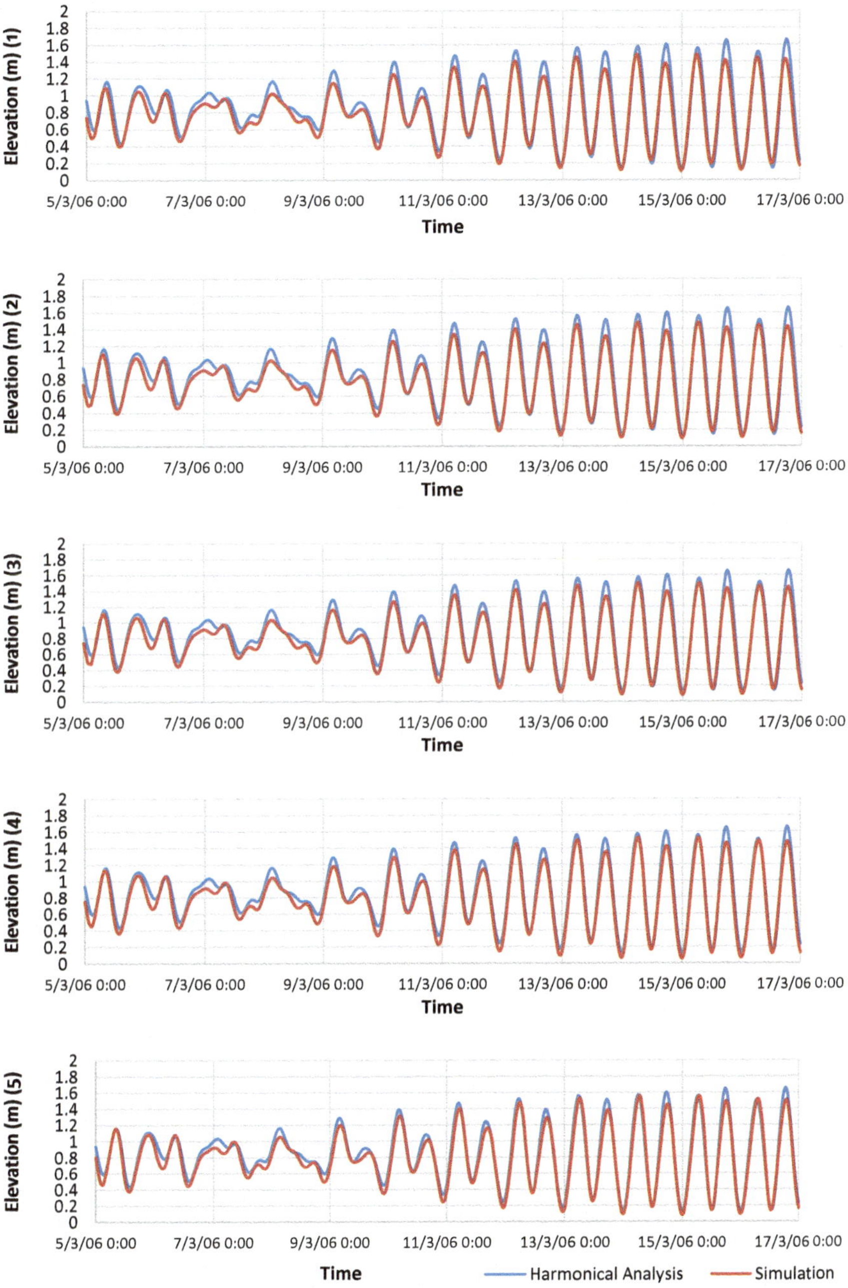

Fig. 11.13 Time series comparative between harmonic analysis (blue) and simulation (red) for tide gauge station (1) Ilha das Palmas, (2) Praticagem, (3) Conceiçãozinha, (4) Ilha Barnabé e (5) Cosipa

Table 11.2 Root Mean Square Error (RMSE) for each tide gauge station

Tide gauge	RMSE (m)
(1) Ilha das Palmas	0.0817
(2) Praticagem	0.0496
(3) Conceiçãozinha	0.0405
(4) Ilha Barnabé	0.0677
(5) Cosipa	0.0755

Due to lack of data, the period of validation covers only 8 days, from March 9th to 17th of 2006, and the model had no adjustment in the validation process. Figure 11.14 shows comparison between flow measurement and simulation.

The validation results show good or acceptable agreement with measured values from flow stations along Santos Estuary. Considering that flow measurements are influenced by meteorological tide, river discharges and salinity, the results show acceptable error (Table 11.3), since the model consists of astronomical tide forcing and long-period river discharges.

Skill score for S07 and S09 flow stations may be lower because the bathymetry has coarser scatter data in this area. Besides, S07 and S09 are also influenced by São Vicente Estuary and Bertioga Estuary, respectively.

Due to high influence of rivers Mogi, Itapanhaú, and Cubatão (considering Henry Borden discharge), S10 and S11 flow stations presents the two lowest Skill score among all. In addition, the model has constant river discharges.

Therefore, the calibration and validation support that this model is reliable to retrieve results of tidal currents velocity, and water elevation from Santos estuary.

11.6 Stability of Santos Estuary Tidal Inlet

11.6.1 Calibration of Area-Prism Relationship Using the Slice Method

Following the Slice Method described in Sect. 11.4.1, Santos estuary tidal inlet was divided into slices, and the specific discharge time series of each slice was retrieved from the hydrodynamic numerical model described in Sect. 11.5.

Then, tidal prism was computed during each tidal period (consecutive flood and ebb water slacks). The larger spring tidal prism for the simulated period summed a volume of 55.1×10^6 m^3. During the tidal period related to the highest tidal prism, slice tidal prism (ΔP) was computed for the 43 slices of Santos estuary tidal inlet, and plotted with their corresponding slices areas (ΔA) (Fig. 11.15).

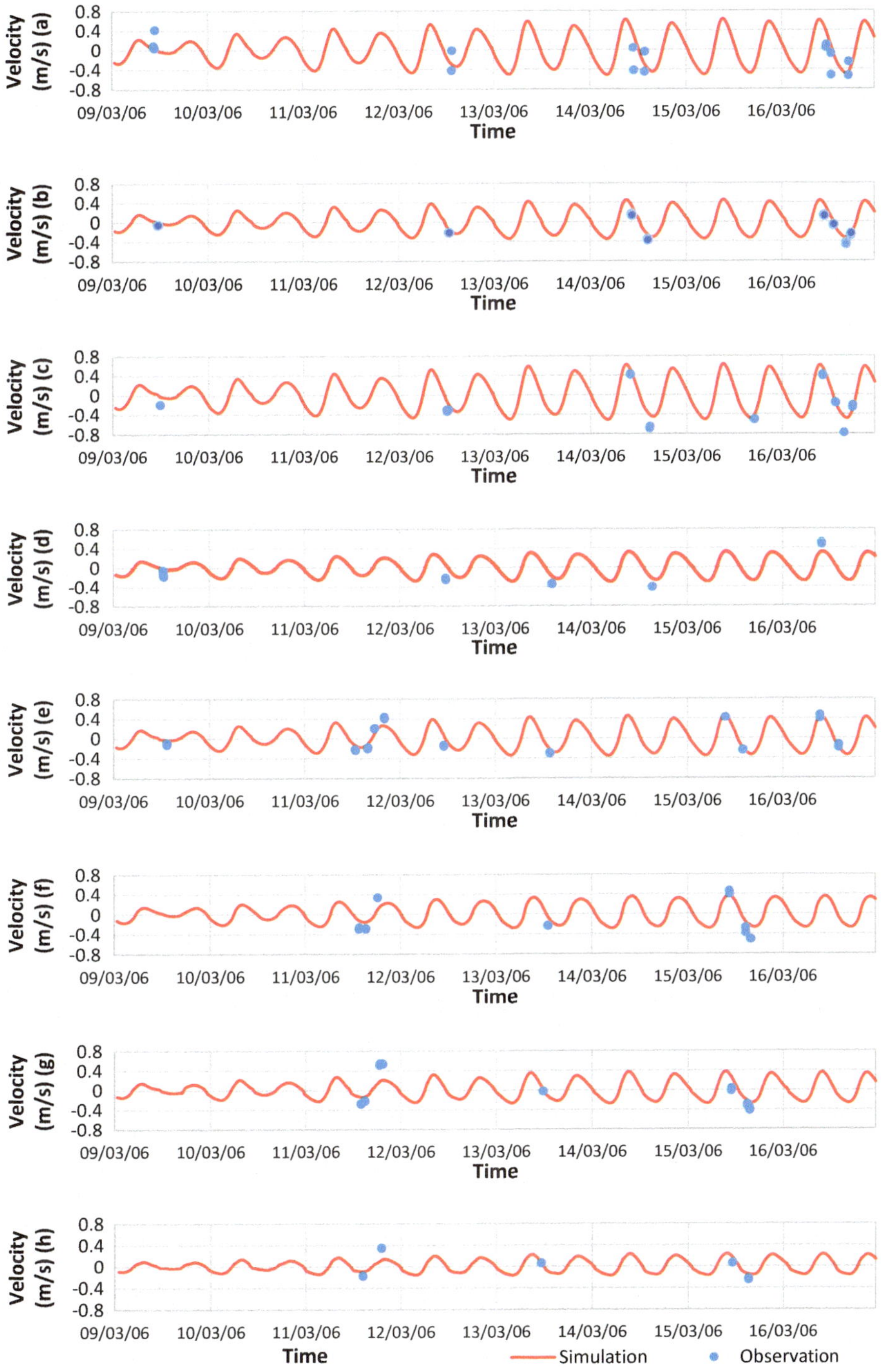

Fig. 11.14 Model validation for flow stations S04 (**a**), S05 (**b**), S06 (**c**), S07 (**d**), S08 (**e**), S09 (**f**), S10 (**g**), S11 (**h**). Comparative between flow measurement (blue) and simulation (red)

Table 11.3 Skill score for current measurements in each section (0 – poor agreement and 1 – good agreement)

Tide gauge	Skill
(a) S04	0.9366
(b) S05	0.9360
(c) S06	0.9279
(d) S07	0.8362
(e) S08	0.9001
(f) S09	0.8431
(g) S10	0.6165
(h) S11	0.6427

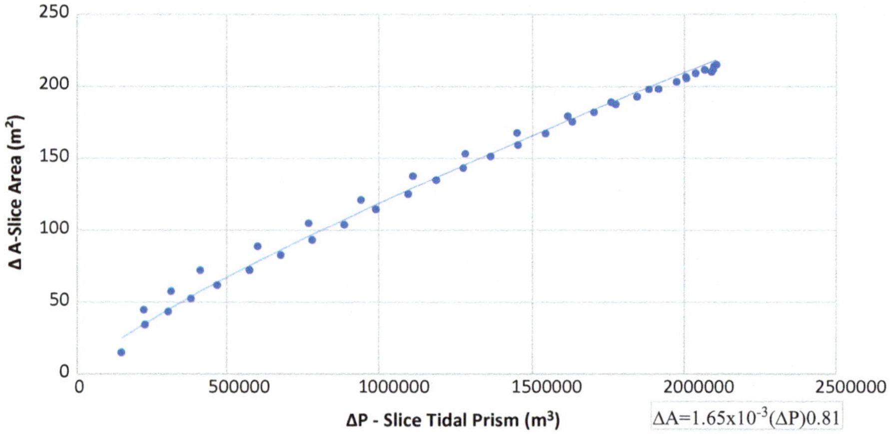

Fig. 11.15 AP relationship for Santos estuary tidal inlet derived from the Slice Method, and potential trend line for scatter data

For the calibration of coefficients C and q, a non-linear optimization model was used. The model consists of calculating the differences between measured (ΔA_j) and computed ($\overline{\varnothing A_j}$) area for each slice, and then minimizing the sum of the square of these differences (Eq. 11.9). The Generalized Reduced Gradient method was applied to solve the optimization model.

$$Min\,Error = \sum_{w}^{j=1} \left(\Delta A_j - \overline{\Delta A_j} \right)^2 \tag{11.9}$$

Then, the coefficients that best fit the scatter data (Fig. 11.15) were $C = 1.65 \times 10^{-3}$ and $q = 0.81$. It is important to emphasize that this adjustment works better for interpolations. Therefore, the Slice Method may be applied using Eq. 11.10, so the total cross-sectional area will be the sum of all slices (Eq. 11.11).

$$\Delta A_j = C\left(\Delta P_j\right)^q \tag{11.10}$$

$$A = \sum_{w}^{j=1} \Delta A_j = \sum_{w}^{j=1} C\left(\Delta P_j\right)^q \tag{11.11}$$

Before evaluating Santos estuary tidal inlet stability, the baseline scenario from 2006 must be reproduced. So, Eq. 11.11 was applied for three different approaches with the empirical coefficients from Jarrett (1976), with the theoretical coefficients from Hughes (2002), and with the coefficients provided by the Slice Method.

The cross-sectional area was computed by the sum of the 43 slices (Eq. 11.11). Each slice was geometrically approximated to a trapezium. Two consecutive depths are the bases (parallel sides) and the slice width (Δl) is the orthogonal leg. So, the real cross-sectional area in 2006 was 6105 m^2.

Thus, the cross-sectional profile and area were estimated using Jarrett (1976) empirical coefficients, Hughes (2002) theoretical coefficients, and the Slice Method coefficients. For Hughes (2002) AP relationship (Eq. 11.3), the coefficient C was calculated using data from Santos estuary, where T = 43,200 s (tidal period), $d_e = 0.11$ mm (median grain size), $W = 385$ m (tidal inlet width), $g = 9.81$ m/s^2 (gravity acceleration), $S = 2.65$ (sediment specific gravity), and $k_a = 1$ (coefficient related to the effects of non-sinusoidal tide).

First, Eq. 11.10 was applied to provide the area of each slice (ΔA_j), and then the slices were summed to estimate the total cross-sectional area (A) (Eq. 11.11). The coefficients used for each approach and the corresponding estimated area with relative error (compared to the real cross-sectional area of 6105 m^2) are listed in Table 11.4.

The Slice Method had the smallest relative error (+0.4%) and the cross-sectional profile was similar to field data from 2006 (Fig. 11.16), because its coefficients were designed minimizing the error of each slice area

When the classical AP relationship (Eq. 11.1) is used to estimate the entire cross-sectional area (A) in function of tidal prism ($P = 55.1 \times 10^6$) with Hughes (2002) and Jarrett (1976) coefficients, the relative errors are higher (Table 11.5) than the application of these coefficients on Eq. 11.11. Thus, slicing the cross-section improved the accuracy of cross-sectional area estimative in both cases.

Table 11.4 Estimative of Santos estuary tidal inlet cross-sectional area using the Slice Method with different coefficients

Coefficient	C	q	Cross-sectional area (m^2)	Relative error (%)
Jarrett (1976)	2.41×10^{-4}	0.93	4,912	−19.5
Hughes (2002)	5.78×10^{-4}	0.89	6,679	+9.4
Slice method	1.65×10^{-3}	0.81	6,132	+0.4

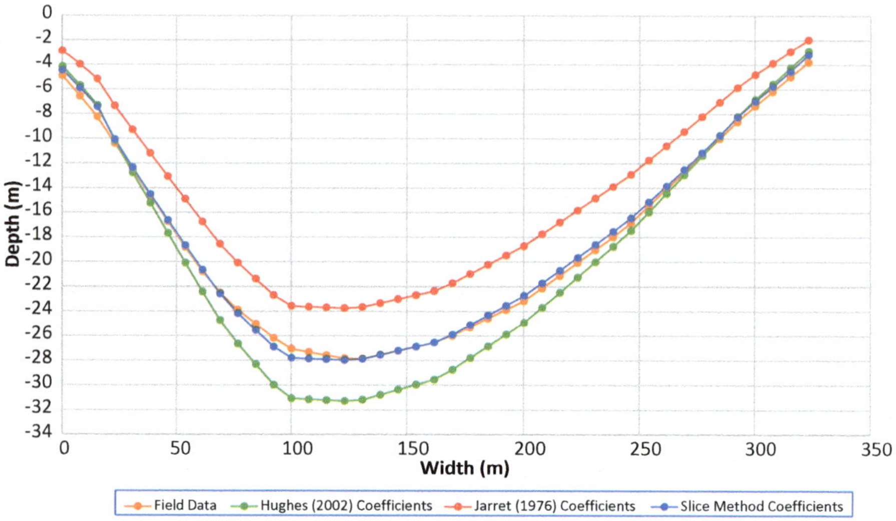

Fig. 11.16 Comparison between the real and estimated cross-sectional profiles in 2006

Table 11.5 Estimative of Santos estuary tidal inlet cross-sectional area using the classical area-prism relationship with Jarrett (1976) and Hughes (2002) coefficients

Coefficient	C	q	Cross-sectional area (m^2)	Relative error (%)
Jarrett (1976)	2.41×10^{-4}	0.93	3,813	−37.6
Hughes (2002)	5.78×10^{-4}	0.89	4,482	−26.6

Therefore, the Slice Method not only provided the cross-sectional profile of Santos estuary tidal inlet, which is not provided by the classic AP relationships, but also had better cross-sectional area estimative than classical AP relationships.

11.6.2 Evaluation of Tidal Inlet Stability

The foreseen scenario is 2014, 8 years later the baseline scenario from 2006. Both scenarios were simulated by hydrodynamic numerical modelling, and the interventions considered in foreseen scenario were land reclamation (construction of EMBRAPORT and BTP) and deepening dredging (from 13 to 15 m). The tidal forcing in sea and riverine boundaries remained the same to assess the isolated effects of anthropogenic interventions.

Following the Slice Method described in Sect. 11.4.1, Santos estuary tidal inlet from 2014 was divided into 43 slices, and the specific discharge time series of each slice were retrieved from the modified hydrodynamic numerical model described in Sect. 11.5. The 43 slice tidal prisms (ΔP) were computed using Eq. 11.7, and Santos estuary cross-sectional area (A) was estimated using Eq. 11.11.

Table 11.6 Summary of Santos estuary tidal inlet cross-sectional area for baseline (2006) and foreseen scenario (2014)

	Cross-sectional area (m^2)		
Scenario (year)	Field data	Slice method	Relative error (%)
Baseline (2006)	6,105	6,132	+0.4
Foreseen (2014)	5,769	6,021	+4.4
Area reduction	5.5%	1.8%	–

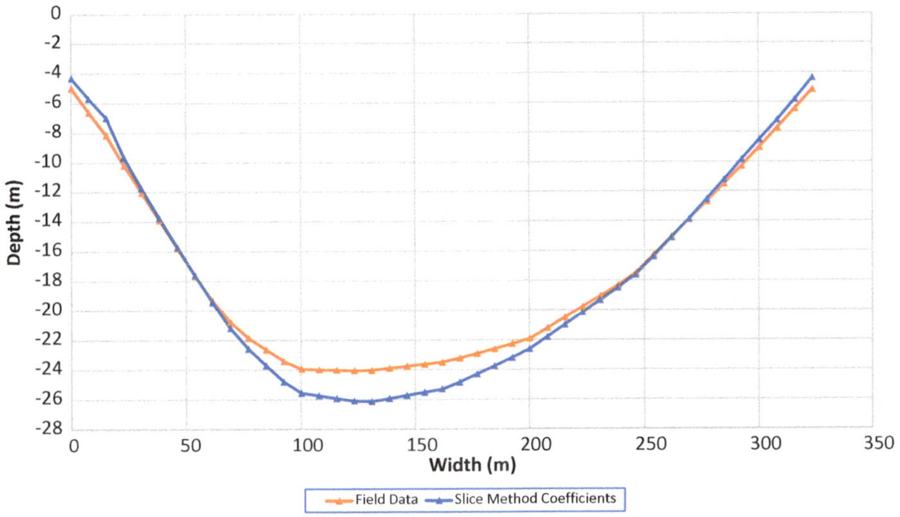

Fig. 11.17 Comparison between real and estimated cross-sectional profiles in 2014

As expected in section 11.3, Santos estuary tidal prism reduced by 2.7% to 53.6 × 10^6 m^3. The effect of this reduction is confirmed by field data and by Slice Method: both approaches show that Santos estuary tidal inlet cross-sectional area reduced (Table 11.6). This result is consistent with other cases, where cumulative land reclamation reduced tidal prism (Cuvilliez et al. 2009; Feng et al. 2015).

Despite deepening dredging tends to increase tidal prism (Oliveira et al. 2006; Malhadas et al. 2009; Picado et al. 2010), Santos estuary tidal prism reduced between 2006 and 2014. Therefore, land reclamation effect on tidal prism reduction overlapped deepening dredging opposite effect.

Regarding tidal inlet cross-sectional area, the Slice Method estimated a lesser reduction (1.8%) compared to the actual reduction (5.5%) from 2006 to 2014 (Table 11.6 and Fig. 11.17). Only land reclamation and deepening dredging in Santos estuary were accounted, so changes and responses from mangrove and interventions in São Vicente and Bertioga estuarine channels were not updated. Not updating these areas, which also affect Santos estuary tidal prism, might explain why the Slice Method cross-sectional area reduction estimative is lower than actual reduction.

Moreover, tidal inlet may take decades to adapt from land reclamation (van de Kreeke 2004; Oost 1995). Despite Santos estuary tidal inlet may be under adaptation, field data and the Slice Method confirm that tidal inlet got shallower and its stability decreased during this short period of 8 years.

11.7 Conclusions

By dividing tidal inlet in several slices, the Slice Method enhanced the estimative of tidal inlet cross-sectional area (Eq. 11.11). Also, the calibration of coefficients C and q were improved by minimizing the sum of the square of the differences between measured (ΔA_j) and computed ($\overline{\varnothing A}_j$) slice areas (Eq. 11.9).

Therefore, the Slice Method provided a reliable estimative of cross-sectional profile, making possible the evaluation of tidal inlet geometry changes. Even using Jarrett (1976) and Hughes (2002) coefficients, the results were more accurate than the application of AP relationship to the entire tidal inlet (Eq. 11.1).

Santos estuary tidal prism reduced from 55.1×10^6 in 2006 to 53.6×10^6 in 2014, and the cross-sectional area reduced 5.5%. These observations support the assumption that tidal inlet stability decreased due to land reclamation. The Slice Method estimative of 1.8% on cross-sectional area reduction could be more accurate if bathymetric data of mangrove and São Vicente and Bertioga estuarine channels were updated. Nevertheless, the Slice Method estimated Santos estuary tidal inlet area with good accuracy, only 4.4% larger. Moreover, the Slice cross-sectional profile output is consistent with field data investigations; both identified that tidal inlet got shallower.

Since land reclamation affects tidal prism, slight changes in estuarine lands may bring major effects in tidal inlet stability. Then, future Environment Impact Assessment of estuarine occupations in Santos estuary must consider the cumulative effects of land reclamation on tidal inlet stability in order to mitigate channel siltation and possible impacts on adjacent shoreline.

Acknowledgements The authors thank DHI (Danish Hydraulic Institute) for providing a time-limited license for thesis of the Mike 21 FM software, and the first and the third authors thank CAPES (Coordenação de Aperfeiçoamento de Pessoal de Nível Superior) for the funding of their scholarship.

References

Brasil. (2007). *Federal Law N 11610/2007: "Plano nacional de dragagem portuária e hidroviária". Brasília.* Available at: http://www.planalto.gov.br/ccivil_03/_ato2007-2010/2007/Lei/L11610.htm.

Brasil. (2013). *Federal Law N 12815/2013: "Plano nacional de dragagem postuária e hidroviária 2". Brasília.* Available at: http://www.planalto.gov.br/ccivil_03/_ato2011-2014/2013/lei/l12815.htm.

Bruun, P. (1978). Development of tidal inlets. In P. Bruun (Ed.), *Stability of tidal inlets theory and engineering, Developments in Geotechnical Engineering* (Vol. 23, pp. 13–38). Amsterdam: Elsevier.

CODESP – Companhia de Docas do Estado de São Paulo. (2015). *Porto de Santos – Relatório Anual 2014*. Available at: http://201.33.127.41/down/relatorio/Relatorio2014.pdf

Cuvilliez, A., Deloffre, J., Lafite, R., & Bessineton, C. (2009). Morphological responses of an estuarine intertidal mudflat to constructions since 1978 to 2005: The seine estuary (france). *Geomorphology, 104*(3), 165–174.

EMBRAPORT. (2013). Um passo importante foi dado em 2009, com a entrada de dois grandes acionistas, ODEBRECHT TRANSPORT e a DUBAI PORTS WORLD. EMBRAPORT – Empresa Brasileira de Terminais Portuários. Available at: http://www.embraport.com/aembraport/historia/. Accessed 22 Feb 2017

Feng, L., He, J., Ai, J., Sun, X., Bian, F., & Zhu, X. (2015). Evaluation for coastal reclamation feasibility using a comprehensive hydrodynamic framework: A case study in Haizhou Bay. *Marine Pollution Bulletin, 100*(1), 182–190.

Garcia, P. D., Araújo, R. N., Silva, G. D. C., Baptistelli, S. C., Alfredini, P. (2002). Preparo de bases batimétricas, de agitação e circulação para o litoral do estado de São Paulo. Boletim Técnico da Escola Politécnica da USP. BT/PMI, Escola Politécnica da USP, 105(1), 1–56.

Gireli, T. Z., & Vendrame, R. F. (2012). Deepening the Port of Santos: A critical analysis. *Revista Brasileira de Recursos Hídricos, 17*(3), 49–59.

Harari, J., & Camargo, R. (1994). Simulação da propagação das nove principais componentes de maré na plataforma sudeste brasileira através do modelo numérico hidrodinâmico. *Boletim Instituto Oceanografico, 42*(1), 35–54.

Hughes, S. A. (2002). Equilibrium cross sectional area at tidal inlets. *Journal of Coastal Research*, 160–174.

Hume, T. M., & Herdendorf, C. E. (1993). On the use of empirical stability relationships for characterising estuaries. *Journal of Coastal Research, 9*(2), 413–422.

INPH. (2007). Relatório INPH n° 018/2007 – Projeto Geométrico da Infra-Estrutura Aquaviária do Porto de Santos – SP – Anexo XI. Rio de Janeiro.

Jarrett, J. T. (1976). *Tidal prism–inlet area relationships, GITI report no. 3*. Fort Belvoir: Coastal Engineering Research Center, US Army Corps of Engineers.

Kraus, N. C. (1998). Inlet cross-sectional area calculated by process-based model. *Coastal Engineering, 1998*, 3265–3278.

Leitão, J. C., Mateus, M., Neves, R. (2008). Calibration of the hydrodynamic model for the Santos Estuary. Technical report.

Malhadas, M. S., Silva, A., Leitão, P. C., & Neves, R. (2009). Effect of the bathymetric changes on the hydrodynamic and residence time in Óbidos Lagoon (Portugal). *Journal of Coastal Research, SI56*, 549–553.

MARMIL – *Centro de Hidrografia da Marinha*. Marinha do Brasil. (2015). Available at: http://www.mar.mil.br/dhn/chm/box-cartas-raster/raster_disponiveis.html. Accessed 30 Sept 2015.

Mehta, A., & Ozsoy, E. (1978). 3 – inlet hydraulics: 3.1 flow dynamics and nearshore transport. In P. Bruun (Ed.), *Stability of Tidal Inlets: Theory and Engineering* (pp. 83–161). Elsevier. (Developments in Geotechnical Engineering, volume 23).

MKR. (2003). Volume I: Capítulos 1 a 7. In EIA – Terminal Portuário EMBRAPORT – Empresa Brasileira de Terminais Portuários.

O'Brien, M. P. (1969). Equilibrium flow areas of inlets on sandy coasts. *Journal of the Waterways and Harbors Division, ASCE, 95*, 43–52.

Oliveira, A., Fortunato, A. B., & Rego, J. R. (2006). Effect of morphological changes on the hydrodynamics and flushing properties of the Óbidos lagoon (Portugal). *Continental Shelf Research, 26*(8), 917–942.

Oost, A. P. (1995). Sedimentological implications of morphodynamic changes in the ebb-tidal delta, the inlet and the drainage basin of the Zoutkamperlaag tidal inlet (Dutch Wadden Sea), induced by a sudden decrease in the tidal prism. In *Tidal signatures in modern and ancient sediments* (vol. 24, pp. 101–119).

Picado, A., Dias, J. M., & Fortunato, A. B. (2010). Tidal changes in estuarine systems induced by local geomorphologic modifcations. *Continental Shelf Research, 30*(17), 1854–1864.

Powell, M. A., Thieke, R. J., & Metha, A. J. (2006). Morphodynamic relationships for ebb and flood delta volumes at Florida's tidal entrances. *Ocean Dynamics, 56*, 295–307.

Roversi, F., Colonna Rosman, P. C., & Harari, J. (2016). Análise da renovação das águas do Sistema Estuarino de Santos usando modelagem computacional. *Ambiente & Água-Na Interdisciplinary Journal of Applied Science, 11*(3), 567–585.

Scazufca, M. (2012). *A primazia do Porto de Santos no cenário portuário nacional no período contemporâneo.* Determinantes logísticos, territoriais e de gestão. Tese (Doutorado) – Universidade de São Paulo, 2012.

Sondotécnica. (1977a). Comportamento Hidráulico e Sedimentológico do Estuário Santista. Relatório Final- Texto.

Sondotécnica. (1977b). Comportamento Hidráulico e Sedimentológico do Estuário Santista. Relatório Final- Desenhos.

Stive, M. J. F., Ji, L., Brouwer, R. L., Van De Kreeke, J., Ranasinghe, R. (2010). Empirical relationship between inlet cross-sectional area and Tidal Prism: A re-evaluation. In J. M. Smith, & P. Lynett (Eds.), *Proceedings of 32nd conference on coastal engineering, Shanghai, China, 2010. Sediment transport and morphology*.

Townend, I. (2005). An examination of empirical stability relationships for UK estuaries. *Journal of Coastal Research, 21*(5), 1042–1053.

van de Kreeke, J. (2004). Equilibrium and cross-sectional stability of tidal inlets: Application to the frisian inlet before and after basin reduction. *Coastal Engineering, 51*, 337–350.

WEDA. (2013). *WEDA 2013 environmental dredging award: Embraport container and bulk terminal – Santos, Brazi"l.* Available at: https://www.westerndredging.org/phocadownload/2013 EnvironmentalAwards/weda%202013%20embraport%20application.pdf.

Part IV
Adaptation

Chapter 12
Vulnerability of Critical Infrastructure Under Climate Change Scenarios: The Case of Santos

Vitor Baccarin Zanetti, Wilson Cabral de Sousa Júnior, and Eduardo Kimoto Hosokawa

Abstract Vulnerability is a concept that may be applied to many things, and is a way to identify points that must be prioritized by risk analysis. To that principle this chapter will explore the application of the SEVICA index (Zanetti VB, de Sousa Júnior WC, De Freitas DM. Sustainability 8:811–823, 2016) to the city of Santos, and try to identify sensible infrastructure that are in vulnerable areas, to extreme events effects, to show that much of the basic infrastructure that would be necessary in the occurrence of a extreme climate event, would in fact be affected by it, amplifying the potential vulnerability of the city habitants. In the study of the city of Santos, that most of the infrastructure in Santos, an important port city in Brazil, is located in areas of high vulnerability, which poses a major threat to the population, and a big challenge to city planners, to adapt the city and try to mitigate the possible impacts of the climate changes.

Keywords City infrastructure · Extreme climate events · Rainfall · Sea level rise

12.1 Introduction

Since the first assessment report of the Intergovernmental Panel on Climate Change (1990), vulnerability and risk assessments have been important approaches to examining climate change-related risks and impacts such as variations in temperature, rainfall, and sea-level rise (e.g., IPCC 1990; Gornitz 1991; Shaw et al. 1998). These assessments have focused mostly on coastal areas, where climate-related stressors are predicted to significantly affect urban populations and infrastructure.

V. B. Zanetti (✉) · W. C. de Sousa Júnior
Water and Environments Department, Civil Engineer Division,
Technological Institute of Aeronautics, São José dos Campos, Brazil

E. K. Hosokawa
Municipal Government of Santos, Secretariat of Urban Development, Santos, SP, Brazil
e-mail: eduardohosokawa@santos.sp.gov.br

L. H. Nunes et al. (eds.), *Climate Change in Santos Brazil: Projections, Impacts and Adaptation Options*, https://doi.org/10.1007/978-3-319-96535-2_12

Noteworthy is the number of indexes that has been developed to assess coastal vulnerability (e.g., Gornitz 1991; Cutter et al. 2003; Balica et al. 2012). This is also true for Brazil (see Alves 2006; Nicolodi and Petremann 2010; Iwama et al. 2014; Furlan et al. 2011). However, rarely do coastal vulnerability indexes consider inland (e.g., landslides and flooding) and ocean (sea-level rise and coastal erosion) hazards in conjunction, which can be particularly relevant in areas where the relief plays an important role in the landscape. To help fill this gap, the Socio-Environmental Vulnerability Index for Coastal Areas (SEVICA) was developed by Zanetti et al. (2016), who used the city of Santos, on the central coast of São Paulo, as a case study.

A study that applied the SEVICA index to assess the vulnerability of the municipality of Santos has already been conducted. To extend that study, we plotted the city's critical public infrastructure onto the vulnerability maps for the Representative Concentration Pathways (RCP) 8.5 scenario from the IPCC AR5 report (IPCC 2014). We then analysed the resulting infrastructure map and the possible impacts of an extreme event on the city.

The analysis is based on the premises that (1) the infrastructure is critical to rescue operations and to assisting the population after the occurrence of an extreme event; and (2) the failure of the infrastructure could lead to an increased number of victims, due to a possible inability of that infrastructure to fully attend to the populace.

12.2 Methods

The Socio-environmental Vulnerability Index for Coastal Areas (SEVICA) was developed by Zanetti et al. (2016) and consists of seven factors: four geophysical and three socio-economic factors (Fig. 12.1). Each of these factors has independent parameters and a distinct composition. Each parameter is ranked from 1 to 5 on the quantitative vulnerability scale. The vulnerability level of a factor is defined by the parameters' mean.

To calculate the SEVICA, all the parameters used to estimate the index factors are first determined for the city area and plotted on thematic maps for each parameter. Those values are then used to calculate the factor maps, using map algebra. The factor maps, in turn, are used to calculate the SEVICA through weighted mean map algebra.

For the infrastructure analysis, we plotted the city's main infrastructure (based on information provided by the city council) over the results from Zanetti et al. (2016) for the RCP 8.5 scenario (IPCC 2014). Using the resulting map, we evaluated the possible impacts based on the position of the infrastructure.

12.3 Infrastructure Vulnerability Analysis

Five main components of public infrastructure were analysed in this study: firefighters, who are the first line of rescue in the city; police, who also act on the first line of defence; military, which in Brazil is often an auxiliary force in the rescue of, and

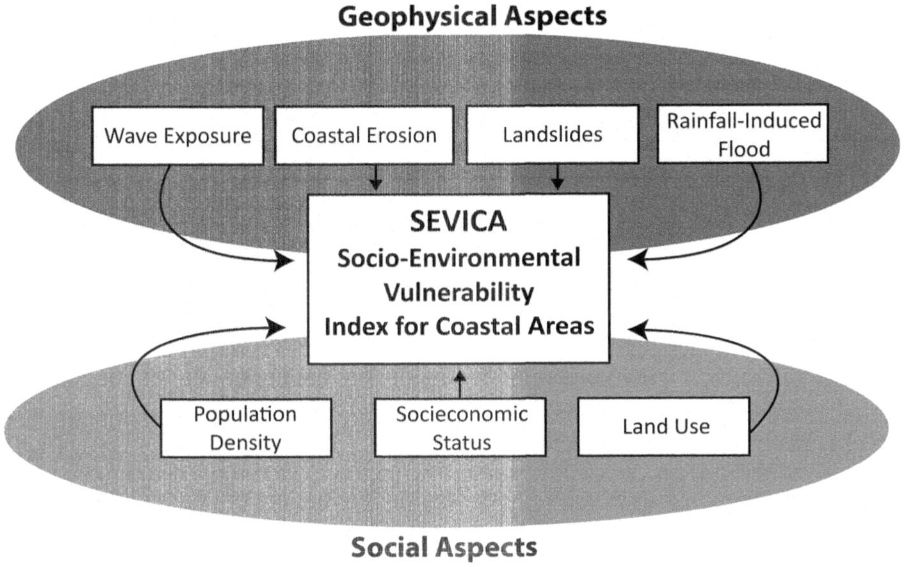

Fig. 12.1 Structure of the Socio-environmental Vulnerability Index for Coastal Areas – SEVICA. (Zanetti et al. 2016)

attendance to, the population in extreme events; public hospitals, which receive all the victims; and gymnasiums, which are used as shelter if residents are displaced from their homes due to extreme events. In a second analysis, critical transportation infrastructure, such as main roads and main public transport terminals, were plotted over the same vulnerability map.

Although Zanetti et al. (2016) discusses the vulnerability map that resulted from the SEVICA for the RCP 8.5 scenario, it is important to also briefly present it here, as it is the basis of our discussion. Figure 12.2 presents the result for the most severe vulnerability scenario in Santos, in which more than 70% of the city is categorised as high or very high vulnerability. In contrast, the area categorised as low or very low vulnerability is extremely small. This general situation poses a big challenge for the local government.

The vulnerability in Santos is directly linked to several factors: The city is located on an island, with predominance of alluvial soils, and most of the area is at altitudes below 5 m. Furthermore, the city is densely populated and the area is densely built, factors that concur to create a very critical scenario.

Figure 12.3 shows that the essential infrastructure of the city is located in areas of high or very high vulnerability, amplifying the critical scenario.

Almost all of the fire stations are in high vulnerability areas, which could affect the first line of response to an extreme event. The situation is aggravated by the fact that fire stations in Brazil also house some of the ambulance services. The SAMU (Serviço de Atendimento Móvel de Urgência, in Portuguese) is the main source of

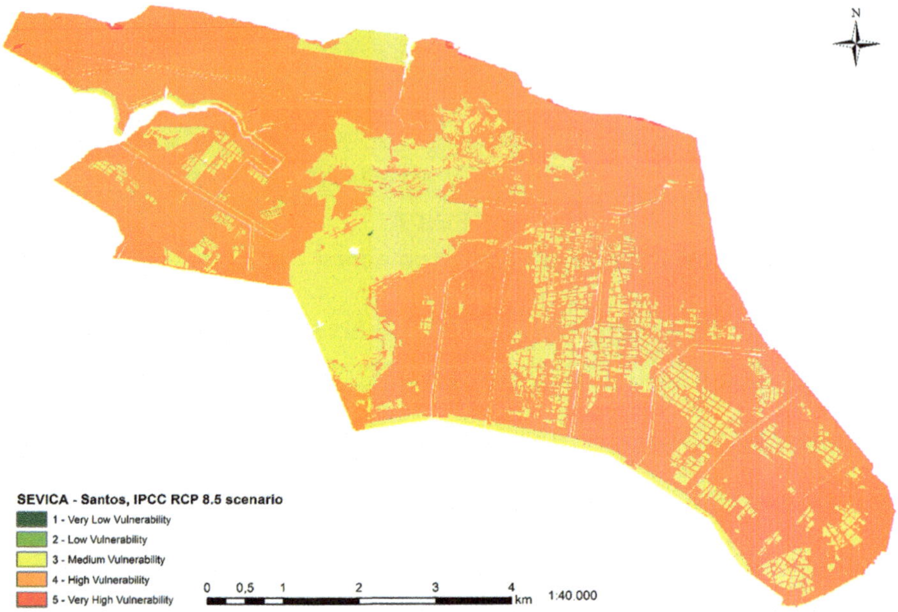

Fig. 12.2 Vulnerability of Santos under IPCC RCP 8.5 scenario (0.38 m sea-level rise projection) based on the SEVICA index

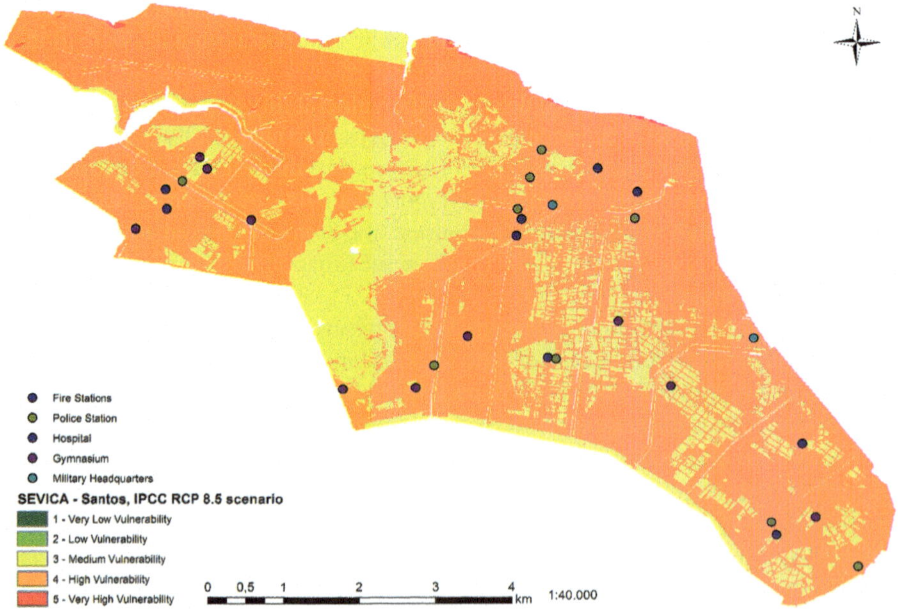

Fig. 12.3 Critical infrastructure plotted over vulnerability map

rescue ambulances in Santos, and one of the SAMU units is in a fire station in the city's historical centre, which is a high vulnerability area.

Like the fire stations, the police and military headquarters are in the high vulnerability areas, which may prevent rapid responses if a disaster hit the city. If these basic services were affected by a severe event, the city's capacity to cope with that event would be reduced to critical levels.

Furthermore, the two main hospitals in the city, Santa Casa and Pronto Socorro Municipal (both in the centre of the map in Fig. 12.3), are in high vulnerability areas. These two facilities are the main source for public medical care and first aid in extreme events, and both are located on the margin of the city's drainage channel, in very low altitude areas, and thus prone to flooding. If both hospitals are affected by a hazardous event, the city's capacity to provide support for victims would be extremely reduced, with disastrous consequences. Other care facilities, such as Emergency Units, might also be affected, and access to them may be totally or partially limited.

Although gymnasiums, which would provide shelter for displaced people, are the least critical structures, most of them are in high vulnerability areas. Moreover, the current poor condition of urban infrastructure such as road pavement and asphalt may prevent access to these shelters.

In Fig. 12.4, the vulnerability map shows the main roads and the public transportation terminal, illustrating the high vulnerability of the urban transportation infrastructure. The main roads are in vulnerable or highly vulnerable areas. Given that individual cars and motorcycles are the main means of transportation in Brazil, the

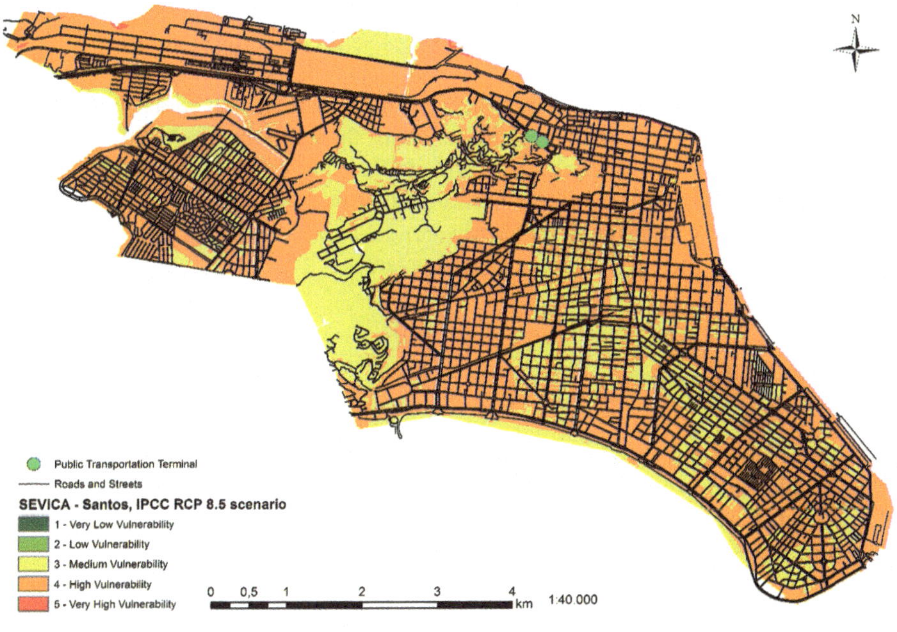

Fig. 12.4 Roads and transport infrastructure plotted over vulnerability map

flooding of major roads can virtually paralyse a city, as vehicular travel throughout the area becomes difficult or even impossible.

Public transport would also be affected, since the main municipal bus terminal is in a high vulnerability area associated with the main drainage channels of the city. The roads next to the bus terminal are one of the city's vulnerability hotspots in terms of flooding. One major event could affect all the bus lines throughout the city. The bus terminal regularly serves a high volume of people, who, in the event of a severe flood, could become trapped temporarily in the building.

The main connection between the city and the mainland is the Anchieta Highway, in the northern part of the island, a region of high vulnerability due to low altitude areas and the presence of chemical plants (namely oil and gas). If this particular highway were affected, the city would be almost isolated from the mainland. With respect to the city infrastructure, the highway is a particularly critical issue because the area surrounding it has historical importance, featuring buildings from the colonial period of Brazil. These buildings are part of the Brazilian Historical Heritage and are protected by restrictive laws that make it difficult to implement adaptation measures or major interventions to mitigate the vulnerability of the area.

The vulnerability of roads is by far the city's greatest problem, because it affects the entire municipality, posing a major challenge to people trying to leave the affected areas and to the rescue teams trying to reach people in need. In a critical scenario, where most of the city would be affected by an extreme event involving heavy rainfall accompanied by abnormally high tides, the city's main roads would be affected directly because they are close to the drainage channels. Therefore, most of the population would not be able to leave the area, and those who might try to escape could become trapped in traffic jams or by flash flooding.

In addition, most of the residential parking lots are underground or below street level, and many of them have only one way to enter and exit. This poses a serious threat, as demonstrated during past flood situations in which fatalities occurred when people tried to rescue their vehicles. Adjustments to the city's building codes could help to diminish this particular threat. New buildings have avoided underground garages, but in the case of buildings that do have them, the challenge for local government is to promote public awareness about the risk of trying to recover material assets in the case of a catastrophic event.

12.4 Conclusion

The study shows that most of the infrastructure in Santos, an important port city in Brazil, is located in areas of high vulnerability, which poses a major threat to the population. City planners must place particular focus on adapting the city infrastructure to the new scenario created by climate change.

Although the present study offers information that city planners can use to improve the resilience of Santos, additional evaluations are necessary to ensure a more resilient city. The results presented here should be used to design further risk analyses, as the city has a limited budget for research of this kind. The study can also serve to inform special city councils' discussions on climatic change and city development.

The current analysis considered only the most critical scenario of SEVICA, but other scenarios should be considered in further evaluations on the theme.

Although this infrastructure analysis is specific to the city of Santos, it can be replicated for other cities. Where data or time are lacking, or where budget constraints do not allow for more comprehensive risk assessments, a study of the impacts of climate change on urban infrastructure is a useful tool for policy makers and provides relevant information concerning urban adaptation to climate change.

References

Alves, H. P. F. (2006). Vulnerabilidade socioambiental na metrópole paulistana: uma análise sociodemográfica das situações de sobreposição espacial de problemas e riscos sociais e ambientais Revista Brasileira de Estudos de População. *Rio de Janeiro, 23*(1), 43–59.

Balica, S. F., Wright, N. G., & Van Der Meulen, F. (2012). A flood vulnerability index for coastal cities and its use in assessing climate change impacts. *Natural Hazards, 64*(1), 73–105. https://doi.org/10.1007/s11069-012-0234-1.

Cutter, S. L., Boruff, B. J., & Shirley, W. L. (2003). Social vulnerability to environmental hazards. *Social Science Quarterly, 84*, 242–261.

Furlan, A., Bonotto, D. M., & Gumiere, S. J. (2011). Development of environmental and natural vulnerability maps for Brazilian coastal at São Sebastião in São Paulo State. *Environmental Earth Sciences, 64*, 659–669.

Gornitz, V. (1991). Global coastal hazards from future sea-level rise. *Palaeogeography, Palaeoclimatology, 89*(4), 379–398.

IPCC – International Panel on Climate Change. (1990). *Climate change: The IPCC scientific assessment* (410 pp). Cambridge: Cambridge University Press.

IPCC – International Panel on Climate Change. (2014). Summary for policy makers. In C. B. Field, V. R. Barros, D. J. Dokken, K. J. Mach, M. D. Mastrandrea, T. E. Bilir, M. Chatterjee, K. L. Ebi, Y. O. Estrada, R. C. Genova, et al. (Eds.), *Climate change 2014: Impacts, adaptation, and vulnerability. Part A: Global and sectoral aspects. Contribution of working group II to the fifth assessment report of the Intergovernmental Panel on Climate Change* (pp. 1–32). Cambridge: International Panel on Climate Changes (IPCC).

Iwama, A. Y., Batistella, M., & Ferreira, L. C. (2014). Geotechnical risks and social vulnerability in coastal areas: Inequalities and climate change. *Ambiente e Sociedade, 17*, 251–274.

Nicolodi, J. L., & Petremann, R. M. (2010). Mudanças climáticas e vulnerabilidade da zona costeira do Brasil: aspectos ambientais, sociais e tecnológicos. *Revista da Gestão Costeira Integrada, 10*(2), 151–177.

Shaw, J., Taylor, R. B., Forbes, D. L., RUz, M.-H., & Solomon, S. (1998). *Sensitivity of the coasts of Canada*. Ottawa: Atlantic Geoscience Centre, Geological Survey of Canada.

Zanetti, V. B., de Sousa Junior, W. C., & De Freitas, D. M. (2016). A climate change vulnerability index and case study in a Brazilian coastal city. *Sustainability, 8*, 811–823.

Chapter 13
Climate Change and Adaptive Capacity in the City of Santos

Fabiano de Araujo Moreira, Shona Paterson, Lucí Hidalgo Nunes, and Mark Pelling

Abstract Knowing the adaptive capacity of a given local – central objective of this work – contributes to understanding the generic capacities existing in a society that allow collective actions and its self-protection to avoid or face changes in multiple systems. The goal is to evaluate the adaptive capacity of the municipality of Santos, São Paulo, Brazil, based on the application of an index with institutional approach, the Adaptive Capacity Index (ACI) in a series of agents from the various sectors who are responsible for dictating the regulations of the municipality, observing the issue by its social aspect, conducting research on adaptation among local communities analyzing the role of institutions, culture and political changes that affect individual and/or collective in the face of adverse effects that may arise due to climate changes. According to the interviewees, there has been a slight progression of ACI components in the last 10 years in the city of Santos due to: (i) lack of integration of risk management organizations; and (ii) domain of the adaptation agenda by the local Civil Defence.

Keywords Adaptive capacity · Risk management · Adaptation · Climate change

F. d. A. Moreira (✉)
Department of Geography, University of Campinas (Unicamp), Campinas, SP, Brazil

S. Paterson
Future Earth Coasts, Cork, Ireland
e-mail: shona.paterson@ucc.ie

L. H. Nunes
Santos, São Paulo, Brazil
e-mail: luci@ige.unicamp.br

M. Pelling
Department of Geography, King's College London, London, UK
e-mail: mark.pelling@kcl.ac.uk

L. H. Nunes et al. (eds.), *Climate Change in Santos Brazil: Projections, Impacts and Adaptation Options*, https://doi.org/10.1007/978-3-319-96535-2_13

13.1 Introduction

The fifth report from the Intergovernmental Panel on Climate Change (2014), upholds the assertion that human actions continue to affect the Earth's energy balance through emission of important greenhouse gases and through land-use change. According to the AR-5, the past reports' scenarios describing CO_2 concentrations, average world temperature, and sea levels are being confirmed, while only CH_4 and N_2O levels are below those predicted.

Since the 2007 report, the IPCC has shown the highest concentration of greenhouse gases in the atmosphere in the last 500,000 years. These gases have increased by 70% in the last four decades alone. According to various models, estimates for the year 2100 show that the average global surface temperature could increase by 1.1–6.4 °C, while the average sea level could rise by 18–59 cm, and oceans could become more acidic. Indeed, the most recent IPCC report, AR5, increased its estimates of the maximum possible sea-level rise relative to the 2007 report, varying from 17 to 82 cm by the year 2100. There could be an increase in cases of temperature and precipitation that according to the current weather patterns are classified as extreme, that would become more common.

Analysing the case of Brazil, Carmo and Silva (2009) warned that most of the population residing in coastal urban agglomerations may suffer, directly or indirectly, some impact from the climate change. The authors explain that there is still a large degree of uncertainty concerning the elevation that sea-level rise can reach, since ocean dynamics involve a complex interrelationship among factors such as ocean thermal expansion and the melting of polar ice caps and large glaciers. Therefore, the models adopted by researchers in the field generate values that vary widely.

Considering the array of issues, the creation of public policies on adaptation and reduction of vulnerability is paramount, as are actions to mitigate the environmental impacts on the coastal urban agglomerations. The present analysis fits within this theme by defining a case study of the municipality of Santos, state of São Paulo, and considering the city's relationship to the Metropolitan Region of Baixada Santista (MRBS), to which it belongs.

Santos occupies a strategic role in the national economy, and like many other coastal areas, has been subject to significant financial speculation as well as environmental deterioration, which may worsen with the advent of climate change and by sea-level rise in particular. These impacts may irreversibly compromise the urban apparatus and increase the population's vulnerability, which is already high in various sectors of the municipality (Silva and Nunes 2016; Zanetti et al. 2016).

In this chapter, we present the results of one of the pillars of the Metropole Project, which was to determine the adaptive capacity of Santos through analysing the factors that affect the planning of necessary adaptation measures and policy changes within the social context of the region. The complete report is part of a working technical series of the Department of Geography, King's College London (Moreira et al. 2017).

The adaptive capacity of different private and public institutions and non-governmental organisations was assessed through interviews designed by a group of researchers at King's College London (Pelling and Zaidi 2013; Paterson and Pelling 2017; Paterson et al. 2017). Denominated the *Adaptive Capacity Index* (ACI), the set of questions was applied individually, preferentially face-to-face, and aimed to understand the institutional framework and the potential for individual actions that constitute adaptive capacity in Santos. The application of the ACI requires information about the socioenvironmental characteristics of the locale and a review of the regulatory foundation for institutional resilience and of the adaptation plans for the country, for the state of São Paulo, and for the municipality.

13.2 Study Context

Santos is the seat of the Metropolitan Region of Baixada Santista (MRBS), which encompasses eight other municipalities, covers 2423 km^2 and is home to around 1,749,343 inhabitants (Fig. 13.1). Santos is the most populous municipality of the MRBS, with almost 420,000 inhabitants (Ibge 2010). Founded in 1546, it is one of the oldest cities in Brazil. The Port of Santos was built during the city's nascence, becoming the most important port in South America and central to the economy of the municipality and of Brazil.

Santos and surroundings were originally covered by the Atlantic Rainforest, but urban expansion caused a comprehensive disturbance of its ecological dynamics, as well as of the local climate and patterns of energy and water consumption. The rainfall volumes in the MRBS are high but variable and mostly concentrate during summer months, a period which also coincides with an increase of tourism activities, thus exposing more people to the risk of floods and landslides.

Santos, like the entire coast of the state of São Paulo, is bordered by the Serra do Mar. mountain range. Of the municipality's total area of 280.7 km^2, around 15% lies on the island of São Vicente, and the rest, on the mainland. In contrast, 99.3% of the population lives in the insular part of Santos because the area on the mainland is under protection and occupation is restricted.

The spatial occupation in Santos is extremely complex, with the harbour activities, commerce and tourism on one side, and an asymmetrical urban occupational pattern on the other side (Zundt 2006). Upscale neighbourhoods are closer to the shoreline, while poor neighbourhoods are situated on the hillsides and in mangrove areas. Furthermore, the conservation reserves and legal protection areas that cover the mainland are sparsely populated; nonetheless, they are under constant pressure.

The environmental problems in the region date back to the very beginning of its occupation and persist today. Some examples are the contamination of soils, air, mangroves and estuarine waters due to effluents from heavy industry located in the adjacent city of Cubatão. This industrial activity also promotes erosive processes on the hillsides as well as deforestation.

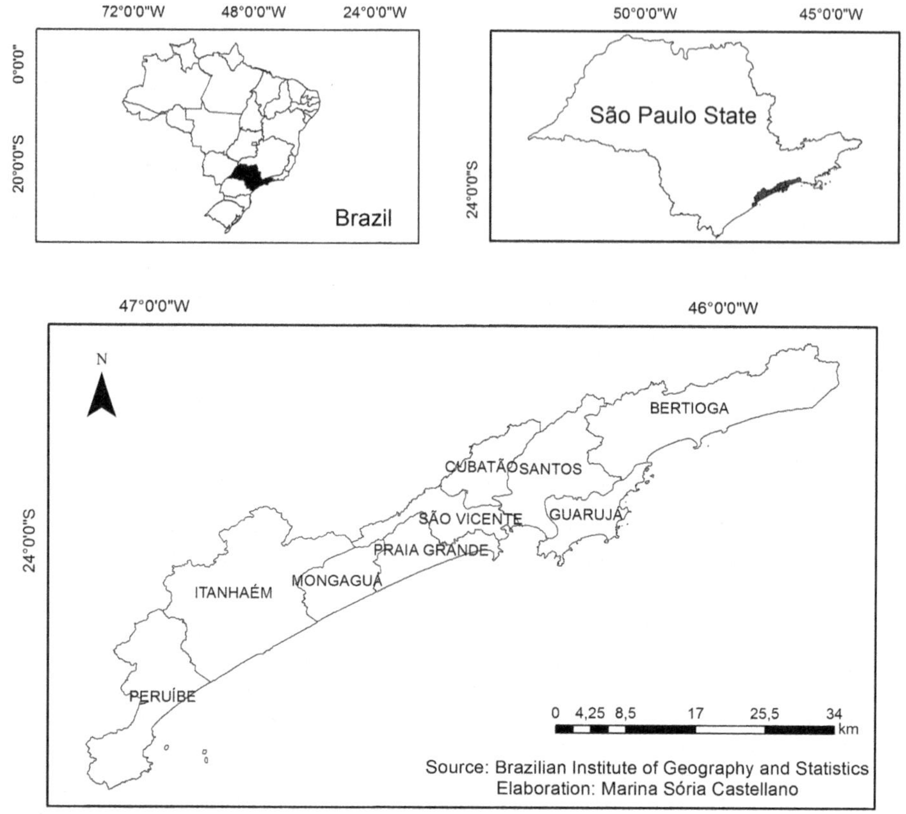

Fig. 13.1 Location of the Metropolitan Region of Baixada Santista, in Brazil

Climate change may increase both environmental susceptibility and social vulnerability, including impacts on infrastructure (see Chap. 12), demanding fast, effective and locally appropriate adaptation measures that consider the unique physical, economic and socio-spatial characteristics of the municipality. There is a heightened awareness of the urgent need for climate change adaptation measures by public authorities, as well as by public and private institutions and the population, that can evolve into effective and sustainable adaptation policies. These policies must address the effects of multiple drivers such as population growth and water demands, as well as health crises and natural disasters as direct impacts of climate change.

In spite of the importance of Santos, the municipal legislation so far does not cover climate change adaptation measures, which are instead regulated by state and national laws (Table 13.1). The state of São Paulo was one of the first Brazilian states to adopt a climate change policy, first in 1995 and then in 2009. In 2015, Santos created a committee to discuss climate change adaptation measures (see

Table 13.1 Timeline of the adaptation policies and laws for Santos

1980	• Establishment of the Santos Municipal System for Civil Defence
1988/99	• Publication of the Preventative Plan for Civil Defence of the State of São Paulo
2000	• Establishment of the Brazilian Forum on Climate Change
2007	• Establishment of the Interministerial Committee on Climate Change and the Executive Group
2009	• Establishment of the National Plan for Adaptation to Climate Change • Establishment of the São Paulo State Policy on Climate Change
2011	• Establishment of the State Programme for Prevention of Natural Disasters and the Reduction of Geological Hazards
2012	• Establishment of the Work Group Adaptation for creation of the National Plan for Adaptation to Climate Change • Publication of the National Policy on Civil Protection and Defence • Publication of the National Plan for Risk Management and Response to Natural Disasters
2015	• Establishment of the Santos Municipal Commission for Adaptation to Climate Change

Chap. 15), which may be a promising step towards regulating legislation on local adaptation to climate change.

The total or partial lack of effective urban planning measures, as well as the difficulties inherent in operationalising public policies effectively and at the same rate as urban pressure, pose major challenges for city planning aimed towards a healthier and safer environment and higher quality of life for all segments of the population.

13.3 ACI

13.3.1 Methodology

As shown above, the country and the state of São Paulo have established initiatives and laws that deal with climate issues, including laws on adaptation to climate change, risk management and disaster prevention, among others. Local adaptive capacity is directly related to the capacity of the public actors and organisations involved in these issues to implement such policies, and to the capacity of public and private organisations to create opportunities for adaptation. Therefore, understanding the possible limitations and barriers at the scale of organisations and agencies is fundamental, particularly given the argument that it is the scale at which society's responses to climate-related impacts will be managed and implemented (Berkhout 2012; Eisenack et al. 2014).

The methodology of the ACI consists of (1) structuring the tool around general qualities of adaptive capacity, derived from the theory; (2) confirmation of the locally relevant categories, through preliminary discussion with a group of interviewees; (3) implementation of the tool; and (4) presentation of the results to the interviewees, enabling the opportunity for verification, challenge and dissemination.

The ACI analysis for the municipality of Santos was conducted through semi-structured interviews with a sample of 24 specialists and representatives from different institutions and organisations that currently populate the arena of environmental risk and climate change management in Santos (Table 13.2). Consequently, this index reflects the performance of risk reduction (or not) related to climate and adaptation, based on the assessments of academics, professionals and officials in the region. Overall, 24 interviews were conducted in the municipality of Santos – 22 in person and 2 remotely via Skype – between 20/10/2015 and 17/02/2016.

Interviews were conducted with representatives of the municipality's key institutions, both private and public, and nongovernmental organisations. They were scheduled in advance and required that the interviewees have sufficient experience and authority to represent their organisation, as well as an understanding of organisational structure and policy context.

The qualitative analysis of the data from the questionnaires focused on eight elements that pervade the issue of adaptive capacity of organisations:

1. "Critical self-reflection", or an organisation's capacity to deal with and learn from their mistakes;
2. "Ability to experiment", which refers to an organisation's capacity for experimentation within its structure;
3. "Ability to learn", which is a question of access to and generation of knowledge;
4. "Ability to plan for the future", which considers whether institutions are able to plan for the short and/or long term, an important factor when thinking about the creation of public policies on issues such as climate change, for which models and predictions may generate uncertainties;
5. "Command over available resources", which refers to the capacity of an organisation to make changes to their structures according to the available resources;
6. "Organisational responsibility", which refers to the recognition of risk management and adaptation as part of the responsibilities of various institutions involved in environmental issues;
7. "Organisational architecture", or how the organisation is structured; and
8. "Levels of capital": financial, human and technical.

The interviews lasted around 1 h each, with the interviewees being asked to attribute a performance value in each question related to the eight elements, on a

Table 13.2 Final organisational function matrix

	Function								
	Preventive plan for civil defence								
	Santos City organisational matrix	Land use/ planning/ management	Environment	Emergency and risk management	Transport	Energy and water	Economy	Social structure	Health
Actors	Government (local, regional, state and federal agencies)	i. Baixada Santista Metropolitan Agency (AGEM)	i. Secretariat of Environment	i. Civil Defence of Santos	i. Santos Traffic Engineering Company (CET- Santos)			i. Secretariat of Tourism	i. Secretariat of Health
		ii. Santos Novos Tempos Programme Management Unit (UGP)	ii. Fishery Institute	ii. Fire Department	ii. Company of Sao Paulo State Docks (CODESP)			ii. Secretariat of culture	ii. Department of Health Surveillance (DEVIG)
		iii. Secretariat of Urban Development							
	Civil society		i. Ecofaxina Institute				i. Association of Construction Entrepreneurs of Baixada Santista (ASSECOB)	i. Fórum da Cidadania	
			ii. Mar Azul Institute					ii. Order of Attorneys of Brazil (OAB)	
			iii. Maramar Institute						
	Private sector		i. Sinergética			i. Piratininga Power and Light Company (CPFL)	i. Mendes Group	i. Industry Social Service (SESI)	
								ii. Trade Social Service (SESC)	

five-point scale (1- Very Limited, 2- Basic, 3- Appreciable, 4- Outstanding, 5- Optimal), for 3 years: 2005, 2010 and 2015. These years were selected based on critical events across all of the Metropole study sites to assess the evolution of each ACI element over time. Interviewees were encouraged to compare the situation of a given element in the most recent period (2015) to its situation during the two earlier periods (2005 and 2010), to indicate concrete actions that have been taken, and resultant changes in the situation of a given element.

It is important to emphasise that the interviewees were asked to respond spontaneously when attributing values to each year while providing examples to support their choice of values. This ensured that the information obtained was a result of the participants' critical self-reflection based on their determination of concrete events and actions.

The combination of both qualitative and quantitative parameters provided greater context and clarity, and acted both to validate the examples given by the interviewees and to highlight potential policy recommendations. Based on these eight selected elements, we observed various dynamics and recurring actions in Santos, as well as problems on some issues, which provided opportunities for discussion and the application of measures necessary to increase the adaptive capacity of the municipality.

13.3.2 Overall ACI Results for Santos

The results for 2005, 2010 and 2015 are presented as overall adaptive capacity sub-component scores (Table 13.3). The overall results of the ACI for Santos show a persistent progressive trend in the last 10 years (Fig. 13.2).

Santos showed an increase in its adaptive capacity in all sub-components between 2005 and 2015, with the greatest increase between 2010 and 2015. This increase resulted from favourable conditions for activities related to risk management, facilitated by government actions following the natural disasters that occurred in 2008 and 2011 in various parts of the country. These occurrences caused numerous fatalities and widespread damage, and instigated the creation of new national and state

Table 13.3 Overall adaptive capacity sub-component scores

	2005	2010	2015
Critical self-reflection	3.03	3.14	3.41
Ability to experiment	2.69	2.69	2.94
Ability to learn	2.91	3.12	3.41
Ability to plan for the future	2.99	3.17	3.43
Command over available resources	2.37	2.59	3.01
Organisational responsibility	3.11	3.33	3.61
Organisational architecture	2.77	2.91	3.18
Levels of capital	2.99	3.19	3.40

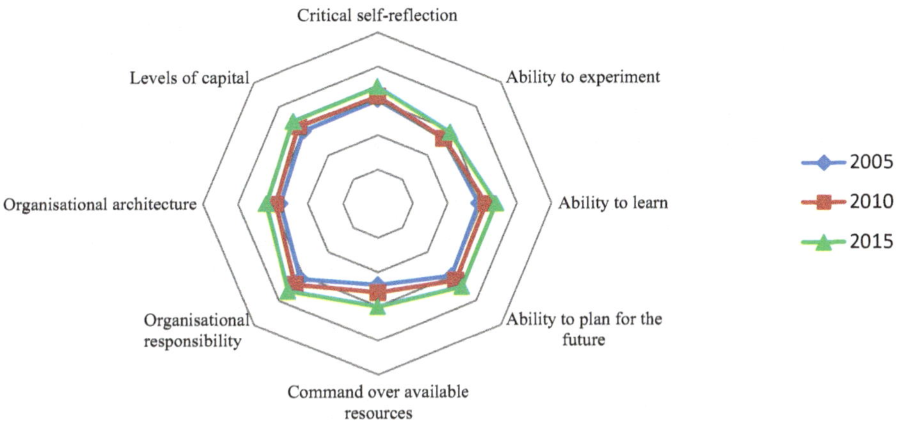

Fig. 13.2 Overall adaptive capacity result for Santos. Note: increasing size of the octagonal rings represents increasing capacity, from very limited (centre) to optimal (outermost ring)

instruments and laws (National Policy on Protection and Civil Defence and the National Plan for Risk Management and Disaster Response; State of São Paulo Programme for Prevention of Natural Disasters and Geological Hazard Mitigation).

However, although the interview respondents indicated close to optimal indices (strong formal capacity for integrated planning and strategising among sectors), in many components the difference between former and current conditions was not significant. This low progress in the components of the ACI, from the perspective of local actors, was explained by the lack of organisational integration of the risk management regime and by the dominance of Civil Defence over the adaptation agenda. This focus on one agency suppresses innovation and new leadership, especially given the many agencies that act in this area. The global economic crisis of 2008 had a considerable impact on the region, particularly after 2014, when all sectors and agencies reported restrictions on resources and funding opportunities. Together, these pressures served to stabilise the adaptive capacity of Santos.

The interviews highlighted the importance of municipal Civil Defence for risk management and climate-related issues. This emphasis reflects the work that has been underway locally since the end of the 1980s, following the establishment of the Preventative Plan for Civil Defence (PPCD) by the government of the state of São Paulo. Instituted in the summer of 1988/1989, this plan is incorporated by research institutions and civil defences of the state of São Paulo and various coastal municipalities, Santos among them. The primary objective of the plan is to avoid loss of human life due to events triggered by rain, which had been a constant occurrence in this region. The PPCD is implemented between November and March, the period of the greatest rainfall in the area, and may be extended to April if rains are expected to continue. During the plan's implementation, the area is under continuous observation, a status that can change to concern or critical according to the meteorological and geotechnical conditions. Each status prescribes different actions that aim to minimise the consequences stemming from the impacts of heavy rainfall (Nunes and Modesto 1992). The impact of precipation comes out as a really important factor in Santos, and the establishment of the PPCD has intensified risk management efforts in the municipality and resulted in fewer disasters and fatalities, and society has recognised these results and the work accomplished by the civil defence.

The civil sector of Santos showed high levels of adaptive capacity in terms of organisational structure and sufficient resources to take risk management actions (Fig. 13.3). Simultaneously, however, it showed restriction of legislation, restructuring of governmental agencies with each new administration – impeding the continuity of existing policies – and a lack of integration across the entire risk management regime, as well as the difficulty of undertaking new experiments in the area and in access to and production of knowledge concerning adaptation to climate change.

The private sector showed high levels of adaptive capacity in many components, with the exception of "command over available resources" (Fig. 13.4). This suggests that the adaptive capacity and the development agenda of Santos are driven by the private sector much more than the civil or government sectors. The privatisation or contracting of previously municipal services led to fragmented communication among different agencies, public service companies and the municipal administration, reducing the municipality's capacity to deal with the effects of climate change.

Government agencies showed high levels of adaptive capacity, with limited or no change over time (Fig. 13.5). This probably reflects the slow pace of such agencies to change over time due to stable and rigid structures. The regional government showed limited change in adaptive capacity over time, with the greatest change occurring at the city level. The interviewees associated these increases, especially those in the 2010–2015 period, with a broad investment in risk management and

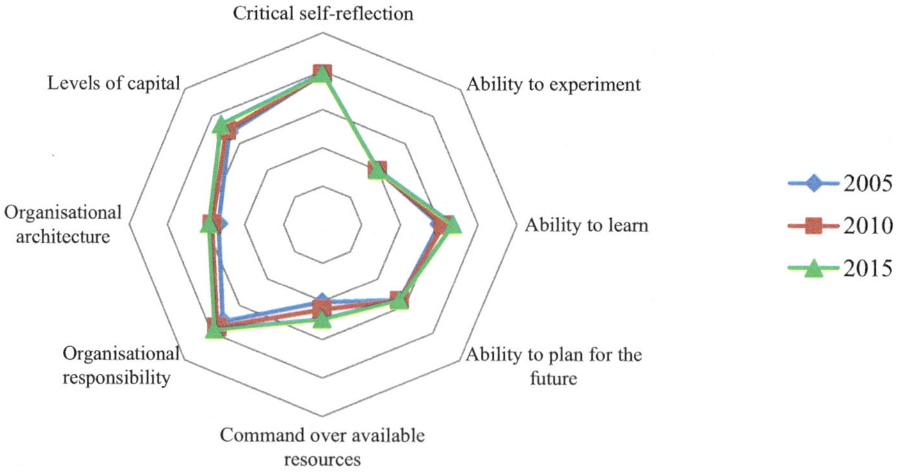

Fig. 13.3 Civil sector adaptive capacity result. Note: increasing size of the octagonal rings represents increasing capacity, from very limited (centre) to optimal (outermost ring)

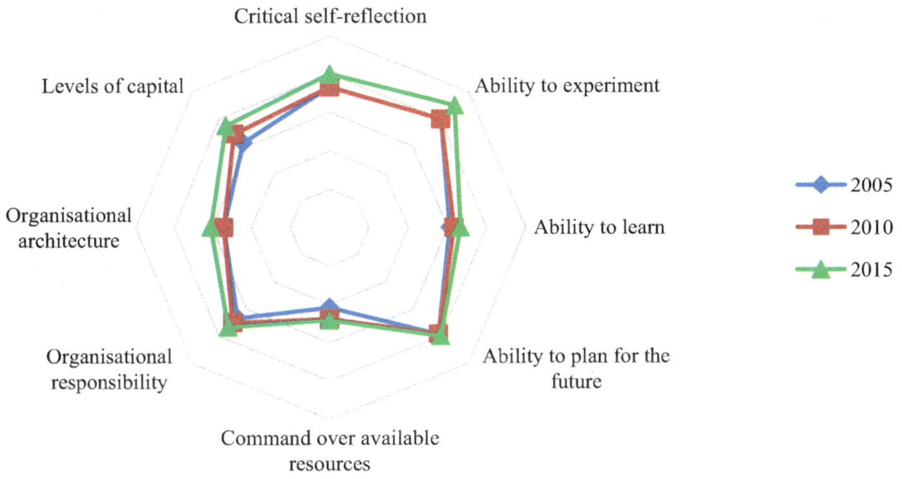

Fig. 13.4 Private sector adaptive capacity result. Note: increasing size of the octagonal rings represents increasing capacity, from very limited (centre) to optimal (outermost ring)

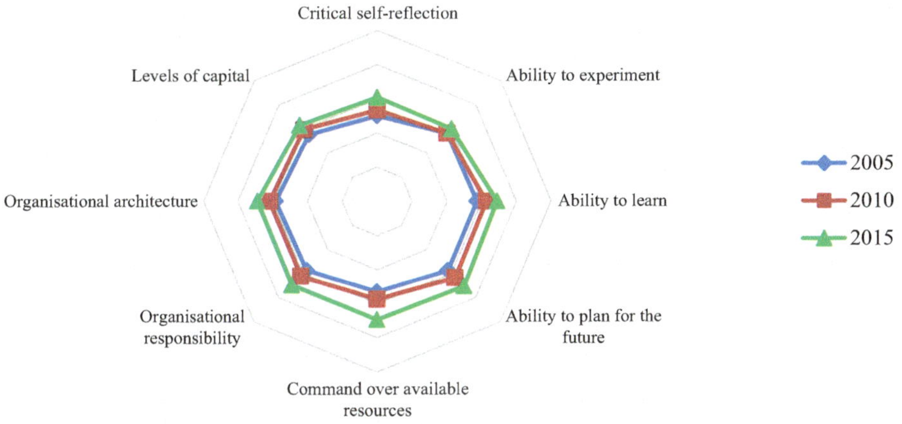

Fig. 13.5 Government sector adaptive capacity result. Note: increasing size of the octagonal rings represents increasing capacity, from very limited (centre) to optimal (outermost ring)

adaptation in response to major natural disasters that occurred in the country between 2008 and 2011. Driven by the willingness of government to create new policy and the establishment of new laws and directives in the following years, specially in 2011 and 2012 (Table 13.1), changes occurred in various central organisations, particularly in the Civil Defence, which created political and bureaucratic space for organisational change. That shows a positive response to lessons learned from the events in 2008 and 2011.

Some of the main conclusions are listed below to summarise the information collected in the interviews for this study:

1. Limited financial resources and complex administrative systems impede the approval of new projects and, therefore, the acquisition of resources from different sources;
2. Lack of awareness of the issue within organisations;
3. Lack of support by the public sector for other sectors;
4. Need for integration among organisations for better planning and action in the area, which is also affected by a lack of continuity across different administrations.

The dominance of a single actor in the public sector – the Civil Defence – in the conception, institutionalisation and implementation of a formal adaptation policy had a limiting effect on municipal adaptive capacity. Reflecting this situation is the lack of recognition of climate change issues by other organisations within the overall risk management regime, as well as the limited integration among Civil Defence and other sectors. The lack of a broader participation in the risk environment to the problem of adaptation and appropriation seems to restrict the adaptive capacity of the city. Consequently, organisations remain static in the face of emerging risks, with rare exceptions.

The lack of financial resources was identified in most interviews as a barrier to risk management, especially as a result of state agencies' and the federal government's inflexibility and slow disbursement of resources to the city. The bureaucracy has been identified as a key area that reduces the adaptive capacity of the municipality, since the actors in the risk management arena opt not to use their time or resources to prepare funding requests for adaptation projects because doing so is burdensome. This has limited experimentation within the city and within key organisations, negatively affecting adaptive capacity and reducing the possibility to strengthen central elements of the city infrastructure.

Another aspect revealed by the study is the use of shadow space, understood as the informal interactions between communities and networks at different scales inside the organisations structures that do not count as official, but may be effective due to their greater agility. Through the analysis of shadow space it is possible to identify "hidden, implicit patterns of behaviour and organisational forms that are hard even to delineate and thus hard to rationally control" (Pelling et al. 2008).

The interviewees in Santos strongly emphasised these informal relationships as a way to circumvent the limitations caused by regular restructuring and modification of job descriptions and responsibilities within government agencies. Such changes result in an environment of instability, in terms of jobs and of organisational structure, as well as a sense of isolation and disconnection within and between hierarchical levels. In such situations, informal relationships between organisations become vital to advance the adaptation agenda throughout the risk management sector. Although the use of informal relationships is considered a complex area to be explored, and is often seen as a source of corruption and inefficiency that requires greater management and control (High et al. 2004), this study observed that the shadow space is partly responsible for a significant change in the organisational and governmental capacity of Santos, enabling the organisations to overcome barriers such as periodic government overhaul, which causes policy discontinuity (Paterson et al. 2017).

13.4 Final Considerations

Analysis of the adaptive capacity index for the municipality of Santos identified some areas where local adaptive capacity can be better developed and increased, and highlighted new opportunities for innovation and integration among the various institutions involved in local risk management.

In light of the situation observed in Santos, we make some recommendations so that local actors at various levels can think about issues of learning, information, division of responsibilities, and networks of interaction:

1. Greater connectivity and comprehension between local efforts and activities motivated by specific sectoral and local concerns, and the structure of risk management planning and development for the city;
2. Greater transparency of decision-making processes at higher levels to increase society's and institutions' knowledge and awareness of the issues;
3. Improve the formal relationships/organisational structure to create a mechanism for collaboration;
4. Promote ownership and leadership of the issue over and above the Civil Defence;
5. Develop voluntary and cooperative partnership between the local governmental bodies as a mechanism to alleviate the impact of decentralising responsibility without resources – an issue that often affects adaptation efforts – which can serve as a potential model for the Municipal Policy on Adaptation to Climate Change.

Beyond the recommendations, it is important to highlight how the decision-making agents of different institutions of the municipality will have central importance and will be directly responsible for the actions undertaken to confront the possible effects of climate change. To the degree that climatic changes intensify and accelerate, the capacity to deal with the impacts and the capacity to adapt to opportunities become critical attributes:

As the natural sciences are becoming better in predicting the potential future environmental impacts of anthropogenic activities, for example in the case of climate change, institutions will increasingly need to be able to rise to the challenge of incorporating new information and becoming more proactive and progressive in coping with the projected impacts of environmental change (Gupta et al. 2010, P. 460).

The case study of Santos raises a series of recommendations and insights in relation to adaptive capacity. However, the capacity of local actors is, in many respects, shaped by their relationship to and interdependence on the broader governance picture, principally the political and economic context in which they operate (High et al. 2004). In this sense, the importance of maintaining and making better use of informal relationships established within (and between) the municipality's organisational structures is evident, as they collaborate for the continuity of plans that are essential to local adaptive capacity, which may be affected by regular changes of administration.

Therefore, it is imperative that this analysis of the adaptive capacity of Santos become another foundational element for local organisations and members of the community, to strengthen the organisational architecture and to identify opportunities for new actions or integration, so that more organisations become involved in these issues and cooperate with the Civil Defence, which is already established and active. This would create an environment in which cooperation and knowledge can enhance and improve the local adaptive capacity.

Acknowledgements The results of this analysis were funded by NERC (NE/L008963/1), NSF (ICER 1342969) and FAPESP (Fapesp Proc. 12/51876-0 and 14/14598-8). The authors thank the respondents for the time given and for exposing and sharing their visions.

References

Berkhout, F. (2012). Adaptation to climate change by organizations. *Wiley Interdisciplinary Reviews: Climate Change, 3*, 91–106.

Carmo, R. L., & Silva, C. A. M. (2009). População em zonas costeiras e mudanças climáticas: redistribuição espacial e riscos. In D. J. Hogan & E. Marandola Jr. (Eds.), *População e mudança climática: dimensões humanas das mudanças ambientais globais*. Campinas/Brasilia: Núcleo de Estudos de População – Nepo, Unicamp/UNFPA.

Eisenack, K., et al. (2014). Explaining and overcoming barriers to climate change adaptation. *Nature Climate Change, 4*, 867–872.

Gupta, J., et al. (2010). The adaptive capacity wheel: A method to assess the inherent characteristics of institutions to enable the adaptive capacity of society. *Environmental Science & Policy, 13*, 459–471.

High, C., Pelling, M., & Rengasamy, S. (2004). *Local agency, adaptation and the shadow system: The institutional architecture of social learning in rural areas of the UK and India*. XI World Congress on Rural Sociology, Trondheim, Norway, 25–30 July 2004.

IBGE – Brazilian Institute of Geography and Statistics. (2010). Census 2010.

IPCC – Intergovernmental Panel on Climate Change. (2014). *Climate change 2014: Impacts, adaptation, and vulnerability*. Cambridge: Cambridge University Press.

Moreira, F. A., Nunes, L. H., Paterson, S. K., et al. (2017). *Adaptive capacity index: Part II. Santos, Brazil* (Working Paper 77. Contested Development Working Paper Series). Department of Geography, King's College London. https://www.kcl.ac.uk/sspp/departments/geography/research/Research-Domains/Contested-Development/workingpapers.aspx. Accessed 15 May 2018.

Nunes, L. H., & Modesto, R. P. (1992). Comportamento pluviométrico nos municípios atendidos pelo Plano Preventivo de Defesa Civil – PPDC. *Revista do Instituto Geológico, 13*, 47–57.

Paterson, S. K., & Pelling, M. (2017). Adaptive capacity: Methodological working paper. Paper 75. Contested Development Working Paper Series Department of Geography, King's College London. https://www.kcl.ac.uk/sspp/departments/geography/research/Research-Domains/Contested-Development/workingpapers.aspx. Accessed 15 May 2018.

Paterson, S. K., Pelling, M., Nunes, L. H., et al. (2017). Size does matter: City scale and the asymmetries of climate change adaptation in three coastal towns. *Geoforum, 81*, 109–119.

Pelling, M., & Zaidi, R. Z. (2013). *Measuring adaptive capacity: Application of an indexing methodology in Guyana* (EPD Working Paper #47). Department of Geography, King's College London.

Pelling, M., High, C., Dearing, J., et al. (2008). Shadow spaces for social learning: A relational understanding of adaptive capacity to climate change within organisations. *Environment and Planning A, 40*(4), 867–884.

Silva, R. B., & Nunes, L. H. (2016). *Vulnerability profiles in a Metropolitan Area of Brazil.* Saarsbrüchen: Lambert Academic Publishing.

Zanetti, V. B., Sousa, W. C., Jr., & Freitas, D. M. (2016). A climate change vulnerability index and case study in a Brazilian Coastal City. *Sustainability, 8*(8), 811. https://doi.org/10.3390/su8080811.

Zundt, C. (2006). Baixada Santista: uso, expansão e ocupação do solo, estruturação de rede urbana regional e metropolização. In J. M. P. Cunha (Ed.), *Novas Metrópoles Paulistas: População, vulnerabilidade e segregação.* Campinas: Núcleo de Estudos de População – Nepo. Unicamp.

Chapter 14
Population Matters: Listening to Past Experiences and Future Aspirations Regarding Risks and Adaptation Actions

Roberto Greco and Lucí Hidalgo Nunes

Abstract Successful adaptation to climate change needs the involvement of the local population. Thus, to identify individual's perception of climate-change-driven sea-level rise, to know whether individuals recognise climate change as a priority and to observe their willingness to bear the costs for implementing adaptation measures, pre- and post-surveys were applied in two stakeholder engagement meetings in Santos, in Sep. and Dec. 2015. Attendees of both engagement meetings were invited to answer a survey on individual and community risk experience, funding preferences for adopting adaptation measures, and their perceptions about barriers to implement adaptation. Results showed different viewpoints and even antagonistic interests: on the one hand those focused on environmental conservation and on the other, interests focused on urban expansion, increase of port operations, and of tourism and commercial activities. A pro-environmental pattern was identified, but some contradictions have also emerged: participants consider that high wind storms is the most preoccupant hazard for their own houses but not to the city as a whole; they also recognise the urgency of the matter, but do not seem willing to support adaptation measures by new taxes or fees.

Keywords Adaptation · Perception · Survey · Stakeholder engagement

14.1 Introduction

Amongst the foundations of Metropole Project is the understanding of connections between stakeholder values, beliefs and preferences regarding risks, adaptation options and funding choices, in view of understanding barriers to adaptation and providing findings to inform regional stakeholders' priorities for adaptation

R. Greco (✉)
Institute of Geosciences, University of Campinas (Unicamp), Campinas, SP, Brazil
e-mail: greco@ige.unicamp.br

L. H. Nunes
Santos, São Paulo, Brazil
e-mail: luci@ige.unicamp.br

L. H. Nunes et al. (eds.), *Climate Change in Santos Brazil: Projections, Impacts and Adaptation Options*, https://doi.org/10.1007/978-3-319-96535-2_14

strategy (Marengo et al. 2017a, b). But developing any kind of adaptation to climate change is not a purely scientific and academic matter, so that to be fully successful, adaptation needs the involvement of the local population, which is the main interested part in the issue, since they can be on the frontline of the impacts (Adger et al. 2009; Ross et al. 2015).

To cope with the effects of climate change requires increased levels of public investment, incentives, and/or leveraging for private investment (IPCC 2014). Thus, it is of great importance to know how to encourage stakeholders and decision makers to support funding mechanisms.

In addition, involving people and raising their awareness concerning climate change require knowing the demands and aspirations of the community, which are based on individual experiences and perceptions of the environment, (Martín-Vide 2001), as well as their experiences with the hazards, their assumptions with respect to the subject and their willingness to bear the costs for implementing adaptation measures through new taxes.

In order to put into practice adaptation measures in the city of Santos that allow the maintenance of the local physical processes, the economic activities and the physical integrity of the population, it is necessary to know whether individuals and local government recognise that climate change is a priority issue and can no longer be postponed. With this purpose, individual's perception of climate-change-driven sea-level rise were analysed by means of pre- and post-surveys applied in two stakeholder engagement meetings in Santos that presented estimates of costs and benefits in a near future in two situations: if adaptation measures are taken and if any attitude towards adaptation to climate change is considered. A final meeting to present to the participants the results of the project occurred in the end of the project.

Attendees of both engagement meetings were invited to answer a survey partly based on NEP (New Ecological Paradigm), used to assess environmental concern in group of people. The survey has questions on individual and community risk experience, funding preferences for adopting adaptation measures in Santos, as well as their perceptions about barriers to implement adaptation. There were also questions about participant's characteristics (age, gender, monthly income etc.).

The success of this stage of the project was guaranteed by the commitment of the entire team of the Project and the support of the local government, which collaborated in a strictly way in order to standardise the procedures, so that the same protocols might be applied in other locations, allowing future comparisons.

14.2 The Participatory Approach

Pre- and post- surveys were applied in two engagement meetings, which also presented scenarios of flooding damage to assets in two sectors of Santos (Southeast and Northwest Zone) from expected 50 to 100 year storms, and estimated benefit/costs of two adaptation actions defined by participants. Scenarios of sea-level rise were obtained by using the COAST model (Coastal Adaptation to Sea-level rise Tool, Catalysis Adaptation Partners 2015).

NEP inspired part of the survey, designed to measure the environmental concern of a given group of people from individual responses to a number of statements, in which respondents are asked to indicate the strength of their agreement or disagreement for each question and the relationship of environmental view to demographic, economic, and behaviour change. Developed by Dunlap and Van Liere in 1978 and revised in the present format in 2000 (Dunlap et al. 2000), NEP has been used in many countries since 1979, so that there is already a basis for its analysis (The Berkshire Encyclopedia of Sustainability 2012). In a study of NEP involving Brazil, the United States and Mexico, it was observed that the three groups view the environment differently, with Brazilians presenting the most ecological notion of environment, as this group sees no need for separation between nature and growth (Bechtel et al. 1999). Therefore, local and cultural aspects are important issues to be considered in environmental studies, and even following the same protocols comparative studies might produce quite distinct results.

The surveys aimed to clarify issues related to personal involvement with climate change, evaluating aspects such as whether individuals consider climate change a priority issue in their lives since its effects tend to be seen as a long-term concern, their risk experience or what would be the barriers which affect how people engage in a more effective way with adaptation to climate change. Based on the personal lived experience of hazardous events, the analysis is anchored on phenomenology, which evaluates how a phenomenon is perceived from an individual experience and how this experience is internalized and interpreted. In this sense, respondents were viewed as "co-researchers".

Prior to the meetings, municipal partners invited potential participants via personal email invitations, planning and community listserves and by verbal outreach from local officials. Both meetings took place at the Associação Comercial de Santos, a well-known historical building in the town centre of Santos.

In order to express the large socioeconomic and socio-spatial diversity of the Santos population (see Chaps. 8 and 9), we tried to involve participants from all areas and social groups of Santos. Nevertheless, it was noted by the Zip code and by the incomes that the participants come from the classes of highest income, which might biased some responses.

Surveys were applied in the beginning of meeting one (30th September, 2015) and in the end of meeting two (1st December, 2015). Because some people attended the second meeting only, in the beginning of meeting two the new participants answered the survey for the first time before presentations. In total there were 42 attendees on meeting one but only 36 questionnaires were answered, while 34 people attended meeting two, but only 21 answered the survey, of which, 9 were new participants.

In meeting one, after responding the survey, the results of the first run of the COAST model were presented to the participants by means of visualisation tools, with the potential losses in two sectors of Santos (Southeast and Northwest zones) due to cumulative damage of flooding to assets in 2050 and 2100 from expected 50 and 100 return period storm surges, if any adaptation measure is adopted (the "no action" scenario). Examples of adaptation to sea-level rise already put in force in other coastal locations were showed to stimulate discussions about the adaptation measures that could be feasible and effective in the case of Santos. In the second part of meeting one, participants were split into two groups, each one discussing adaptation measures specific for one of the zones under evaluation, and then groups moved to discuss the other area, so that both groups evaluated the two areas: Southeast and Northwest Zones (Fig. 14.1).

In the end of the meeting one, two options were chosen by vote for each zone. Between the two meetings, it was estimated the benefit/cost rates of the adaptation actions chosen by attendees in the first meeting. These results were presented in the second meeting and showed how cost effective the measures chosen by participants are for avoiding damages to real estate (see Chap. 1 and Marengo et al. 2017a, b). It was stressed that no endorsement of any candidate action was implied by this public process, which aimed to involve citizens and local government to start discussing feasible alternatives to avoid the harmful effects of climate change; thus other alternatives may arise in the future.

The surveys had 20 statements distributed in 5 sections. The first one, named "your experience with hazards" had three questions and respondents assigned whether they had experienced or not to specific natural hazards in the city (Santos) and /or in their residence. Hazards were: storm surge, extended flooding, high winds in storms, rising sea level and coastal or beach erosion. In the second session, about potential adaptation actions, respondents chose when the local government should implement (if so) 16 activities. Possibilities were now, in 10, 25, 100 years, never or

Fig. 14.1 Group discussion, meeting one, 30th Sep. 2015. (Photo: Eduardo K. Hosokawa)

unsure. Section 3 had three questions about possible funding sources to implement new hazard protection efforts, while section 14.4 had three questions concerning perspectives about adaptation and the environment. For answering some of the questions of sections 1, 3 and 4 respondents used a Likert scale, indicating his/her strength of agreement. Finally, demographic questions (gender, family income, age, school level, etc.) were on section 5. Annexes 1 and 2 show the sessions and questions in detail.

Through the application of the surveys it was sought to evaluate how stakeholder attitudes and judgments concerned to the impacts of climate change altered through the participatory engagement process. Responses were anonymous and held in strict confidence. Data were reported in aggregate manner, but because participants provided their ZIP codes and date of birth, the answers of the same attendee in meetings one and two could be identified and compared. This was important for evaluating whether or not each participant changed his/hers initial opinion after being aware of the projections of sea-level rise delivered by COAST tool. By the ZIP code it was possible to verify that some participants do not reside in Santos, but in nearby cities.

In addition, a satisfaction survey was applied at the end of each survey, answered by 19 people in meeting one and 16 participants in meeting two.

Table 14.1 shows the questions that are the same in both surveys and from where it is possible to make comparisons before and after the meeting two:

Table 14.1 Questions which remained the same in meeting one (30th Sep. 2015) and two (1st Dec. 2015)

Section 1. Your experience with hazards
2. How concerned are you that the following natural hazards might seriously and negatively affect your town in the next 10 years in terms of physical and economic damage?
3. Thinking about the next 10 years, how concerned are you that these natural hazard may seriously and negatively affect your primary household in terms of physical and economic damage?
Section 2. Questions about potential adaptation actions
4. There are a variety of actions a town or district might implement to reduce the potential for physical and economic damage caused by climate-related hazards
Section 3. Questions about possible funding sources
5. Do you agree or disagree with the following statement: implementing projects to reduce potential impacts of climate-related hazards in our community should be a local or regional government priority, even if it will require a slight increase in taxes or new fees?
6. Please consider the following funding options that local government and agencies could use/offer and tell us whether you think they are acceptable
7. If a referendum was put on the 2016 ballot to create a Community Resiliency Bond (a long-term loan) that would generate a fund by 2036 to support multiple adaptation projects, how likely would you be vote for it?
Section 4. Perspectives about adaptation and the environment
8. Some people in your community might NOT want to support local government adaptation plans. What do you think are some the most common reasons for NOT supporting plans?
9. Are there other reasons why people in your community might NOT support local government adaptation plans? Please tell us your thoughts

14.3 Survey Results and Discussion

The following statements concern to answers provided by participants that attended both meetings: 60% were men; 50% were between 45 and 64 years-old, 39% between 26 and 44 years-old, and 11% between 21 and 25 years-old. Further, 36% have children or grandchildren under the age of 18 years old and the majority of participants (86%) were white, 11% declared themselves mixed, while 2% were Indigenous.

All participants earn more than 36.000,00 reais (Brazilian currency) per year[1] and 70% earn more than 90.000,00 reais per year, of which 25% earn more than 180.000,00 reais, and 4% more than 300.000,00 reais.

Participants have a high educational attainment: 90% reported they had attained at least a bachelor's degree and almost 60% declared they hold a master or a doctoral degree. Almost 70% of attendees are owner of their houses.

In the sample 35% of the participants are employers of the municipality, 10% of State institutions while 23% are researchers from local universities or other research institutions. There were also delegates of environmental non-governmental organizations, representative of trade union, company owners, professionals from various fields such as engineers and geologists, militaries and journalist.

Concerning their reported political preference, almost 18% declared a right or central right preference, 46% declared a left or central left preference, and 36% declared others.

The participants were also tested by the New Ecological Paradigm (NEP) Scale, used to assess the environmental concern in a group of people. A pro-environmental attitude would have agreement with the odd-numbered items and disagreement with the even–numbered items (Merril et al. manuscript). The authors pointed out that some of the non-pro-environmental statements about the human role in modifying the environment are the ones that might be consistent with support for climate risk adaptation; in this case, item 4 belief in the ability to adapt could be consistent with the belief that the earth's systems are fragile and thus, adaptive action would be necessary.

Table 14.2 shows the results, clustered in three classes (agree, undecided and disagree). Participants demonstrate a pro-ecological perspective, with the exceptions of items 4 and 6. It is of note that the main purpose of the meetings was to discuss possible adaptive measures for Santos, but by the answers for item 4 – quite related to this purpose – we see that the majority of respondents seem not to believe in adaptive measures.

In relation to the experience with hazards (storm surge, extended flood, rising sea level and coastal or beach erosion), each type of hazard was evaluated individually and the collective responses to a specific hazard were converted to a percentage, indicating the most common experiences in relation to a specific hazard. A third of the participants had personal experience with high winds in storms, but around 70%

[1] Currency rate at the time: 1US$ = approximately 3.854 Brazilian Real.

Table 14.2 Results of Question 10 based on NEP (new ecological paradigm)

Q10	Disagree	Undecided	Agree
1. We are approaching the limit of the number of people the earth can support	32%	4%	64%
2. Humans have the right to modify the natural environment to suit their needs	68%		32%
3. When humans interfere with nature it often produces disastrous consequences	27%		73%
4. Human ingenuity will insure that we do not make the earth unliveable	27%	34%	39%
5. Humans are seriously abusing the environment	5%		95%
6. Earth has plenty of natural resources if we just learn how to develop them	14%	2%	84%
7. Plants and animals have as much right as humans to exist	2%	5%	93%
8. The balance of nature is strong enough to cope with impacts of modern industrial nations	84%		16%
9. Despite our special abilities, humans are still subject to the laws of nature	9%		91%
10. The so-called "ecological crisis" facing humankind has been greatly exaggerated	80%	2%	18%

and 80% of them stated they are aware that the city of Santos is impacted by storm surge, extended flooding, rising sea level and coastal or beach erosion. About half of the participants expressed concern about impacts of extended flooding, high winds in storms and rising sea level to their own households within the next 10 years. Considering the impacts in Santos in the next 10 years the main concern of more of two-thirds of the participants is about extended flooding, coastal or beach erosion, rising sea level, storm surge and high winds in storms.

Although participants have pointed out wind storm as the hazard that more commonly affect their houses and that this hazard is their main concern for the next 10 years, when they consider the city as a whole this hazard appears as the least concern. This could be linked to the economic and social position of the participants, who live in areas of the city which are less affected by the natural hazards that hit the city as a whole more frequently and strongly.

There is a variety of actions that a city can implement to reduce the potential of physical and economic damages caused by climate-related hazards. Thus, it was proposed to the stakeholders a list of actions they believe that might be effective for Santos and asked what activities they think the local government should implement, and when. In parentheses are the answers of the participants before they receive information of what could happen in Santos in 2050 and 2100 without adaptive measures and in the second, the answers given by the participants after the presentation of the results of COAST. The main actions that should be implemented by the government related to land use and policy change are to restrict new buildings in highly vulnerable locations (84%, 100%) and to require that the new constructions are made above a certain level, to be defined by the responsible entity (56%, 53%). Among the more nature-based actions are to conserve existing natural areas such wetlands and mangroves to protect coastal sectors (93%, 89%) and to restore/increase natural areas to protect the coast (73%, 74%). Actions related to infrastructure improvements are basically to use innovative/green technology to reduce flooding due to increased rains (permeable surfaces or other storm water management systems) (76%, 74%). The answers did not reveal any substantial change but showed some discredit regarding the pro-environment solutions presented. Interestingly, the greatest change was observed in relation to restrict new buildings in highly vulnerable locations: it is possible that the greater adhesion of the participants to this possibility is due to their belief that the vulnerable areas of Santos are restricted to the poorest neighbourhoods: by the income and the information provided by the Zip code this is not the origin of the participants. Therefore, they probably do not consider the areas where they live susceptible to severe hazards.

While some actions do not require significant expenditures but political involvement, some others need resources; in the latter case, respondents consider them for medium and long term of 10 or 25 years. Considering the responses given in surveys one and two, the solutions would be: elevate or harden coastal transportation infrastructure – roads, bridges (61%, 53%); climate proof ongoing infrastructure improvements and development efforts (69%, 56%); create a plan to purchase vulnerable land and structures from residents (52%, 31%); relocate water and sewage services to more remote locations away from the coast (64%, 68%), and relocate vulnerable public facilities such as water and wastewater treatment plants (57%, 44%). There was less support for solutions in the second survey responses, with one single exception.

The implementation of these actions to reduce potential impacts of climate-related hazards requires economic resources. Thus, the participants were asked if these measures should be a local or state government priority, taken even if the measures require a slight increase in taxes or new fees. The majority of them agreed in the first meeting (90%) but disagreed in the second meeting (43%).

Results of which funding options would be more acceptable for population showed that participants prefer to develop a special local assessment which applies to properties in areas designated as highly vulnerable (75%, 89%), issue a bond (long-term borrowing) to finance public infrastructure improvements (66%, 68%) and create a low-interest loan program for flood proofing and elevating residences (65%, 47%). Other options were less accepted, like to raise the local sale tax slightly (21%, 42%), to add a flood resiliency surcharge on the monthly water utility bill (ex: specific to storm water drain improvements) (20%, 26%) and to create a new city-wide resiliency fund based on property taxes (19%, 21%). Overall, there was low support for the creation of new fees to finance preventive measures.

Most respondents (68%, 74%) stated that they would vote favourably to create a community resilience obligation (a long-term loan) that would generate funds to support various adaptation projects in 2036, if there was a referendum in 2016.[2]

[2] This popular consultation did not occur.

Participants believe that the main reason for the lack of community support for local adaptation plan is the lack of knowledge/understanding of future hazards and local consequences (82%, 79%). The scenarios of future impacts presented during meetings one and two were intended to provide useful information, but after the participants received information that was supposed important to their opinions, there was less support for this type of initiative.

New fees and taxes that would be needed to fund adaptations actions could be another important reason (54%, 74%) for a lack of support. This point appeared in the comment space of the survey, in which it was possible to observe that citizens do not trust in public administration and politicians due to corruption.

Analysing the responses given by the participants after they have received information of the project, we noted there was a general increase in concern with all natural hazards. Concerns with the impacts of storm surge in the respondent's house presented the highest increase, but the concern of this type of hazard in the city decreased. Concern over coastal or beach erosion has increased significantly.

According to the participants, the actions that the city of Santos should implement now to reduce the potential damages of climate-related hazards are:

- Land Use Policy Changes: create a plan to purchase vulnerable land and structures from residents (36%, 63%); create a plan to purchase vulnerable land and structures from small businesses residents (18%, 42%); restrict new building in highly vulnerable locations (75%, 100%); restrict rebuilding in highly vulnerable areas after major damage has occurred (75%, 100%). The answers demonstrated clearly that participants believe that land use change measures have to be taken, in special those that restrict some current uses.
- Infrastructure improvements: relocate vulnerable public facilities such as water and wastewater treatment plants (39%, 56%); build new or higher seawalls (43%, 26%); raise the height of canal flood gates (26%, 17%).
- Nature-Based enhancements maintained high appreciation but there was some decrease: conserve existing natural areas (such as wetlands and mangroves) to protect coastal sectors (93%, 89%), restore/increase amount of natural areas (such as wetlands and mangroves) to protect coastal sectors (73%, 74%) and preserve and restore mangrove - measure proposed by participants in meeting one (77%)

The more acceptable options to collect funds to implement the measures needed to avoid or minimise the negative effects of climate-change-driven sea-level rise are slightly raise of the local sales tax (from 21% to 42%) and develop a special assessment which applies to properties in areas designated as highly vulnerable (75%, 89%).

Question 8 is related to the respondent's opinion about what would be the reasons why some people might not want to support local government adaptation plans and what would be the most common reasons for this refuse. Table 14.3 presents the positive answers given before the participants have contact with the results of the

Table 14.3 Results of Question 8 (possible reasons for the little support to the municipal plan)

Q8	Before Coast	After Coast
(a) Lack of knowledge/understanding of future hazards and local consequences	82%	79%
(b) Adaptation actions will need funding – people are generally opposed to new fees and taxes	54%	74%
(c) Climate change is a distant issue. Other social/economic issues are more important now	45%	32%
(d) Distrust the media and news reports	2%	5%
(e) Uncertain about scientific data – no one really knows how bad it will get	36%	21%
(f) Local government doesn't have technical expertise to solve the problems	14%	21%
(g) Denial. People don't want to believe their homes will be impacted/ don't want to move	37%	32%
(h) Businesses are concerned about the impact on real estate investment	30%	16%
(i) Concerns that tourism businesses and jobs will decline	7%	0%

COAST platform (survey 1), and after they were aware of the results shown by the COAST platform (survey 2). Attendees revealed they have a lack of knowledge related to the impacts of future hazards, but the process Metropole Information provided was useful for their improvement; they agreed people are opposed to new taxes and fees. Especially in the responses given in survey 2 we can verify that participants do not see climate change as a distant issue or an unimportant matter, but they agree about the uncertainties of scientific information and that local government do not have technical expertise. They trust the information received by the media and reports and they see the general concern about the matter as low.

The following analyses are related to a set of questions which were specific of the second survey. The participants considered that the technical information was presented in a clear and comprehensible way (88%). They expressed confidence in the technical information that was passed on to them (82%) and in the information on costs and biases presented (69%). After participating at the meetings, they found that their knowledge of local risks and impacts due to long-term events increased (81%) as well as their knowledge in the different adaptation options (76%).

They were almost unanimous in believing that local government will need to implement some of the adaptation options raised in the meetings (94%) and with no exception they agreed with the general opinions regarding adaptation options expressed by the group in which they were inserted.

The main aspects that influenced stakeholders' opinion to support or not the option presented in the second meeting are, in their words, the trustful and clear information presented by the researches of Metropole Project, a comprehensive representation of the diverse interests of the Santos society, their personal experience with hazards, international data on climate change and recent events that affected the city of Santos.

Negative aspects raised were the disbelief in the public authority, lack of interest and lack of information, and inaccurate approximations of some information provided by the project.

In order to understand how the information provided would affect participants' future decisions and actions, they were asked if the information provided during the process would be used by them (or their organisations) in the next three months. In a list of options the preferred were: share information with community members (65%), up-date and incorporate this information in the existing plans (59%), organise a meeting with the leader in order to discuss these topics and the next steps to be implemented (53%), organise meetings among institutions to discuss this topic, and determine what the next steps would be (47%). The less accepted options were to contact politicians in order to know more about existing plans and ask them to be focused in this topic (18%), and carry out an internal revision of the city's Master Plan (18%).

14.4 Conclusions

Issues involving long-term vision such as adapting to climate change entail an iterative association involving individuals, practitioners, academics and local government.

In order to raise awareness to deliver adaptation actions it is necessary that people realise that climate change is a key issue to them and must be seen as a priority. But climate change is still a controversial issue and involves many uncertainties, like causes, future effects, best measures to cope with it, suitable measures to be taken etc. In addition, climate change involves distinct and even competitive interests to be accommodated.

After the two engagement meetings of Metropole Project the municipality of Santos was severely affected by storm surges that caused enormous destruction and damages in Ponta da Praia (in April, October and especially August of 2016). Although it is a mere speculation, it is possible that if these hazardous events occurred before or at the same time of the survey's application, responses could be influenced and more favourable to the adoption of adaptation measures. On the other hand, although during the period of the two engagement meetings (September and December of 2015) the economic and political situation in Brazil was giving clear signs of instability, the situation worsened enormously throughout 2016 and 2017, a fact that could have influenced negatively responses concerning forms of financing measures to adapt to climate change, especially because this aspect is seen as something distant and out of people's everyday lives.

Even though the researchers of the Metropole Project were a bit frustrated by the relatively low attendance at both meetings when compared to the number of invitations sent, participants showed positive attitudes and genuine enthusiasm in taking part of the process. Many of them returned to the second meeting, enhancing their interest. In addition, the local media covered the meetings, especially in the case of the first meeting. However, the news coverage was a little sensationalist, since the emphasis was on the "no action" scenarios, that is, what can happen in Santos in 2050 and 2100 if no adaptation measure is taken, although during the presentation researchers made it clear that the scenarios of disruption showed for 2050 and 2100 can be changed if effective adaptation measures are adopted in the municipality: this was the real aim of Metropole Project.

Results showed different viewpoints, consistent to the complexity of a city like Santos, which involves diverse and even antagonistic interests: on the one hand those focused on environmental conservation and on the other, interests focused on urban expansion, increase of port operations, and of tourism and commercial activities, in a place of great natural fragility (see Chaps. 8, 9, 11, and 12). Besides, there was a relative lack of representativeness of certain social groups and thus, of their interests, despite the fact that the Metropole team intended to aggregate the socioeconomic diversity of the municipality.

Although the small number of answers does not allow many conclusions, it is possible that the group of participants do not see themselves vulnerable to some hazards, which could justify some collective responses – in special, that most of participants in survey one and all of them in survey two are favourable for restricting new buildings in highly vulnerable locations. However, in the one hand the land appreciation in the Southeast Zone has increased even more in the last 15 years, attracting to the area large real estate developments and high income population, a process that has diminished thanks to the economic crisis but is still under way; on the other hand, the Southeast Zone (where some of the participants live) is very vulnerable to storm surges and beach erosion (see Chaps. 6 and 11), which could limit plans for more investments in the area, which seems very far from reality. Thus, it is possible that there is a misinterpretation of vulnerability, which for some people would be limited to the poor neighbourhoods of Santos, like the ones of the Northwest Zone, where new constructions should be limited.

Considering the NEP questions, it was possible to identify a pro-environmental pattern. But participants showed some reluctance as to the forms of financing the necessary actions to adaptation. This and other aspects could be interpreted in the light of the current political moment in which Brazilian society passes, ruled by a great discredit at all levels of government, which causes a natural rejection to any possible financing through public money for the distrust of bad use of money because of high corruption. Besides the political crisis, the economic crisis is also an impediment to discussing new taxes or rates to be used for adaptation measures, which poses a new challenge, in addition to those imposed by climate change.

Participants do not seem willing to support adaptation measures, although the whole process was aimed at assessing possible adaptation measures in Santos.

It was observed changes of the participants in relation to the theme. They acknowledge their improvement in the understanding of the issue as well as the urgency of the matter. In addition, during the second meeting, the local government created the Commission for Adaptation to Climate Change (see Chap. 15).

Thus, the most important legacy of this participatory process, which is a central part of the Metropole Project, is to strengthen a democratic structure in which different actors participate in the governance process, helping to build a safer future for all.

Acknowledgements Samuel Merrill, Jack Kartez, Catherine J. Reynolds, Karen Langbehn and Frank Muller-Karger (US Metropole Project), for the support given to the analyses and for providing the manuscript entitled: Identifying Stakeholder Values and Priorities for Adaptation Planning and Finance: A Florida Case Study (not published).

Annexes

Annex 1 Sections of the survey

Survey 1	Survey 2
Section 1. Your experience with hazards	Section 1. Your experience with hazards
Section 2. Questions about potential adaptation actions	Section 2. Questions about potential adaptation actions
Section 3. Questions about possible funding sources	Section 3. Questions about possible funding sources
Section 4. Perspectives about adaptation and the environment	Section 4. Perspectives about adaptation and the environment
Section 5. Demographic questions	Final section. Your perspective about the meeting

Annex 2 Questions of the surveys

Survey 1	Survey 2
	Your information
	Your zip code
	Day and months of your birth
	How you define your political position?
Section 1. Your experience with hazards	**Section 1. Your experience with hazards**
1. Which of the following natural hazards that seriously and negatively affected your household or town in the past 10 years have you experienced?	
2. How concerned are you that the following natural hazards might seriously and negatively affect your town in the next 10 years in terms of physical and economic damage?	2. How concerned are you that the following natural hazards might seriously and negatively affect your town in the next 10 years in terms of physical and economic damage?
3. Thinking about the next 10 years, how concerned are you that these natural hazard may seriously and negatively affect your primary household in terms of physical and economic damage?	3. Thinking about the next 10 years, how concerned are you that these natural hazard may seriously and negatively affect your primary household in terms of physical and economic damage?
Section 2. Questions about potential adaptation actions	**Section 2. Questions about potential adaptation actions**
4. There are a variety of actions a town or district might implement to reduce the potential for physical and economic damage caused by climate-related hazards	4. There are a variety of actions a town or district might implement to reduce the potential for physical and economic damage caused by climate-related hazards
	Plus 4 options proposed by participants at the 1th meeting
Section 3. Questions about possible funding sources	**Section 3. Questions about possible funding sources**
5. Do you agree or disagree with the following statement: implementing projects to reduce potential impacts of climate-related hazards in our community should be a local or regional government priority, even if it will require a slight increase in taxes or new fees?	5. Do you agree or disagree with the following statement: implementing projects to reduce potential impacts of climate-related hazards in our community should be a local or regional government priority, even if it will require a slight increase in taxes or new fees?
6. Please consider the following funding options that local government and agencies could use/offer and tell us whether you think they are acceptable	6. Please consider the following funding options that local government and agencies could use/offer and tell us whether you think they are acceptable
7. If a referendum was put on the 2016 ballot to create a Community Resiliency Bond (a long-term loan) that would generate a fund by 2036 to support multiple adaptation projects, how likely would you be vote for it?	7. If a referendum was put on the 2016 ballot to create a Community Resiliency Bond (a long-term loan) that would generate a fund by 2036 to support multiple adaptation projects, how likely would you be vote for it?
Section 4. Perspectives about adaptation and the environment	**Section 4. Perspectives about adaptation and the environment**

(continued)

Annex 2 (continued)

Survey 1	Survey 2
8. Some people in your community might NOT want to support local government adaptation plans. What do you think are some the most common reasons for NOT supporting plans?	8. Some people in your community might NOT want to support local government adaptation plans. What do you think are some the most common reasons for NOT supporting plans?
9. Are there other reasons why people in your community might NOT support local government adaptation plans? Please tell us your thoughts	9. Are there other reasons why people in your community might NOT support local government adaptation plans? Please tell us your thoughts
10. People have different views about managing and adjusting to the environment around us	
Section 5. Demographic questions	**Final Section. Your perspective about the meeting**
11. What is your home zip code or postal code?	
12. Please circle the month and day you were born	
13. What is your gender?	
13. How old are you?	
14. Do you have children or grandchildren under the age of 18 living with you?	
15. What was your total household income last year (the income of yourself and everyone who contributes to your household and lives with you)?	
16. Do you currently own a home or condo?	
17. In politics today, do you consider yourself a Republican, Democrat or Independent or other? (Please circle one or write in.)	
18. What is the highest level of school you have completed?	
19. Finally, which of the following describes your role in your community at this meeting?	
	Final section. Your perspective about the meeting
	20. In order to continue to improve the process of choose of the community related to the adaptation option, express yourself about the information's presented and discussed today
	21. Which factors (for instance believes, previous information's) most influence your opinion in order to support or not an option of adaptation presented in this meeting?
	22. Please help us to understand what you or your organization may do with this information's in the next three months

References

Adger, W. N., Dessai, S., Goulden, M., et al. (2009). Are there social limits to adaptation to climate change? *Climatic Change, 93*, 335–335. https://doi.org/10.1007/s10584-008-9520-z.

Bechtel, B. T., Verdugo, B. C., & Pinheiro J de, Q. (1999). Environmental belief systems. United States, Brazil and Mexico. *Journal of Cross-Cultural Psychology, 30*, 122–128. https://doi.org/10.1177/0022022199030001008.

Catalysis Adaptation Partners, LLC. (2015). COastal Adaptation to Sea-level rise Tool (COAST) version 3.0. Catalysis Adaptation Partners, LLC, South Portland, ME, licensed through GEI Consultants, Inc.

Dunlap, R. E., Van Liere, K. D., Mertig, A. G., et al. (2000). Measuring endorsement of the new ecological paradigm: A revised NEP scale. *Journal of Social Issues, 56*, 425–442.

IPCC Intergovernmental Panel on Climate Change. (2014). Climate change 2014: Synthesis report. Contribution of working groups I, II and III to the fifth assessment report of the intergovernmental panel on climate change.

Marengo, J. A., Nunes, L. H., Souza, C. R. G., et al. (2017a). A globally deployable strategy for co-development of adaptation preferences to sea-level rise: The public participation case of Santos, Brazil. *Natural Hazards, 88*, 39–53. https://doi.org/10.1007/s11069-017-2855-x.

Marengo, J. A., Nunes, L. H., Souza, C. R. G., et al. (2017b). Vulnerability in Brazilian coastal communities: An integrated framework to analyze local decision making and adaptation to sea-level rise in Santos, Sao Paulo – Brazil. In: V. Marchezini, B. Wisner, & L. Londe, et al. (Eds.) *Reduction of vulnerability to disasters: From knowledge to action* (pp. 397–408). https://preventionroutes.weebly.com/capiacutetuloschapters.html. Accessed 12 Sept 2017.

Martín-Vide, J. (2001). Algunas reflexiones y ejemplos del valor de la percepción ambiental en la planificación territorial y de actividades. *Revista de Desenvolvimento Econômico, 3*, 61–65.

Ross, E., Shaw, S., Rissik, D., et al. (2015). A participatory systems approach to understanding climate adaptation needs. *Climatic Change, 129*, 27–42. https://doi.org/10.1007/s10584-014-1318-6.

The Berkshire Encyclopedia of Sustainability. (2012). Measurements, indicators, and research methods for sustainability. *New Ecological Paradigm (NEP) scale, 6*, 260–262.

Chapter 15
Adapting to a Changing Climate: An Operational Space for Local Adaptation Committee in Santos Coastal Area

Débora M. de Freitas, Danielle Almeida de Carvalho, and Eduardo Kimoto Hosokawa

Abstract Coastal living countries are the frontline of climate change impacts presenting vulnerability to natural disasters and other climate hazards. Specially, in developing countries, there is greater vulnerability and environmental risks, related to populous irregular settlements common in coastal areas. The impacts of climate change will materialize over the next century, however by taking action now it is possible to adapted and avoid risks. At the local scale, the impacts are more significant so local government policies play an important role in preventing and managing risks. This chapter summarizes recent research findings in planning and management about climate change adaptation on coastal areas and presents a study case in the coastal city of Santos, Brazil. Based on decision makers and local community perceptions, this chapter analyses the creation of a local policy and points out the importance of a governance structure to manage climate change related decisions. Finding an adequate operational space for adapting to a changing climate in coastal complex regions such as Santos requires local and regional strategies. In this sense, local committees can act as enablers to ensure institutional support for climate adaptation as a priority issue in the political and planning agendas.

Keywords Policy · Governance · Decision-making · Climate adaptation · Local government · Planning

D. M. de Freitas (✉) · D. A. de Carvalho
São Paulo State University (UNESP), Institute of Biosciences, Coastal Campus, São Vicente, SP, Brazil
e-mail: debora.freitas@unesp.br; danielle.almeida@unesp.br

E. K. Hosokawa
Municipal Government of Santos, Secretariat of Urban Development, Santos, SP, Brazil
e-mail: eduardohosokawa@santos.sp.gov.br

L. H. Nunes et al. (eds.), *Climate Change in Santos Brazil: Projections, Impacts and Adaptation Options*, https://doi.org/10.1007/978-3-319-96535-2_15

15.1 Introduction

Coastal zones are attractive to humans for both aesthetic and practical reasons. This appreciation for the coast, however, poses a grave risk for coupled coastal environments and the people concentrated in these narrow geographical spaces (Baztan et al. 2015). This fact not also amplifies historical issues, such as coping with conflicting interests over the diverse uses of the coastal environment, but it also unfolds new challenges to emerging events such as climate variability and change. Coastal living countries, such as Brazil and Australia among others, are the frontline of climate change impacts presenting greater vulnerability to natural disasters and other climate hazards.

Despite the obvious impacts in coastal settlements and nearshore infrastructure, climate change will lead to considerable risks to assets and productivity in the coastal zone. For example, it will lead to increased flooding of communities, greater damage costs, more frequent expenditure on repair, and likely interruptions to the supply of essential services. The challenge for infrastructure owners and decision makers is identifying when assets are likely to fail, and when action is required to ensure delivery of the services that we take for granted (Commonwealth of Australia 2009; Rosenzweig et al. 2011). The risks from climate change in the coastal zone are large, increasing, and in some areas will be felt in the near term. While the impacts of climate change will materialize gradually over the next century and beyond, by taking action now we have time to develop a well-considered; well-managed, and staged adaptation plan to reduce risks and avoid creating new risks. There is a large risk in the coastal zone to buildings and other infrastructure constructed in the past when the implications of climate change were not well known (Zanetti et al. 2016).

The coastal environment presents particular challenges of scale in space and time with broad disconnects between cause and effect, and between benefits and impacts. It is jurisdictionally and socially complex with multiple jurisdictions and properties and biophysically complex because of linkages within and between coastal environments and offshore waters (Kenchington et al. 2012). Additionally, there is the variability and uncertainties of climate change related events add an extra compounded layer on top of that.

Climate change is a critical issue that has been receiving massive attention globally because its impacts are deemed to have adverse effects on natural systems and human societies. Impacts resulting from climate change happen, regardless of the cause of the changes. Worldwide, coastal communities are challenged by the increase in frequency and magnitude of natural events associated with climate variability and change. During the last decade, several significant events such as storm surges, cyclones and sea-level rise have caused substantial damage to coastal infrastructure and communities. For example, continuous erosion and frequent storm tide inundation threaten coastal settlements and infrastructure by making then more vulnerable to landslides and flooding, respectively (Inouye et al. 2015; Iwama et al. 2016; Sakai et al. 2013). These issues demand special attention in developing countries, where populous irregular settlements (such as slums) are common in coastal areas, experiencing greater vulnerability to natural disasters and other environmental risks (Inouye et al. 2015).

When coastal flooding occurs along low lying, highly populated, and developed coastlines, the impacts can be devastating with wide ranging social, economic, and environmental consequences. Several significant events in the last decade, like Hurricane Katrina in New Orleans in 2005; Cyclone Xynthia on the French Atlantic coast in 2010; Hurricane Sandy in New Jersey in 2012; Typhoon Haiyan in the Philippines in 2013; and the recent events in the UK; have dramatically emphasized the high vulnerability of densely populated coastlines to extreme events (Haigh et al. 2015). The growing economic damages caused by a combined exposure of coastal assets to sea-level rise with climate change into the future. It has been estimated that exposure will follow population growth. In coastal countries like Australia, for instance, coastal assets at risk from the combined impact of inundation and shoreline recession include between 5800 and 8600 commercial buildings (estimated value ranging from $58 to $81 billion), between 27,000 and 35,000 km of roads and rail, (estimated value between $51 and $67 billion), and between 187,000 and 274,000 residential buildings (valued between $51 and $72 billion dollars) (Commonwealth of Australia 2009).

Changes are being identified and measured right now, and as the climate changes, adaptations are necessary. Many adaptations are already ongoing, and a combination of information, communication, dissemination, and planning involving all community sectors will have better changes to result in successful adaptation to future climate change (Simões et al. 2017). The IPCC reports, for instance, are heavily based on modelling projections and estimations to assess the scientific, technical and socio-economic information relevant for the understanding of the risk of human-induced climate change. In this chapter the term 'climate-adapted' or 'climate adaptation' is used interchangeably to refer to any changes in actions or planning undertaken to reduce the risks associated with observed or projected climate change (Norman et al. 2013).

Although globally significant for political decisions and goals setting, understand and application of globally based models at the regional and local governance levels where most of the on-ground decisions are made represents a formidable challenge (Fünfgeld 2015). There is now a growing awareness that most climate adaptation measures will need to take place at the local scale, making local government a significant player in this field (Fünfgeld 2015; Füssel 2007; Simões et al. 2017). Nevertheless, climate change and its related impacts are neither well understood nor taken into consideration at local and regional planning scales in developing countries. Brazil is an example of such problem. At the local scale, there are pressures related to settlement for construction and operational assets associated with port expansion and the development of oil and gas industry mega-infrastructure. At the state and national scales, there are political and economic pressures imposed by development of mega-projects, such as Pre-Salt oil exploration offshore, which does not consider local management needs and priorities on land use planning (Teixeira 2013).

Research in the southeast of Brazil also indicates that human settlement, mainly driven by tourism and mega-infrastructure developments, has induced urban sprawl towards the most vulnerable areas, making these populations and associated infrastructure more susceptible to the impacts of climate change (Inouye et al. 2015; Teixeira 2013). Fragmented and sectored licensing processes do not consider the cumulative and synergistic environmental effects resulting in inadequate public policies. Local coastal management initiatives show intense rearrangement, but changes are mostly focused in the ecological-economic zoning and creation of marine protected areas, and climate change mitigation and adaptation strategies are not a priority in the planning agenda (Simões et al. 2017).

Global sustainability requires the right balance between the need for finer scales and the need for ensembles with stakeholders demands on tangible decision making measures. Nevertheless, recent studies point that current approaches to coastal adaptation planning are likely to be beset by: complex processes and uncertain understandings; diverse forms of legitimate knowledge; multiple sources of fragmented information; constrained and conflicting and planning horizons and timeframes; sensitivities regarding the release of information; and, the sense that coastal adaptation is political (Simões et al. 2017).

In Brazil, as other coastal living countries, effective governance of coastal areas is challenged by: complexity of natural coastal systems; diverse uses of coastal areas; diverse jurisdictions and administrative bodies with coastal responsibilities (e.g. shipping/ports, planning, biodiversity management, fishing, recreation); diverse ways of understanding and appreciating coasts; and different perspectives on how it should be governed, managed, and used. It is now evident that climate science uptake requires both accessible knowledge systems and flexile structures and processes that facilitate policy actions in support of climate change adaptation (Kenchington et al. 2012).

Conventional systems of government have not been very successful in resolving coastal management problems (Stocker et al. 2012). This lack of progress is partially attributable to inadequate representation in governance processes of the variety of knowledge present on the coast. In particular, there has been a struggle to engage effectively with climate science and its implications. The current structure of coastal governance such as in Australia, for instance, largely reflect two centuries of policy and practice of British and subsequently Australian governments over several generations of immigrant settlement. There are still gaps at the interfaces between thinking, learning, knowing, governance and successful implementation (Stocker et al. 2012).

Governance is a highly contested concept with a plethora of meanings (Jordan 2008). Governance is a broader concept than government; it involves state and non-state actors in decision-making (Adger and Jordan 2009). Governance "can be formally institutionalised or expressed through subtle norms of interaction or even more indirectly by influencing the agendas and shaping the contexts in which actors contest decisions and determine access to resources" (see Pag. 2 in Lebel et al. 2006). For example, the implementation of coastal adaptation policies and actions,

Fig. 15.1 Sandbags (left picture) and reconstruction works (right picture) in response to accelerated beach erosion in Kingscliff, New South Wales /Australia, 2011. (Photo Débora M. De Freitas)

such as retreating from or protecting the coast, is fundamentally a governance issue (Adger and Jordan 2009).

The risk of coastal flooding and erosion will, or already is, demanding the construction of man-made defences to protect coastal assets and people due to sea-level rise (Balica et al. 2012; Klein et al. 1999) (Fig. 15.1).

It is important, however, to raise awareness of the need to preserve the natural coastal environment and to ensure that its management is sustainable way. As pointed by Beavers et al. (2016), it is necessary first to understand the physical processes that shaped and will continue affecting coastline as a way to better deal with engineering or intervention works. Understanding and documenting this range of complex interactions in political and social responses to climate change, and reflecting on potential strategies to adapt to challenges posed by climate change, is a formidable task posed in decades to come (Allison and Bassett 2015).

15.2 Operational Spaces for Local Adaptation

Climate variability and change, including sea-level rise trends and possible changes in weather patterns, will likely add additional pressure on coastal communities affecting regions differently. The burden of the responsibility for managing these risks resides with local government who is usually in the frontline of small scale and on-ground issues. Most coastal adaptation legislation around the world are still recent and the existing regulatory instruments to prohibit urban development in flood-prone areas (Fletcher et al. 2013). When it comes to respond and to delivery functional adaptation strategies, there is no difference. Protect, accommodate and retreat strategies are common adaptation options worldwide with measures such as hard defences, the removal of buildings and the restoration of natural processes (Fletcher et al. 2013).

Scholars worldwide have contributed to this field of research in a variety of ways, while proposing a series of questions to support the argument that robust climate change adaptation decisions should be evidence based (see Table 15.1).

Politically wise, in contexts where climate adaptation policy is well advanced such as in Australia for instance, regulatory and institutional landscape is designed to be dynamic and flexible, so main existing statutory arrangements do have the capacity to support climate adaptation planning and policy (Fletcher et al. 2013). However, there are some pitfalls, for instance almost all of the statutory and

Table 15.1 Example of key research questions of high relevance to coastal adaptation and policy

Key questions	References
How different vulnerable groups react and adapt to a changing climate?	Macintosh et al. (2013)
How coastal governance might be affected by new climate change policies?	
Would a possible review of current coastal planning mechanisms change the planning and development patterns of the coastal areas?	
How can policy makers affect behaviour to promote more socially-desired outcomes decisions affected by climate change regulations on the coastal zone?	
How individual land use decisions can be affected by climate change regulations on coastal land zoning?	
What are the side-effects of climate change regulations on land use pattern and market mechanisms on coastal development?	
Are existing infrastructure and new investment in the sectors resilient to a changing coast?	Balica et al. (2012)
Can governance systems allocate the costs of, and responsibilities for adaptation under scientific uncertainty?	Fletcher et al. (2013)
How should local authorities prioritize spending on coastal adaptation from their already financially stretched budgets?	CREW (2012)
If no public infrastructure is threatened as a result of coastal damage, who pays for the adaptation measures?	
What implications do such plans have on local business, infrastructure and the historic and natural environment?	
Which are the key organizational drivers that support adaptive capacity in a coastal zone socio-ecological system?	Dutra et al. (2011)
How climate change adaptation takes place in a complex multilevel system of governance?	Fidelman et al. (2013)
How the gap between generic frameworks and situation-specific tools can be addressed in estuarine and coastal marine ecosystems for efficient adaptation strategies?	Sheaves et al. (2016)
How climate change has been considered in the design, implementation, and management and monitoring of Marine Protected Areas (MPAs)?	Hopkins et al. (2016)
How climate change evidence can most effectively raise risk awareness and inspire community adaptation in a context of flood risk in coastal zones?	Lieske et al. (2014)

institutional arrangements relate to climate adaptation apply to new developments not including existing dwellings and infrastructure, except in certain circumstances such as in-fill developments, or post disaster reconstruction (Fletcher et al. 2013). In southeastern Australia, research showed that current coastal planning regulations used to assess future development address only one part of the urban future (Norman et al. 2013) and that the planning system in place still do not fully support or regulate 'retrofitting' of the existing built environment for a climate-adapted future and mechanisms for this risk response are required (Macintosh et al. 2013). As in other countries, Australia is an example of an operational space where a more collaborative regional approach with wide representation may be the best pathway forward for adapting to climate change on the coast. In this case, the "devolution of coastal planning and climate change decisions to local government will not be sustainable in the immediate or longer term as actions on major coastal infrastructure and development requires support and investment from state and national governments" (see Pag. 7 in Norman et al. 2013).

In Brazil, the National Plan for Adaptation, released in 2016, includes as a strategy for action in Brazil's coastal zone to "insert the climate lens in Coastal Management" to "internalize aspects related to sea-level rise in the management and promotion instruments in the coastal zone; develop, implement or re-adjust land use plans; management of coastal space considering the need to adapt coastal ecosystems" (PNA 2016). This strategy is the first national legislative initiative to integrate coastal management and climate change impacts on coastal zone in Brazil, and more, it could support the development of climate change plans in regional and local level.

15.3 Local Adaptation Committee in Santos Coastal Area

This case study aimed to address two major objectives: (1) assess the extent to which existing statutory frameworks, associated institutions and policy processes support or impede local adaptation planning and practice, and (2) make a significant contribution to the development and implementation of a strategic local adaptation policy framework. Understanding critical assets and services at risk in a changing coast requires further knowledge on the potential to local political framework to cope with new demands. The case of the Municipal Committee for Climate Change Adaptation of the city of Santos is an example of that.

With a population around 400,000 inhabitants from which 99% living in the low-lying insular area (39.4 km^2), Santos is a region highly vulnerable to climatic events. Local government buildings, most of the port infrastructure and public services such as hospitals, public schools, public transport stations, and police stations are located in the insular region too. Additionally, the region hosts the largest port of

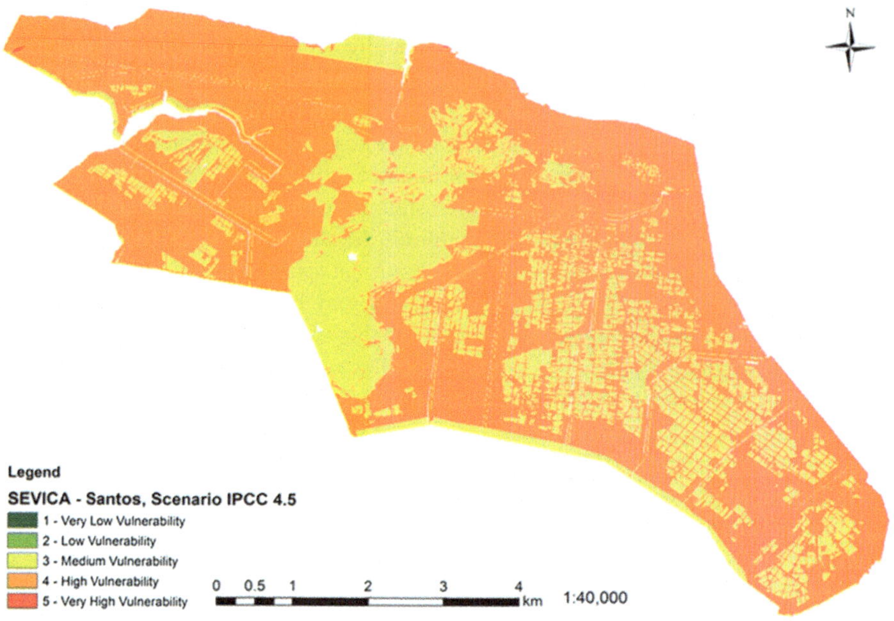

Fig. 15.2 Santos vulnerability map in the scenario of IPCC 4.5 (projection of sea level increase in 0.33 m) according to the Socio-Environmental Vulnerability Index for Coastal Areas –SEVICA. (With permission of the authors Zanetti et al. 2016, p. 8)

Latin America, the Port of Santos. Therefore, are showed in a research study, most of territory is located in lowland region with an average altitude of less than 10 m (Fig. 15.2), thus highly or potentially vulnerable to sea-level rise and flooding (Zanetti et al. 2016). High vulnerabilities areas were classified as low-lying areas of less than 1 m above sea level; so exposed to the sea-level rise risk. In fact, the study found that about 70% of the city's territory is in high vulnerability soil (Fig. 15.2), with residential built on alluvial soils, naturally susceptible to erosion.

Ponta da Praia (the portion located in the down-right corner of Fig. 15.2) is one of the noblest neighborhoods in Santos and houses a series of city infrastructures', such as the headquarters of the 6th Military Police Battalion; Fishing Museum; Municipal Aquarium, the most visited tourist attraction in the city; Museum of the Sea; Sports clubs (Regatas, Vasco da Gama and Saldanha) and part of the Port infrastructure.

According to the Magini et al. (2007) studies, Ponta da Praia neighborhood historically (Fig. 15.3), suffers erosion due to action of waves, which even destroyed urban infrastructure such as sidewalks and streets.

The frequency and intensity of storms and high tides reaching this area has increased lately. In 21 August 2016, the coast of Santos was reached by a storm surge that surpassed expectations of scientists and local Civil Defense by the intensity of the waves. According to Civil Defense, high tide and high pressure winds were combination of such intense event. For 11 years, storm surge of such intensity

Fig. 15.3 Historical aerial imagery from 1962 to 2014 showing the progressive erosion at Ponta da Praia suburb. (Image kindly provided by the Prefecture of Santos)

was not registered; with wind gusts of 40 km/h and tidal elevation of up to 2.6 m.[1] The main coast road, the Coast Avenue, and other nearby were invaded by the water, demonstrating the strength of this storm surge. Numerous damages were caused to the residents, condos and enterprises of the neighborhood; as well as damage to the urban infrastructure, such as the destruction of part of the famous Santos walls and the Fisherman's deck, a tourist point of the city (Fig. 15.4).

Climate change impacts are one of the greatest challenges to be faced in our society, and Santos coastal region is no exception. While measures to mitigate the climate change effects have been widely discussed and motivated major global meetings, adaptation measures are increasingly gaining ground in the discussions. At the local level, the climate change impacts, as extreme events, are more intense (Carter 2011), so local governments have been the protagonists of adaptation measures.

In December 2015, Decree No. 7293 the municipality of Santos established the Municipal Committee for Adaptation to Climate Change (MCACC) who was responsible for elaborating the Municipal Plan for Climate Change in Santos. At this point, it is important to highlight that the initiative to create the Municipal Plan for Climate Change emerged from local interests and was driven by the influence of

[1] Information taken from the local online newspaper "Portal G1 Santos" – "Ressaca inunda avenida da praia em Santos e provoca transtornos"; published in 21/08/2016 http://g1.globo.com/sp/santos-regiao/noticia/2016/08/ressaca-inunda-avenida-da-praia-em-santos-e-causa-bloqueios-no-canal-6.html. Accessed 10 Mar 10 2017.

Fig. 15.4 Damages caused by intense storm surge in the city of Santos. Both pictures illustrate the destruction of the Santos walls on a part of the beach that have been suffering erosion, and it have been intensified with current climatic variations. (Photo: Danielle Almeida de Carvalho)

the project funded by the São Paulo Research Foundation (FAPESP), the Metropole Project.[2] Within the Metropole's scope, studies demonstrated that direct economic impacts on real estate market would affect the area if adequate adaptation and mitigation measures to face sea-level rise were not taken. Local political leaders were engaged in the project and made possible the creation of the MCACC. Therefore, the creation of the plan was a demand that arose locally and not instituted by the State or Federal government.

15.3.1 A Local Operational Space for Climate Adaptation in Santos

This initiative on adaptation to the climate change impacts by the municipality of Santos was unprecedented in south coast cities in Brazil. As coastal erosion and flooding became more frequent, the creation of MCACC came as a response to local concerns about the city's future scenario. MCACC is coordinated by the Municipal Secretary of Urban Development and has a vice-coordinator indicated by vote. The MCACC is composed of secretariats representatives of the municipality of Santos as follow: Mayor's Office (GAB); Municipal Secretariat of Urban Development (SEDURB); Municipal Secretariat of Communication (SECOR); Municipal Secretariat of the Environment (SEMAM); Municipal Secretariat of Public Services (SESERP); Municipal Secretariat of Infrastructure and Buildings (SIEDI); Municipal Secretariat of Port and Maritime Affairs (SEPORT); Department of Civil Defense of the Municipal Security Department; Municipal Secretariat for Economic

[2] Full title: An integrated framework to analyze local decision-making and adaptive capacity to large-scale environmental change: Community case studies in Brazil, the United Kingdom and the United States.

Development and Innovation (SEDES); Santos Technology Park Foundation (FPTS) and Municipal Health Secretariat (SMS).

The communication process within the organizational structure of the Committee occurs through monthly meetings among the members. Despite efforts including publication about meetings' agenda in the Official Gazette of Santos, same secretariats do not participate creating communication gaps between. Low representation of secretariats has turned difficult to discuss and decide on certain procedures and issues. This has been making the decisions taken less representative of the totality of compartmented visions and unbalanced influence on climate change decisions and adaptation strategies the city should implement. As the major driver of regional and national economy, Port of Santos, for instance, has high influence over local-based decisions. Decision makers also reported the strong role played by real estate market influencing urban planning and infrastructure development in the region.

According to Decree No. 7.923, which establishes the MCACC, committee members must be indicated by each secretariat convened and must develop this activity combined with the day-to-day of the public servant tasks. At this point, two problems emerge: (i) accumulation of functions and tasks, which can lead to absence or lack of representativeness; and (ii) lack of technical capacity, whereas the members convened by their respective sectors do not necessarily have a technical understanding of climate change. In this situation, the introduction of agreements and further support from all spheres of governance, NGOs, private institutions and universities could minimize the accumulation of functions and improve technical capacity.

Substantial support and/or long-term agreement could further support to solve the lack of technical capacity problem identified in Santos case study. Participation of experts could improve technical capacity of the Committee responsible for developing the local adaptation plan. In addition, supporting MCACC members to cope with climate change would also enhance the technical-scientific knowledge of decision makers, such as offering courses and workshops. Thus, the translation of climate change global impacts into local impacts would be clearer for the decision makers understanding. So, the decision-making process would be more comprehensive and realistic regarding local particularities and, especially, would provide greater team cohesion.

As mentioned by Moser and Ekstrom (2010), insufficient resources are among the main difficulties faced by decision makers to implement climate change adaptation measures. In the case of Santos, the MCACC does not receive any kind of support from federal or state level. Local planning faces therefore an additional task, which means, finding or proposing new municipal funding such as income generation through the imposition of payments for excessive emission of greenhouse gases or funding from international agencies (Department of the Environment and Water Resources 2007; Department for Environment, Food and Rural Affairs 2010; Hunter et al. 2010).

At the local level, there is also space to further develop understanding of the complexity and uncertainties imposed the climate change impacts. Here, there is an open field to strength communication between scientific-academic environment and decision makers on climate risk uncertainties tends to grow, while socioeconomic processes are regulators and determinants of the climate change impacts (Norman et al. 2013; Pidgeon and Butler 2009; Pidgeon and Fischhof 2011).

The lack of effective coordination between sectors, departments and levels of government represents a great challenge but also an ongoing opportunity to ensure coherence in decision making related to the management of natural resources, as in the case of climate change (Fidelman et al. 2013; OECD 2002). Horizontal (institutions on the same level of social organization, e.g. sectoral policies) and vertical (strategies for adaptation at different levels of social and administrative organization, e.g. state and municipality) interplay between institutions provides an adequate framework for the success of the adaptive measures to be adopted (Young 2002). The local Committee is composed by representatives from different sectors of the local government, which ensures the horizontal interplay if it is efficiently implemented. However, the initiative of the city of Santos was a bottom-up approach and there is deficient in vertical communication between local policies and regional or national efforts.

In regional scale, a potential agent for the elaboration of strategies to mitigate and adapt to climate change in the south coastal cities is AGEM (Metropolitan Agency of São Paulo South Coast) an autonomous entity of the Government of the São Paulo State. The AGEM council is composed of representatives of the municipalities that integrate the south coast of São Paulo. The city of Santos can represent the local government that could influence the regional adaptation to climate change; so that the other cities of the region could follow in direction for the development of climate change local plans. Therefore, adaptive capacity can be built locally and regionally, if properly conducted (Fig. 15.5).

The implementation of adaptation policies requires not just a functional interplay between sectors and institutions but also a qualified communication directed to lay people targeting people engagement. In this way, adaptation strategies could endure in long-term horizons. Aiming the creation of an operational space for risk perception and management for climate change in the city of Santos, we found gaps in communication between local government and local population. Then, we follow this chapter showing some results from interviews applied directed to local trade owners in a vulnerable district of Santos.

15.3.2 Communication Perception, Communication and Engagement

During the event on August 21, 2016, action of waves and winds in a strong storm surge reached the region of the Ponta da Praia neighborhood leaving economic damages and losses to the residents, local business e public infrastructure (Fig. 15.6). Sea-level rise; coastal and beach erosion; strong storm surges; changes in the seasons; extreme events; weather abrupt changes and intense winds were pointed by

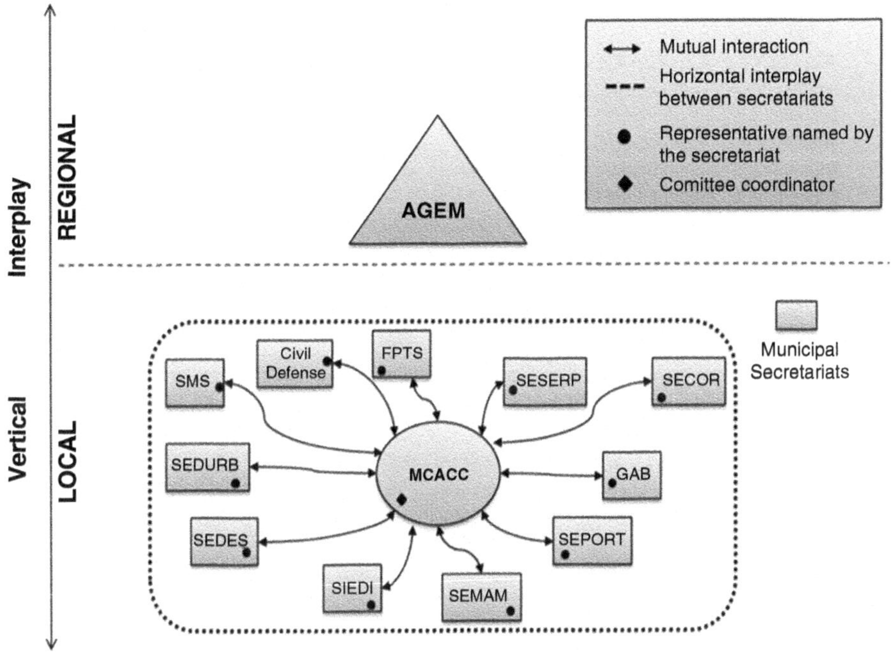

Fig. 15.5 A better organizational structure purposed by this study to climate change Committee (MCACC) in the city of Santos. Horizontal interplay, in local level, could be improved if each secretariat could name your own "climate change responsible", a person with abilities and technical capacity related to climate change subjects (indicates with black circle) to the Committee. So, it could enhance the communication between local decision makers. Vertical interplay is possible between the local government and the Metropolitan Agency (AGEM) that manages the metropolitan region in south coast cities of São Paulo, such as the city of Santos

the trade owners as the major climate change impacts. Association between the intense storm surges in Ponta da Praia region and climate change, and storm surges in the region were recognized more frequently in recent years.

Damages of trade owners were related to loss of products, which included damages in appliances, loss of part or all of inventory and others. In addition, infrastructure losses (i.e. losses related to damages in the electrical installation) and losses related to the impossibility of offering services, days that the merchants stopped profiting because they did not have adequate conditions to open the establishment, were other damages. Local business owners in Ponta da Praia attribute the City Hall, in first place, followed by Port of Santos as principal responsible for rebuild damages and compensate for material losses mainly by construction of containment works, repayment and development of risk prevention programs.

Individual initiatives for adaptation are common in the absence of policies on climate change (Laukkonen et al. 2009). Some initiatives were taken by local traders in Santos to reduce damages and risks; such as modification of electrical installations and relocation to upper store part; installation of steps in appliances; use of

Fig. 15.6 Infrastructure damages caused by the intense storm surge event on 21 August 2016 at Ponta da Praia, Santos. (Photos Danielle Almeida de Carvalho)

Fig. 15.7 Examples of a watergates-floodgate built by local trade owners of a cosmetics store (left) and a restaurant to prevent water intrusion in the property located at Ponta da Praia, Santos (2016). (Photos Danielle Almeida de Carvalho)

sandbags as a barrier to water intrusion; and walls and watergates (Fig. 15.7). They also point up possible solutions and proposal to deal with impacts caused by storm surges and flooding events, including: installation of floodgates to control water level in the city's waterways; prevention and warning program for the population; construction of barriers to block water flow, and long-term studies of the impacts of Port dredging.

Overall, at this stage, local traders demonstrated that they did not trust local institutions for risk aid. Absence of effective risks and disasters communication increases the vulnerability of the city. The mapping of populations and properties located in risk areas to recurrent natural disasters of climate change is necessary, coupled with a communication network and public service in case of emergencies in disasters. Nevertheless, social communication about climate change is still very incipient in Santos. Exploratory research conducted by the authors of this chapter showed that, depending on the explicit scenario, it may influence the inertia or engagement of the local community. That means, catastrophic scenarios or controversial uncertainties in the challenges to climate variability can lead to a paralytic and inert state of impotence in the face of calamitous disasters (Pidgeon and Fischhof 2011). On the other hand, efficient communication to population ensures awareness of impacts and adaptation measures, building and improving the city's adaptive capacity (Adger et al. 2005; Beatley 2009). Online newspaper and television are the most common sources of information consulted by trade owners to know what is happening in the region, and followed by the internet and television news.

Therefore, political participation of the communities in adaptation decision making process is essential for reaching the goals and objectives raised the Local Adaptation Plan in Santos. Research pointed that community knows more than anyone the weaknesses and opportunities of the place where they live, and this information coupled with the knowledge of experts and decision makers should create a collaborative scenario for risk reduction and resilience of the city (Beatley 2009). However, this is very challenging given that most part of the traders located at Ponta da Praia are do not integrate any local association or civil organization. This fact turns difficult to organize this specific vulnerable population for future representations in the context of political participation in decision making.

15.4 Conclusions

The elaboration of the Municipal Plan for Climate Change in the municipality of Santos is a possibility of integration between coastal management (CM) and adaptation. Decision makers understand that CM and adaptation to climate variability relate to: (i) land use and occupation in the coastal zone; (ii) the socioeconomic pathways of the region; and (iii) protection of coastal environments, such as mangroves.

Possible influence of the Port of Santos dredging in the coastal dynamics could exacerbate storm surges scenarios. Although, there are not specific studies measuring the influence of Port dredging on wave dynamics; one of the negative environmental impacts of dredging is the alteration of hydraulic, circulation, flow and water mixing patterns, as well as salinity and turbidity (United Nations 1992; Castro and Almeida 2012). Probably, climatic and meteorological variations on the coastal zone of Santos have been modified by the Port dredging. Therefore, the development of studies, such as precipitation regime, winds, beach dynamics and others is essential and urgent. In this way, local planners could be aware of region vulnerabilities and could decide the most qualified measures to the current and future city scenario.

Understanding critical assets and services at risk in a changing coast is ultimately an ethical, moral, equity and intergenerational issue. Current approaches to coastal adaptation planning are likely to be beset by: complex processes and uncertain understandings; diverse forms of legitimate knowledge; multiple sources of fragmented information; constrained and conflicting and planning horizons and timeframes; sensitivities regarding the release of information; and, the sense that coastal adaptation is political. Nevertheless, the economic risks from climate impacts are also important and often neglected in the public discourse (Steffen et al. 2014).

Adaptive measures more adopted by countries, regions and localities have been those of short-term, incremental and non-transformative changes. Crucially, finding an adequate operational space for adapting to a changing climate in coastal complex regions such as Santos requires local and regional strategies. In this sense, local committees can act as enablers to ensure institutional support for climate adaptation as a priority issue in the political and planning agendas.

References

Adger, W. N., & Jordan, A. (2009). Sustainability: Exploring the processes and outcomes of governance. In: *Governing sustainability* (pp. 3–31). Cambridge: Cambridge University Press.

Adger, W. N., Arnella, N. W., & Tompkins, E. L. (2005). Successful adaptation to climate change across scales. *Global Environmental Change, 15*, 77–86.

Allison, E. H., & Bassett, H. R. (2015). Climate change in the oceans: Human impacts and responses. *Science, 350*, 778–782.

Balica, S. F., Wright, N. G., & van der Meulen, F. (2012). A flood vulnerability index for coastal cities and its use in assessing climate change impacts. *Natural Hazards, 64*, 73–105.

Baztan, J., Chouinard, O., Jorgenses, B., et al. (2015). *Coastal zones: Solutions for the 21st century* (1st ed.). Amsterdam: Elsevier.

Beatley, T. (2009). *Planning for coastal resilience: Best practices for calamitous times*. Washington, DC: Island Press.

Beavers, R. L., Babson, A. L., & Schupp, C. A. (2016). *Coastal adaptation strategies handbook* (NPS 999/134090). Washington, DC: National Park Service.

Carter, J. G. (2011). Climate change adaptation in European cities. *Current Opinion in Environmental Sustainability, 3*, 193–198.

Castro, S. M., & Almeida, J. R. (2012). Dragagem e conflitos ambientais em portos clássicos e modernos: Uma revisão. *Society & Nature, 24*(3), 519–534.

Centre of Expertise for Waters – CREW. (2012). *Coastal flooding in Scotland: A guidance document for coastal practitioners*. http://www.crew.ac.uk/sites/default/files/sites/default/files/publication/Coastal%20Flooding_300812_Final.pdf. Accessed 14 Feb 2017.

Commonwealth of Australia. Department of Climate Change and Energy Efficiency. (2009). *Climate change risks to Australia's coast – A first pass national assessment*. https://www.environment.gov.au/system/files/resources/fa553e97-2ead-47bb-ac80-c12adffea944/files/cc-risks-full-report.pdf. Accessed 26 Oct 2017.

Department of Environment, Food and Rural Affairs – DEFRA. (2010). *Adapting to climate change: A guide for local councils*. United Kingdom Government. https://assets.publishing.

service.gov.uk/government/uploads/system/uploads/attachment_data/file/218798/adapt-local-councilguide.pdf. Accessed 15 May 2017.

Department of the Environment and Water Resources. (2007). *Climate change adaptation actions for local government.* Australian Greenhouse Office. Australian Government. http://www.lincoln.ac.nz/PageFiles/6702/4230_localgovernment_s13808.pdf. Accessed 27 Aug 2017.

Dutra, L. X. C., Ellis, N., Boschetti, F., et al. (2011). The use of the Healthy Waterways management strategy evaluation framework for active learning and decision- making: A first assessment. In *CSIRO*. Australia: Dutton Park.

Fidelman, P. I. J., Leitch, A. M., & Nelson, D. R. (2013). Unpacking multilevel adaptation to climate change in the Great Barrier Reef, Australia. *Global Environmental Change, 23*, 800–812.

Fletcher, C. S., Taylor, B. M., Rambaldi, A. N., et al. (2013). *Costs and coasts: An empirical assessment of physical and institutional climate adaptation pathways.* Gold Coast: National Climate Change Adaptation Research Facility https://www.nccarf.edu.au/sites/default/files/attached_files_publications/Fletcher_2013_Costs_and_coasts.pdf. Accessed 14 Aug 2017.

Fünfgeld, H. (2015). Facilitating local climate change adaptation through transnational municipal networks. *Current Opinion in Environmental Sustainability, 12*, 67–73.

Füssel, H. M. (2007). Adaptation planning for climate change: Concepts, assessment approaches and key lessons. *Sustainability Science, 2*, 265–275.

Haigh, I. D., Matthew, P., Wadey, M. P., et al. (2015). A user-friendly database of coastal flooding in the United Kingdom from 1915–2014. *Nature, 2*, 150021.

Hopkins, C. R., Bailey, D. M., & Potts, T. (2016). *Perceptions of practitioners: Managing marine.*

Hunter, P., Burkitt, Z., & Trangmar, B. (2010). Local government adapting to climate change: Managing infrastructure, protecting resources, and supporting communities. In R. A. C. Nottage, D. S. Wratt, J. F. Bornman, & K. Jones (Eds.), *Climate change adaptation in New Zealand: Future scenarios and some sectoral perspectives* (pp. 122–133). Wellington: New Zealand Climate Change Centre.

Inouye, C. E. N., Sousa, W. C., Jr., De Freitas, D. M., & Simões, E. (2015). Modelling the spatial dynamics of urban growth and land use changes in the North Coast of São Paulo, Brazil. *Ocean and Coastal Management, 108*, 147–157.

Iwama, A. Y., Batistella, M., Ferreira, L. C., et al. (2016). Risk, vulnerability and adaptation to climate change: An interdisciplinary approach. *Ambiente Soc, 19*(2), 93–116.

Jordan, A. (2008). The governance of sustainable development: Taking stock and looking forwards. *Environment and Planning C: Government and Policy, 26*, 17–33.

Kenchington, R., Stocker, L., & Wood, D. (Eds.). (2012). *Sustainable coastal management and climate adaptation – Global lessons from regional approaches in Australia.* Victoria: CSIRO Publishing.

Klein, R. J. T., Nicholls, R. J., & Mimura, N. (1999). Coastal adaptation to climate change: Can the IPCC technical guidelines be applied? *Mitigation and Adaptation Strategies for Global Change, 4*, 239–252.

Laukkonen, J., Blanco, P. K., Lenhart, J., et al. (2009). Combining climate change adaptation and mitigation measures at the local level. *Habitat International, 33*, 287–292.

Lebel, L., Anderies, J. M., Campbell, B., et al. (2006). Governance and the capacity to manage resilience in regional social-ecological systems. *Ecology and Society, 11*(1), 19.

Lieske, D. J., Wadeb, T., & Ronessa, L. A. (2014). Climate change awareness and strategies for communicating the risk of coastal flooding: A Canadian maritime case example. *Estuarine, Coastal and Shelf Science, 140*, 83–94.

Macintosh, A., Foerster, A., & McDonald, J. (2013). *Limp, leap or learn? Developing legal frameworks for climate change adaptation planning in Australia* (p. 262). Gold Coast: National Climate Change Adaptation Research Facility https://www.nccarf.edu.au/sites/default/files/attached_files_publications/Macintosh_2013_Spatial_planning_instruments_adaptation_Final.pdf. Accessed 23 June 2017.

Magini, C., Harari, J., & Abessa, D. M. S. (2007). Circulação Recente de Sedimentos Costeiros Nas Praias de Santos Durante Eventos de Tempestade: Dados Para a Gestão de Impactos Físicos Costeiros. *Geociências – UNESP, 26*, 349–355.

Moser, S. C., & Ekstrom, J. A. (2010). A framework to diagnose barriers to climate change adaptation. *Proceedings of the National Academy of Sciences of the United States of America, 107*, 22026–22031.

Norman, B., Steffen, W., Webb, R., et al. (2013). *South East Coastal Adaptation (SECA): Coastal urban climate futures in SE Australia from Wollongong to Lakes Entrance*. National Climate Change Adaptation Research Facility. 130 pp. https://www.nccarf.edu.au/sites/default/files/attached_files_publications/Norman_2013_SECA_Coastal_urban_climate_futures.pdf. Accessed 17 Mar 2017.

Organisation for Economic Co-Operation and Development – OECD. (2002). Governance for sustainable development: Five OECD case studies. OECD, Paris. Introduction: 9–35. https://www.ulb.ac.be/ceese/nouveau%20site%20ceese/documents/oecd%20governance%20for%20sustainable%20development%205%20case%20studies.pdf. Accessed 5 July 2017.

Pidgeon, N. F., & Butler, C. (2009). Risk analysis and climate change. *Environmental Politics, 18*, 670–688.

Pidgeon, N., & Fischhof, B. (2011). The role of social and decision sciences in communicating uncertain climate risks. *Nature Climate Change, 1*, 35–41.

Plano Nacional de Adaptação à Mudança do Clima – PNA: Estratégias setoriais e climáticas – Vol. II. (2016). Portaria MMA n°150 de 10 de maio de 2016. http://www.mma.gov.br/images/arquivo/80182/LIVRO_PNA_Plano%20Nacional_V2.pdf. Accessed 15 June 2016.

Rosenzweig, C., Solecki, W. D., Blake, R., et al. (2011). Developing coastal adaptation to climate change in the New York City infrastructure-shed: process, approach, tools, and strategies. *Climatic Change, 106*, 93–127.

Sakai, R. O., Cartacho, D. L., Arasaki, E., et al. (2013). Extreme events assessment methodology coupling debris flow, flooding and tidal levels in the coastal floodplain of the San Paulo North Coast (Brazil). *International Journal of Geosciences, 4*(5B), 30–38.

Sheaves, M., Sporne, I., Dichmont, C. M., et al. (2016). Principles for operationalizing climate change adaptation strategies to support the resilience of estuarine and coastal ecosystems: An Australian perspective. *Marine Policy, 68*, 229–240.

Simões, E., de Sousa Junior, W. C., De Freitas, D. M., et al. (2017). Barriers and opportunities for adapting to climate change on the North Coast of São Paulo, Brazil. *Regional Environmental Change, 17*(6), 1739–1750.

Steffen, W., Hunter, J., & Hughes, L. (2014). *Counting the costs: Climate change and coastal flooding*. Climate Council of Australia. http://www.climatecouncil.org.au/uploads/56812f1261b168e02032126342619dad.pdf. Accessed 13 May 2017.

Stocker, L., Kennedy, D., Kenchington, R., & Merrick, K. (2012). *Sustainable coastal management? Sustainable coastal management and climate adaptation, global lessons from regional approaches in Australia* (pp. 29–50p). Australia: CSIRO Publishing.

Teixeira, L. R. (2013). Megaprojetos no litoral norte paulista: o papel dos grandes empreendimentos de infraestrutura na transformacão regional. Tese Doutorado em Ambiente e Sociedade, University of Campinas.

United Nations. (1992). Assessment of the environmental impact of port development – A guide book for EIA of port development. Economic and Social Commission for Asia and the Pacific. Chapter II: Environmental impacts for port development: 6–19. http://www.unescap.org/sites/default/files/pub_1234_fulltext.pdf. Accessed 26 Nov 2016.

Young, O. R. (2002). Institutional interplay: The environmental consequences of cross-scale interactions. In E. Ostrom, T. Dietz, N. Dolsak, P. C. Stern, S. Stonich, & E. U. Weber (Eds.), *The drama of the commons* (pp. 263–292). Washington, DC: National Academy Press.

Zanetti, V. B., de Sousa Junior, W. C., & De Freitas, D. M. (2016). A climate change vulnerability index and case study in a Brazilian coastal city. *Sustainability, 8*, 811.

Printed by Printforce, the Netherlands